**Operator Theory
Advances and Applications
Vol. 88**

**Editor**
**I. Gohberg**

Editorial Office:
School of Mathematical
Sciences
Tel Aviv University
Ramat Aviv, Israel

Editorial Board:
J. Arazy (Haifa)
A. Atzmon (Tel Aviv)
J.A. Ball (Blackburg)
A. Ben-Artzi (Tel Aviv)
H. Bercovici (Bloomington)
A. Böttcher (Chemnitz)
L. de Branges (West Lafayette)
K. Clancey (Athens, USA)
L.A. Coburn (Buffalo)
K.R. Davidson (Waterloo, Ontario)
R.G. Douglas (Stony Brook)
H. Dym (Rehovot)
A. Dynin (Columbus)
P.A. Fillmore (Halifax)
C. Foias (Bloomington)
P.A. Fuhrmann (Beer Sheva)
S. Goldberg (College Park)
B. Gramsch (Mainz)
G. Heinig (Chemnitz)
J.A. Helton (La Jolla)
M.A. Kaashoek (Amsterdam)

T. Kailath (Stanford)
H.G. Kaper (Argonne)
S.T. Kuroda (Tokyo)
P. Lancaster (Calgary)
L.E. Lerer (Haifa)
E. Meister (Darmstadt)
B. Mityagin (Columbus)
V.V. Peller (Manhattan, Kansas)
J.D. Pincus (Stony Brook)
M. Rosenblum (Charlottesville)
J. Rovnyak (Charlottesville)
D.E. Sarason (Berkeley)
H. Upmeier (Marburg)
S.M. Verduyn-Lunel (Amsterdam)
D. Voiculescu (Berkeley)
H. Widom (Santa Cruz)
D. Xia (Nashville)
D. Yafaev (Rennes)

Honorary and Advisory
Editorial Board:
P.R. Halmos (Santa Clara)
T. Kato (Berkeley)
P.D. Lax (New York)
M.S. Livsic (Beer Sheva)
R. Phillips (Stanford)
B. Sz.-Nagy (Szeged)

# The Asymptotic Behaviour of Semigroups of Linear Operators

Jan van Neerven

Birkhäuser Verlag
Basel · Boston · Berlin

Author's address:                                as of May 1, 1996

Mathematisches Institut                          Delft Technical University
Universität Tübingen                             Department of Mathematics
Auf der Morgenstelle 10                          P.O. Box 356
72076 Tübingen                                   2600 AJ Delft
Germany                                          The Netherlands

1991 Mathematics Subject Classification 47D03, 47D06, 35K22, 93D20

A CIP catalogue record for this book is available from the Library of Congress, Washington D.C., USA

Deutsche Bibliothek Cataloging-in-Publication Data
**Neerven, Jan** van:
The asymptotic behaviour of semigroups of linear operators /
Jan van Neerven. – Basel ; Boston ; Berlin ; Birkhäuser, 1996
  (Operator theory ; Vol. 88)
  ISBN 3-7643-5455-0 (Basel ...)
  ISBN 0-8176-5455-0 (Boston)
NE: GT

This work is subject to copyright. All rights are reserved, whether the whole or part of the material is concerned, specifically the rights of translation, reprinting, re-use of illustrations, recitation, broadcasting, reproduction on microfilms or in other ways, and storage in data banks. For any kind of use the permission of the copyright holder must be obtained.

© 1996 Birkhäuser Verlag, P.O. Box 133, CH-4010 Basel, Switzerland
Printed on acid-free paper produced from chlorine-free pulp. TCF ∞
Cover design: Heinz Hiltbrunner, Basel
Printed in Germany
ISBN 3-7643-5455-0
ISBN 0-8176-5455-0

9 8 7 6 5 4 3 2 1

To my family
Silvania and Matthijs

# Preface

Over the past ten years, the asymptotic theory of one-parameter semigroups of operators has witnessed an explosive development. A number of long-standing open problems have recently been solved and the theory seems to have obtained a certain degree of maturity. These notes, based on a course delivered at the University of Tübingen in the academic year 1994-1995, represent a first attempt to organize the available material, most of which exists only in the form of research papers.

If $A$ is a bounded linear operator on a complex Banach space $X$, then it is an easy consequence of the spectral mapping theorem

$$\exp(t\sigma(A)) = \sigma(\exp(tA)), \quad t \in \mathbb{R},$$

and Gelfand's formula for the spectral radius that the uniform growth bound of the family $\{\exp(tA)\}_{t\geq 0}$, i.e. the infimum of all $\omega \in \mathbb{R}$ such that $\|\exp(tA)\| \leq Me^{\omega t}$ for some constant $M$ and all $t \geq 0$, is equal to the spectral bound $s(A) = \sup\{\operatorname{Re}\lambda : \lambda \in \sigma(A)\}$ of $A$. This fact is known as Lyapunov's theorem. Its importance resides in the fact that the solutions of the initial value problem

$$\frac{du(t)}{dt} = Au(t), \quad t \geq 0,$$
$$u(0) = x,$$

are given by $u(t) = \exp(tA)x$. Thus, Lyapunov's theorem implies that the exponential growth of the solutions of the initial value problem associated to a bounded operator $A$ is determined by the location of the spectrum of $A$.

Already long ago it was realized that the corresponding statement is no longer true for unbounded operators $A$. The analogue for the one-parameter family $\{\exp(tA)\}_{t\geq 0}$ and the operator $A$ in the unbounded case is a strongly continuous one-parameter family of bounded operators $\mathbf{T} = \{T(t)\}_{t\geq 0}$ on a Banach space $X$ and its infinitesimal generator $A$. Even in Hilbert spaces, examples are known of semigroups whose uniform growth bound, denoted by $\omega_0(\mathbf{T})$, is strictly larger than the spectral bound $s(A)$.

Since usually one is only given the operator $A$ via some initial value problem, it is desirable to be able to deduce asymptotic properties of the solutions $u(t) = T(t)x$ of the initial value problem from information about $A$. The failure of Lyapunov's theorem means that the location of the spectrum of $A$ no longer provides sufficient information.

This is the starting point of the asymptotic theory of semigroups. As in the bounded case, the presence of a spectral mapping theorem leads to equality $\omega_0(\mathbf{T}) = s(A)$, and many authors have tried to find additional conditions on the semigroup or on its generator under which a spectral mapping theorem holds. A well-known condition of this type is eventual continuity with respect to the uniform operator topology; this covers, e.g., compact semigroups and holomorphic semigroups. Recently, some important new spectral mapping theorems have been found, among which we mention the weak spectral mapping theorem for non-quasianalytic groups and the spectral mapping theorem of Latushkin and Montgomery-Smith.

In some situations, sufficient conditions for equality $\omega_0(\mathbf{T}) = s(A)$ can be given in the absence of a spectral mapping theorem. For example, it has been known for some time that equality holds for positive semigroups on the spaces $L^1(\mu)$, $L^2(\mu)$, and $C_0(\Omega)$. Recently, the problem whether the same is true for positive semigroups on $L^p(\mu)$ was solved affirmatively by Weis.

Another approach to the failure of Lyapunov's theorem is the introduction of more subtle concepts to describe the asymptotic behaviour of the semigroup and the spectrum of its generator. Besides the uniform growth bound and the spectral bound, two quantities appeared to play a crucial role: the growth bound $\omega_1(\mathbf{T})$ of the orbits of the semigroup originating from the domain $D(A)$ of $A$, and the abscissa $s_0(A)$ of uniform boundedness of the resolvent of $A$. Two classical results in this direction are Gearhart's theorem that $\omega_0(\mathbf{T}) = s_0(A)$ for semigroups in Hilbert space and Neubrander's theorem that $\omega_1(\mathbf{T}) = s(A)$ for positive semigroups. Very recently, Weis and Wrobel proved that $\omega_1(\mathbf{T}) \leq s_0(A)$ holds for arbitrary semigroups; an elementary orbitwise proof was found shortly after by the author.

It is also interesting to ask for sufficient conditions for uniform exponential stability ($\omega_0(\mathbf{T}) < 0$) and exponential stability ($\omega_1(\mathbf{T}) < 0$) in terms of a priori weaker conditions on the semigroup. Many results of this type have been obtained in recent years by, among others, Datko, Pazy, Rolewicz, Weiss, Falun Huang, and the author.

Finally, one can ask for conditions guaranteeing uniform stability of the semigroup, i.e. strong convergence to zero of the semigroup or certain of its orbits. Among the most important results in this field are the stability theorem Arendt, Batty, Lyubich and Vũ stating that a bounded semigroup is uniformly stable if the peripheral spectrum of its generator is countable and contains no residual spectrum, and the semigroup version of the Katznelson-Tzafriri theorem. In this book we present generalizations of these theorems to individual orbits; indeed, we have tried to develop a theory of individual strong convergence to zero in a systematic way. We present both the harmonic analysis approach and the Laplace transform approach to this subject.

We have organized these notes in decreasing order of satisfactory asymptotic behaviour of the semigroup: after proving the basic elementary results of the subject in Chapter 1, in Chapter 2 we discuss the various spectral mapping theorems, and in Chapters 3, 4, and 5 we discuss uniform exponential stability, exponential stability, uniform stability, and the relationship between stability of the semigroup and the behaviour near the imaginary axis of the resolvent.

In some sense, these notes can be seen as a follow-up to parts of the book [Na], where most of what was known around 1985 about asymptotics of semigroups is presented. We have taken some effort to keep the overlap to a minimum. Whereas [Na] presents a detailed theory for positive semigroups, the asymptotic theory for general semigroups on Banach spaces was at that time relatively undeveloped. In the past decade, it was the Banach space theory in which most progress occurred. Thus, we decided to include only those results about positive semigroups that were proved after the appearance of [Na]. Accordingly we do not discuss Perron-Frobenius theory and the asymptotic behaviour of special classes of positive semigroups, such as irreducible semigroups and lattice semigroups.

Also, we decided not to include ergodic theory of semigroups. This is mainly to keep this volume at moderate size; a fairly recent account is given in the book [Kr]. For the same reason, we do not discuss the various generalizations of semigroups and related concepts, such as integrated semigroups, regularized semigroups, $C$-semigroups, and cosine families, nor do we attempt to present all results in the greatest possible generality. For instance, many of the results of Chapter 5 can be proved in the setting of strongly continuous representations on $\mathcal{L}(X)$ of subsemigroups of locally compact abelian groups; in particular, there are analogues for the powers $\{T^n\}_{n\in\mathbb{N}}$ of a single bounded operator $T$. Similarly, various results in Chapter 3 have analogues for the powers of a single operator; whenever this is the case the proof is usually slightly simpler than in the semigroup case and we have restricted ourselves to some comments in the notes at the end of each chapter.

At this point, I would like to thank a number of people. First of all, Rainer Nagel for his encouragement to write these notes and his valuable comments and advice. Some material in these notes has not been published yet; I thank Wolfgang Arendt, Charles Batty, Zdzislaw Brzeźniak, David Greenfield, Sen-Zhong Huang, Frank Räbiger, Vũ Quôc Phóng, Lutz Weis, and Volker Wrobel for their kind permission to present their results here. Charles Batty read parts of the manuscript and provided helpful comments. My own research on the subject started during a fourteen months' stay at California Institute of Technology. I am greatly indebted to Wim Luxemburg for his warm hospitality and constant interest; the time spent with him was most enjoyable. This work was written while I was supported by an Individual Fellowship in the Human Capital and Mobility Programme of the European Communities.

# Contents

**Chapter 1.** *Spectral bound and growth bound* . . . . . . . . . . . 1
   1.1. $C_0$–semigroups and the abstract Cauchy problem . . . . . . 2
   1.2. The spectral bound and growth bound of a semigroup . . . . . 8
   1.3. The Laplace transform and its complex inversion . . . . . . . 15
   1.4. Positive semigroups . . . . . . . . . . . . . . . . . . . . 19
   Notes . . . . . . . . . . . . . . . . . . . . . . . . . . . . 23

**Chapter 2.** *Spectral mapping theorems* . . . . . . . . . . . . . 25
   2.1. The spectral mapping theorem for the point spectrum . . . . 26
   2.2. The spectral mapping theorems of Greiner and Gearhart . . 31
   2.3. Eventually uniformly continuous semigroups . . . . . . . . 35
   2.4. Groups of non-quasianalytic growth . . . . . . . . . . . . 42
   2.5. Latushkin - Montgomery-Smith theory . . . . . . . . . . 64
   Notes . . . . . . . . . . . . . . . . . . . . . . . . . . . . 70

**Chapter 3.** *Uniform exponential stability* . . . . . . . . . . . . 73
   3.1. The theorem of Datko and Pazy . . . . . . . . . . . . . . 74
   3.2. The theorem of Rolewicz . . . . . . . . . . . . . . . . . 81
   3.3. Characterization by convolutions . . . . . . . . . . . . . 83
   3.4. Characterization by almost periodic functions . . . . . . . 90
   3.5. Positive semigroups on $L^p$-spaces . . . . . . . . . . . . . 95
   3.6. The essential spectrum . . . . . . . . . . . . . . . . . . 107
   Notes . . . . . . . . . . . . . . . . . . . . . . . . . . . . 111

**Chapter 4.** *Boundedness of the resolvent* . . . . . . . . . . . . 113
   4.1. The convexity theorem of Weis and Wrobel . . . . . . . . 114
   4.2. Stability and boundedness of the resolvent . . . . . . . . . 115
   4.3. Individual stability in $B$-convex Banach spaces . . . . . . . 121
   4.4. Individual stability in spaces with the analytic RNP . . . . . 128
   4.5. Individual stability in arbitrary Banach spaces . . . . . . . 132
   4.6. Scalarly integrable semigroups . . . . . . . . . . . . . . 137
   Notes . . . . . . . . . . . . . . . . . . . . . . . . . . . . 145

**Chapter 5.** *Countability of the unitary spectrum* . . . . . . . . 149
   5.1. The stability theorem of Arendt, Batty, Lyubich, and Vũ . . 149
   5.2. The Katznelson-Tzafriri theorem . . . . . . . . . . . . . 163

5.3. The unbounded case . . . . . . . . . . . . . . . . . 168
5.4. Sets of spectral synthesis . . . . . . . . . . . . . . 175
5.5. A quantitative stability theorem . . . . . . . . . . . 179
5.6. A Tauberian theorem for the Laplace transform . . . . . 188
5.7. The splitting theorem of Glicksberg and DeLeeuw . . . . 199
Notes . . . . . . . . . . . . . . . . . . . . . . . . . 209

**Appendix**
A1. Fractional powers . . . . . . . . . . . . . . . . . . 215
A2. Interpolation theory . . . . . . . . . . . . . . . . . 216
A3. Banach lattices . . . . . . . . . . . . . . . . . . . 219
A4. Banach function spaces . . . . . . . . . . . . . . . . 221

**References** . . . . . . . . . . . . . . . . . . . . . 225

**Index** . . . . . . . . . . . . . . . . . . . . . . . . 235

**Symbols** . . . . . . . . . . . . . . . . . . . . . . . 237

# Chapter 1

## Spectral bound and growth bound

In this chapter we introduce the main concepts of the asymptotic theory of $C_0$-semigroups. The most basic results of the theory are presented, along with several examples which motivate the developments in the later chapters.

In Section 1.1 we start with a general discussion of the abstract Cauchy problem for unbounded linear operators on a Banach space and its relation to $C_0$-semigroups.

In Section 1.2 we define the four main quantities relating to the asymptotic behaviour of a $C_0$-semigroup $\mathbf{T}$ and its generator $A$: the spectral bound $s(A)$, the abscissa $s_0(A)$ of uniform boundedness of the resolvent of $A$, the uniform growth bound $\omega_0(\mathbf{T})$, and the growth bound $\omega_1(\mathbf{T})$. We also present Zabczyk's example of a $C_0$-semigroup on a Hilbert space for which the spectral bound is strictly smaller than the growth bound and the uniform growth bound.

In Section 1.3 we derive a complex inversion formula for the Laplace transform of a $C_0$-semigroup and we prove the Pringsheim-Landau theorem for the Laplace transform of positive functions.

In Section 1.4 we prove some elementary results about the asymptotic behaviour of positive $C_0$-semigroups on Banach lattices and present the example of Arendt of a positive $C_0$-semigroup on $L^p \cap L^q$, $1 \le p < q < \infty$, for which the spectral bound is strictly smaller than the uniform growth bound.

At this point we fix some notations and make some conventions. Since one of our main tools is spectral theory, all Banach spaces and Banach lattices are *complex*. Some results can be stated for real Banach spaces as well, in which case they can usually be derived from the complex case by means of complexification; we leave this to the reader to verify at the particular instances.

For a bounded or unbounded operator $A$ on a Banach space $X$ with domain $D(A)$, we define the *resolvent set* $\varrho(A)$ as the set of all $\lambda \in \mathbb{C}$ for which $\lambda - A$ is invertible, i.e. there exists a bounded operator $B$ with $Bx \in D(A)$ for all $x \in X$ such that $(\lambda - A)Bx = x$ for all $x \in X$ and $B(\lambda - A)x = x$ for all $x \in D(A)$. The set $\varrho(A)$ is an open subset of the complex plane and the family $\{(\lambda - A)^{-1} : \lambda \in \varrho(A)\}$ is usually referred to as the *resolvent* of $A$. It is common to write $R(\lambda, A)$ instead of $(\lambda - A)^{-1}$. The map $\lambda \mapsto R(\lambda, A)$ is a holomorphic $\mathcal{L}(X)$-valued mapping on $\varrho(A)$. For an $x \in X$, the holomorphic $X$-valued mapping $\lambda \mapsto R(\lambda, A)x$ is called the *local resolvent* of $A$ at $x$.

The complement $\sigma(A) := \mathbb{C} \setminus \varrho(A)$ is called the *spectrum* of $A$. The *spectral* bound $s(A)$ of an unbounded operator $A$ is the quantity $\sup\{\operatorname{Re}\lambda : \lambda \in \sigma(A)\}$. If $s(A) < \infty$, then the *peripheral spectrum* of $A$ is the set $\{\lambda \in \sigma(A) : \operatorname{Re}\lambda = s(A)\}$ and the *unitary spectrum* is the set $\{\lambda \in \sigma(A) : \operatorname{Re}\lambda = 0\} = \sigma(A) \cap i\mathbb{R}$. If $T$ is a bounded operator, then the *spectral radius* $r(T)$ of $T$ is the quantity $\sup\{|\lambda| : \lambda \in \sigma(T)\}$, the *peripheral spectrum* is the set $\{\lambda \in \sigma(T) : |\lambda| = r(T)\}$, and the *unitary spectrum* is the set $\{\lambda \in \sigma(T) : |\lambda| = 1\} = \sigma(T) \cap \Gamma$, where $\Gamma$ is the unit circle in the complex plane.

The dual space of a Banach space $X$ will be denoted by $X^*$ and the duality pairing between $X^*$ and $X$ by $\langle \cdot, \cdot \rangle$. The space of all bounded linear operators on a Banach space $X$ is denoted by $\mathcal{L}(X)$.

Unless otherwise stated, all vector-valued integrals in this book are in the sense of Bochner; we refer to the book [DU] for the properties of this integral. At several occasions we shall use vector-valued analogues of results from classical analysis; usually their proofs are straightforward generalizations of their classical counterparts and we refer to the book [HP] for the details.

Finally, $\mathbb{N} = \{0, 1, 2, ...\}$, $\mathbb{Z} = \{..., -1, 0, 1, ...\}$, $\mathbb{R}$ is the field of real numbers, $\mathbb{R}_+ = [0, \infty)$, and $\mathbb{C}$ is the field of complex numbers. The unit circle $\{z \in \mathbb{C} : |z| = 1\}$ is denoted by $\Gamma$. The characteristic function of a set $E$ is denoted by $\chi_E$.

## 1.1. $C_0$–semigroups and the abstract Cauchy problem

Many equations of mathematical physics can be cast in the abstract form

$$(ACP) \quad \begin{aligned} \frac{du}{dt}(t) &= Au(t), \quad t \geq 0, \\ u(0) &= x, \end{aligned}$$

where $A$ is a linear, usually unbounded, operator with domain $D(A)$ on a Banach space $X$. Usually, $X$ is a Banach space of functions suited for the particular problem and $A$ is a partial differential operator.

This abstract initial value problem (ACP) is usually referred to as the *abstract Cauchy problem* associated to $A$. If we want to stress that $x$ is the initial value we will write $(ACP_x)$. A *classical solution* of (ACP) is a continuously differentiable function $u : [0, \infty) \to X$ taking its values in $D(A)$ which satisfies (ACP). A continuous function $u : [0, \infty) \to X$ is a *mild solution* of (ACP) if there exists a sequence $(x_n) \subset D(A)$ such that for each $n$ the problem $(ACP_{x_n})$ has a classical solution $u(\cdot, x_n)$ with $\lim_{n \to \infty} u(t, x_n) = u(t)$ locally uniformly for $t \geq 0$. Clearly, a classical solution is a mild solution, but a mild solution need not be classical.

In this book we undertake a detailed study of the asymptotic behaviour of the classical and mild solutions of the abstract Cauchy problem (ACP). The natural framework to carry out this investigation is the theory of $C_0$–semigroups.

**Definition 1.1.1.** A family $\mathbf{T} = \{T(t)\}_{t\geq 0}$ of bounded linear operators acting on a Banach space $X$ is called a $C_0$-*semigroup* if the following three properties are satisfied:

(S1) $T(0) = I$, the identity operator on $X$;
(S2) $T(t)T(s) = T(t+s)$ for all $t, s \geq 0$;
(S3) $\lim_{t\downarrow 0} \|T(t)x - x\| = 0$ for all $x \in X$.

It is easy to see that the maps $t \mapsto T(t)x$ are continuous for $t \geq 0$. The *infinitesimal generator* of $\mathbf{T}$, or briefly the *generator*, is the linear operator $A$ with domain $D(A)$ defined by

$$D(A) = \{x \in X : \lim_{t\downarrow 0} \frac{1}{t}(T(t)x - x) \text{ exists}\},$$

$$Ax = \lim_{t\downarrow 0} \frac{1}{t}(T(t)x - x), \quad x \in D(A).$$

The generator is always a closed, densely defined operator. The domain is **T**-invariant, i.e. $T(t)x \in D(A)$ for all $x \in D(A)$ and $t \geq 0$, and we have $AT(t)x = T(t)Ax$. Moreover,

$$\frac{d}{dt}T(t)x = AT(t)x, \quad x \in D(A),$$

which shows that for $x \in D(A)$ the problem (ACP) has a classical solution given by $u(t) = T(t)x$. The following theorem shows that this essentially characterizes generators of $C_0$-semigroups and thereby justifies the introduction of this notion. We say that the abstract Cauchy problem associated with a linear operator $A$ is *well-posed* if for each initial value $x \in D(A)$ there exists a unique classical solution $u(\cdot) = u(\cdot, x)$ of (ACP).

**Theorem 1.1.2.** *Let $A$ be a linear operator with domain $D(A)$ on a Banach space $X$. Then the following assertions are equivalent:*

(i) *$A$ is the generator of a $C_0$-semigroup $\mathbf{T}$;*
(ii) *The abstract Cauchy problem associated with $A$ is well-posed and $\varrho(A) \neq \emptyset$;*
(iii) *The abstract Cauchy problem associated with $A$ is well-posed, $A$ is densely defined, and whenever $(x_n)$ is a sequence in $D(A)$ converging with respect to the norm of $X$ to an $x \in D(A)$, then the corresponding classical solutions $u(\cdot, x_n)$ converge to $u(\cdot, x)$, locally uniformly on $[0, \infty)$.*

*In this situation the classical solutions are given by $u(\cdot, x) = T(\cdot)x$, and for all $x \in X$ there is a unique mild solution given by the same relation $u(\cdot, x) = T(\cdot)x$.*

The condition in (iii) expresses continuous dependence on the initial value. We shall not prove this theorem but rather consider it as the starting point of our investigations. A detailed proof can be found in the book [Na].

It is easy to see that $\int_0^t T(s)x\,ds \in D(A)$ for all $t \geq 0$. In fact, a direct application of the definition of the generator shows that

$$A\left(\int_0^t T(s)x\,ds\right) = T(t)x - x, \quad \forall x \in X. \tag{1.1.1}$$

If $x \in D(A)$ we further have

$$A\left(\int_0^t T(s)x\,ds\right) = \int_0^t T(s)Ax\,ds. \tag{1.1.2}$$

The conditions (S1), (S2), and (S3) have an obvious interpretation in terms of the abstract Cauchy problem. First, one observes that the operators $T(t)$ are the 'solution operators', $T(t)x$ being the solution of (ACP) at time $t$ corresponding to the initial value $x$. Thus, (S1) expresses that nothing has happened after zero time, (S2) expresses that the solution at time $t+s$ with initial value $x$ is the same as the solution at time $t$ with initial value $T(s)x$, and (S3) expresses the continuity of the solutions as a function of $t$.

Thus, our object of study will be the asymptotic behaviour of the orbits $t \mapsto T(t)x$ of a $C_0$-semigroup $\mathbf{T}$. More precisely, we restrict ourselves to study conditions under which the orbits are stable, i.e. converge to zero for $t \to \infty$. We distinguish the following three types of stability.

**Definition 1.1.3.** Let $\mathbf{T}$ be a $C_0$-semigroup on a Banach space $X$, with generator $A$. Then $\mathbf{T}$ is said to be:

- *uniformly exponentially stable*, if there exist constants $M > 0$ and $\omega > 0$ such that $\|T(t)\| \leq Me^{-\omega t}$ for all $t \geq 0$;
- *exponentially stable*, if there exist constants $M > 0$ and $\omega > 0$ such that $\|T(t)x\| \leq Me^{-\omega t}\|x\|_{D(A)}$ for all $t \geq 0$ and $x \in D(A)$;
- *uniformly stable*, if $\lim_{t\to\infty} \|T(t)x\| = 0$ for all $x \in X$.

Here, $\|x\|_{D(A)} := \|x\| + \|Ax\|$ denotes the *graph norm* of $x$ with regard to $A$.

A $C_0$-semigroup $\mathbf{T}$ is uniformly exponentially stable if and only if for each $x \in X$ and $x^* \in X^*$ there exist constants $M = M_{x,x^*} > 0$ and $\omega = \omega_{x,x^*} > 0$ such that $|\langle x^*, T(t)x\rangle| \leq Me^{-\omega t}$, $t \geq 0$. Similarly, $\mathbf{T}$ is exponentially stable if and only if for each $x \in D(A)$ and $x^* \in X^*$ there exists constants $M = M_{x,x^*} > 0$ and $\omega = \omega_{x,x^*} > 0$ such that $|\langle x^*, T(t)x\rangle| \leq Me^{-\omega t}$, $t \geq 0$. Let us prove the first of these assertions. Denoting by $H_n$ the set

$$H_n := \{(x, x^*) \in X \times X^* : |\langle x^*, T(t)x\rangle| \leq n e^{-n^{-1}t} \text{ for all } t \geq 0\}$$

we see that $\cup_{n\in\mathbb{N}} H_n = X \times X^*$. Therefore, by Baire's theorem, at least one of the $H_n$ has non-empty interior, say $H_{n_0}$. This means that there are $x_0 \in X$ and $x_0^* \in X^*$, and $\epsilon > 0$ such that $|\langle y^*, T(t)y\rangle| \leq n_0 e^{-n_0^{-1}t}$ for all $t \geq 0$ whenever $\|x_0 - y\| \leq \epsilon$ and $\|x_0^* - y^*\| \leq \epsilon$. Hence for all $x \in X$ and $x^* \in X^*$ of norm $\leq \epsilon$,

$$|\langle x^*, T(t)x\rangle| = |\langle x_0^* - (x_0^* - x^*), T(t)(x_0 - (x_0 - x))\rangle|$$
$$\leq 4n_0 e^{-n_0^{-1}t} \leq 4\epsilon^{-2} n_0 e^{-n_0^{-1}t}\|x\|\|x^*\|.$$

The first assertion follows from this. The second is proved similarly, using the Banach space $(D(A), \|\cdot\|_{D(A)})$ instead of $X$.

Uniform exponential stability implies exponential stability and uniform stability, but none of the other implications generally holds. In Section 1.4 we shall give an example of an exponentially stable $C_0$-semigroup which is not uniformly stable (and hence not uniformly exponentially stable). The following example shows that a uniformly stable semigroup need not be exponentially stable (and hence not uniformly exponentially stable), even if the underlying space is a Hilbert space and the generator is bounded. Let $X = l^2$, the space of all complex sequences $x = (x_n)_{n \geq 1}$ such that $\|x\| := (\sum_{n=1}^{\infty} |x_n|^2)^{\frac{1}{2}} < \infty$. Define the operator $A$ on $X$ by $A(x_n) = (-n^{-1} x_n)$. Clearly, $A$ is bounded and the $C_0$-semigroup $\mathbf{T} = \{e^{tA}\}_{t \geq 0}$ generated by $A$ is given by $T(t)(x_n) = (e^{-\frac{t}{n}} x_n)$. Thus, $\|T(t)\| \leq 1$ for all $t \geq 0$. For each $n$, let $P_n$ be the projection in $X$ onto the first $n$ coordinates. Fix $x \in X$ and $\epsilon > 0$ arbitrary. Choosing $n_0$ so large that $\|(I - P_{n_0})x\| \leq \epsilon$, we have

$$\|T(t)x\| \leq \|T(t)P_{n_0}x\| + \|T(t)\| \|(I - P_{n_0})x\| \leq \|T(t)P_{n_0}x\| + \epsilon.$$

Since $T(t)x \to 0$ coordinatewise, it follows that $\limsup_{t \to \infty} \|T(t)x\| \leq \epsilon$. This proves that $\mathbf{T}$ is uniformly stable. On the other hand, $\mathbf{T}$ is not (uniformly) exponentially stable (which is the same in this case since $D(A) = X$). Indeed, for any $\omega > 0$ we can choose $n_0$ so large that $0 < n_0^{-1} < \omega$. Put $x_{n_0} := (0, 0, ..., 0, 1, 0, ...)$ with the 1 at the $n_0$-th coordinate. Then, for all $M > 0$ there exists a $t_0 > 0$ such that for all $t > t_0$, $\|T(t)x_{n_0}\| = e^{-\frac{t}{n_0}} > Me^{-\omega t}$.

We continue with some useful fact about $C_0$-semigroups that will be used throughout this book. The first of these is the *Hille-Yosida theorem*, which characterizes the generators of $C_0$-semigroups among the class of all linear operators. Before stating it, we make the simple observation that $C_0$-semigroups are always exponentially bounded. In fact, the uniform boundedness theorem and (S3) imply that the norms $\|T(\cdot)\|$ are uniformly bounded in some neighbourhood of 0, and then (S2) easily implies the existence of constants $\omega \in \mathbb{R}$ and $M > 0$ such that $\|T(t)\| \leq Me^{\omega t}$ for all $t \geq 0$. Note that we automatically have $M \geq 1$.

**Theorem 1.1.4.** *Let $A$ be a linear operator on a Banach space $X$, and let $\omega \in \mathbb{R}$ and $M \geq 1$ be constants. Then the following assertions are equivalent:*

(i) *$A$ is the generator of a $C_0$-semigroup $\mathbf{T}$ satisfying $\|T(t)\| \leq Me^{\omega t}$ for all $t \geq 0$;*

(ii) *$A$ is closed, densely defined, the half-line $(\omega, \infty)$ is contained in the resolvent set $\varrho(A)$ of $A$, and we have the estimates*

$$\|R(\lambda, A)^n\| \leq \frac{M}{(\lambda - \omega)^n}, \qquad \forall \lambda > \omega, \quad n = 1, 2, ... \qquad (1.1.3)$$

Here, $R(\lambda, A) := (\lambda - A)^{-1}$ denotes the resolvent of $A$ at $\lambda$. If one of the equivalent assertions of the theorem holds, then actually $\{\operatorname{Re} \lambda > \omega\} \subset \varrho(A)$ and

$$\|R(\lambda, A)^n\| \leq \frac{M}{(\operatorname{Re} \lambda - \omega)^n}, \qquad \forall \operatorname{Re} \lambda > \omega, \quad n = 1, 2, ... \qquad (1.1.4)$$

Moreover, for $\operatorname{Re}\lambda > \omega$ the resolvent is given explicitly by

$$R(\lambda, A)x = \int_0^\infty e^{-\lambda t} T(t) x\, dt, \qquad \forall x \in X. \tag{1.1.5}$$

We shall mostly need the implication (i)⇒(ii), which is the easy part of the theorem. In fact, one checks directly from the definitions that

$$R_\lambda x := \int_0^\infty e^{-\lambda t} T(t) x\, dt$$

defines a two-sided inverse for $\lambda - A$. The estimate (1.1.4) and the identity (1.1.5) follow trivially from this.

A useful consequence of (1.1.3) is that

$$\lim_{\lambda \to \infty} \|\lambda R(\lambda, A)x - x\| = 0, \qquad \forall x \in X. \tag{1.1.6}$$

This is proved as follows. Fix $x \in D(A)$ and $\mu \in \varrho(A)$, and let $y \in X$ be such that $x = R(\mu, A)y$. By (1.1.3) we have $\|R(\lambda, A)\| = O(\lambda^{-1})$ as $\lambda \to \infty$. Therefore, the *resolvent identity*

$$R(\lambda, A) - R(\mu, A) = (\mu - \lambda) R(\lambda, A) R(\mu, A) \tag{1.1.7}$$

implies that

$$\lim_{\lambda \to \infty} \|\lambda R(\lambda, A)x - x\| = \lim_{\lambda \to \infty} \|R(\lambda, A)(\mu R(\mu, A)y - y)\| = 0.$$

This proves (1.1.6) for elements $x \in D(A)$. Since $D(A)$ is dense in $X$ and the operators $\lambda R(\lambda, A)$ are uniformly bounded as $\lambda \to \infty$ by (1.1.3), (1.1.6) holds for all $x \in X$.

We close this section with some elementary facts concerning resolvents and an application to restrictions and quotients of semigroups.

**Proposition 1.1.5.** *Let $A$ be a closed linear operator on a Banach space $X$. Then for all $\lambda \in \varrho(A)$ we have*

$$\|R(\lambda, A)\| \geq \frac{1}{\operatorname{dist}(\lambda, \sigma(A))}.$$

*Proof:* This will be an immediate consequence of the fact that always $\|R(\lambda, A)\| \geq r(R(\lambda, A))$, once we prove that

$$\sigma(R(\lambda, A)) = \frac{1}{\lambda - \sigma(A)}.$$

But it is trivial to check that $(\lambda - \mu)(\lambda - A) R(\mu, A)$ is a two-sided inverse for $(\lambda - \mu)^{-1} - R(\lambda, A)$ whenever $\mu \in \varrho(A)$, which proves the inclusion $\subset$. Similarly, $(\lambda - \mu)^{-1} R(\lambda, A)((\lambda - \mu)^{-1} - R(\lambda, A))^{-1}$ is a two-sided inverse for $\mu - A$ whenever $(\lambda - \mu)^{-1} \in \varrho(R(\lambda, A))$, which proves the inclusion $\supset$. ////

**Proposition 1.1.6.** *Let $A$ be a closed operator on a Banach space $X$. Let $\Omega_0 \subset \varrho(A)$ be an open set and let $\Omega_1$ be an open connected set containing $\Omega_0$. If, for all $x \in X$ and $x^* \in X^*$, the map $F_{x,x^*} : \Omega_0 \to X$ defined by $F_{x,x^*}(z) = \langle x^*, R(z,A)x \rangle$ can be holomorphically extended to $\Omega_1$, then $\Omega_1 \subset \varrho(A)$ and for all $x \in X$ and $x^* \in X^*$ we have $F_{x,x^*}(z) = \langle x^*, R(z,A)x \rangle$ on $\Omega_1$.*

*Proof:* Fix $z_0 \in \Omega_0$ arbitrary. Put $\Omega_2 = \Omega_1 \cap \varrho(A)$ and let $\Omega$ be the connected component of $\Omega_2$ containing the point $z_0$. Assume, for a contradiction, that $\Omega$ is properly contained in $\Omega_1$. Then there is a point $\zeta \in \partial\Omega \cap \Omega_1$. Since $\Omega$ is open, $\zeta \notin \Omega$ and hence $z \notin \varrho(A)$. By Proposition 1.1.5 it follows that $\lim_{z_n \to \zeta} \|R(z_n, A)\| = \infty$ whenever $(z_n)$ is a sequence in $\Omega$ such that $z_n \to \zeta$. By the uniform boundedness theorem, there exist $x_0 \in X$ and $x_0^* \in X^*$ such that $\lim_{z_n \to \zeta} |\langle x_0^*, R(z_n, A)x_0 \rangle| = \infty$. But by uniqueness of analytic continuation we have $\langle x_0^*, R(z,A)x_0 \rangle = F_{x_0,x_0^*}(z)$ on $\Omega$, and it follows that $F_{x_0,x_0^*}(z)$ cannot be extended at the point $\zeta$. Since $\zeta \in \Omega_1$ we have arrived at a contradiction. ////

If $\mathbf{T}$ is a $C_0$-semigroup on a Banach space $X$ and $Y$ is a $\mathbf{T}$-invariant closed subspace of $X$, then the restrictions of the operators $T(t)$ to $Y$ define a $C_0$-semigroup $\mathbf{T}_Y$ on $Y$. Its generator $A_Y$ is precisely the part of $A$ in $Y$, i.e. $D(A_Y) = \{y \in D(A) \cap Y : Ay \in Y\}$ and $A_Y y = Ay$, $y \in D(A_Y)$. The semigroup $\mathbf{T}$ also induces a semigroup $\mathbf{T}_{X/Y}$ on the quotient space $X/Y$ by the formula $T_{X/Y}(t)(x+Y) := (T(t)x) + Y$. The strong continuity of $\mathbf{T}$ implies the strong continuity of $\mathbf{T}_{X/Y}$ and we have $\|T_{X/Y}(t)\| \leq \|T(t)\|$ for all $t \geq 0$. The generator $A_{X/Y}$ of $\mathbf{T}_{X/Y}$ is given by $D(A_{X/Y}) = D(A) + Y$ and $A_{X/Y}(x+Y) = Ax + Y$, $x \in D(A)$.

We denote by $\varrho_\infty(A)$ the connected component of $\varrho(A)$ containing the right half plane $\{\operatorname{Re}\lambda > s(A)\}$.

**Proposition 1.1.7.** *In the above situation, $\varrho_\infty(A) \subset \varrho(A_Y) \cap \varrho(A_{X/Y})$.*

*Proof:* First let $\lambda \in \varrho(A)$, $\operatorname{Re}\lambda > \omega_0(\mathbf{T})$. Then also $\operatorname{Re}\lambda > \omega_0(\mathbf{T}_Y)$ and $\operatorname{Re}\lambda > \omega_0(\mathbf{T}_{X/Y})$, and therefore $\lambda \in \varrho(A_Y) \cap \varrho(A_{X/Y})$. By the representation of the resolvent as the Laplace transform of the semigroup, it is evident that $Y$ is invariant under $R(\lambda, A)$ and that $R(\lambda, A_Y)$ is the restriction of $R(\lambda, A)$ to $Y$. Also, the quotient of $R(\lambda, A)$ modulo $Y$ is well-defined as a bounded operator on $X/Y$ and it is clear from the description of $A_{X/Y}$ that this quotient is a two-sided inverse for $\lambda - A_{X/Y}$. It follows that $R(\lambda, A_{X/Y})$ is the quotient modulo $Y$ of $R(\lambda, A)$.

Let $\mathcal{L}(X,Y)$ denote the closed subspace of $\mathcal{L}(X)$ consisting of all operators leaving $Y$ invariant. Consider the natural maps $r : \mathcal{L}(X,Y) \to \mathcal{L}(Y)$ and $\pi : \mathcal{L}(X,Y) \to \mathcal{L}(X/Y)$ defined by restriction and taking quotients: $(rT)y := Ty$, $y \in Y$, and $(\pi T)(x+Y) := Tx + Y$, $x \in X$.

By what we proved above, we have $rR(\lambda, A) = R(\lambda, A_Y)$ and $\pi R(\lambda, A) = R(\lambda, A_{X/Y})$ for all $\operatorname{Re}\lambda > \omega_0(\mathbf{T})$. But the left hand sides in these identities admit holomorphic extensions to $\varrho_\infty(A)$. Therefore we can apply Proposition 1.1.6 to conclude that $\varrho_\infty(A) \subset \varrho(A_Y) \cap \varrho(A_{X/Y})$ and that the resolvents are given by the extensions of $rR(\lambda, A)$ and $\pi R(\lambda, A)$, respectively. ////

## 1.2. The spectral bound and growth bound of a semigroup

In this section we introduce the growth bounds $\omega_0(\mathbf{T})$ and $\omega_1(\mathbf{T})$ and the spectral bounds $s(A)$ and $s_0(A)$ and discuss some of their elementary properties.

In Section 1.1 we observed that every $C_0$-semigroup $\mathbf{T}$ is exponentially bounded. Therefore, it makes sense to define the *uniform growth bound* $\omega_0(\mathbf{T})$ of $\mathbf{T}$ by

$$\omega_0(\mathbf{T}) := \inf\{\omega \in \mathbb{R} : \exists M > 0 \text{ such that } \|T(t)\| \leq Me^{\omega t}, \; \forall t \geq 0\}.$$

The identity (1.1.4) shows that the spectrum of the generator of a $C_0$-semigroup is always contained in some left half-plane. Therefore, it makes sense to define the *spectral bound* $s(A)$ of $A$ by

$$s(A) := \sup\{\operatorname{Re}\lambda : \lambda \in \sigma(A)\}.$$

Since (1.1.5) holds for all $\omega > \omega_0(\mathbf{T})$, we have:

**Proposition 1.2.1.** *If $\mathbf{T}$ is a $C_0$-semigroup on a Banach space $X$, with generator $A$, then $s(A) \leq \omega_0(\mathbf{T})$.*

It is often useful to have an expression for $\omega_0(\mathbf{T})$ directly in terms of the norms $\|T(t)\|$ or the spectral radii $r(T(t))$:

**Proposition 1.2.2.** *Let $\mathbf{T}$ be a $C_0$-semigroup on a Banach space $X$. Then for all $t_0 > 0$ we have*

$$\omega_0(\mathbf{T}) = \frac{\log r(T(t_0))}{t_0} = \lim_{t \to \infty} \frac{\log \|T(t)\|}{t}. \tag{1.2.1}$$

*Proof:* By Gelfand's theorem for the spectral radius,

$$\lim_{n \to \infty} \frac{1}{n} \log \|T(nt_0)\| = \log r(T(t_0)). \tag{1.2.2}$$

Thus, the limit $\lim_{n \to \infty}(nt_0)^{-1} \log \|T(nt_0)\|$ exists, and the first identity will be proved if we can show that the limit equals $\omega_0(\mathbf{T})$.

If $M > 0$ and $\omega \in \mathbb{R}$ are such that $\|T(t)\| \leq Me^{\omega t}$ for all $t \geq 0$, then

$$\limsup_{n \to \infty}(nt_0)^{-1} \log \|T(nt_0)\| \leq \omega.$$

By taking the infimum over all $\omega \in \mathbb{R}$ for which such $M$ can be found, it follows that $\lim_{n \to \infty}(nt_0)^{-1} \log \|T(nt_0)\| = \limsup_{n \to \infty}(nt_0)^{-1} \log \|T(nt_0)\| \leq \omega_0(\mathbf{T})$.

On the other hand, if $\omega < \omega_0(\mathbf{T})$ there is a sequence $\tau_k \to \infty$ such that $e^{-\omega \tau_k t_0}\|T(\tau_k t_0)\| \geq 1$ for all $k$. Indeed, if such a sequence does not exist, then

$\limsup_{\tau\to\infty} e^{-\omega\tau t_0}\|T(\tau t_0)\| \leq 1$ and hence $\sup_{\tau\geq 0} e^{-\omega\tau t_0}\|T(\tau t_0)\| < \infty$. But then $\omega_0(\mathbf{T}) \leq \omega$, a contradiction.

Let $M = M_{t_0} := \sup_{0\leq s \leq 1} \|T(st_0)\|$ and put $n_k := [\tau_k]$, the integer part of $\tau_k$. The semigroup property (S2) implies that $\|T(n_k t_0)\| \geq M^{-1}\|T(\tau_k t_0)\|$ for all $k$. Therefore,

$$e^{-\omega n_k t_0}\|T(n_k t_0)\| \geq M^{-1} e^{-|\omega|t_0} e^{-\omega \tau_k t_0}\|T(\tau_k t_0)\| \geq M^{-1} e^{-|\omega|t_0}.$$

It follows that

$$\frac{\log \|T(n_k t_0)\|}{n_k t_0} \geq \omega + \frac{\log(M^{-1} e^{-|\omega|t_0})}{n_k t_0}.$$

Since $\omega < \omega_0(\mathbf{T})$ was arbitrary, this proves that $\lim_{n\to\infty}(nt_0)^{-1} \log\|T(nt_0)\| = \limsup_{n\to\infty}(nt_0)^{-1}\log\|T(nt_0)\| \geq \omega_0(\mathbf{T})$. This concludes the proof of the first identity in (1.2.1).

Next, we fix $t > 1$ and choose the integer $n \geq 1$ such that $n < t \leq n+1$. Noting that

$$\frac{\log(M^{-1}\|T(n+1)\|)}{n+1} \leq \frac{\log(\|T(t)\|)}{t} \leq \frac{\log(M\|T(n)\|)}{n},$$

where $M = M_1$ is as above, we see that the second identity in (1.2.1) is a consequence of (1.2.2).    ////

If $A$ is bounded, then $A$ generates the $C_0$–semigroup $\mathbf{T} = \{e^{tA}\}_{t\geq 0}$ and the spectral mapping theorem of the Dunford calculus implies that $\sigma(e^{tA}) = e^{t\sigma(A)}$. Therefore by Proposition 1.2.2,

$$e^{t\omega_0(\mathbf{T})} = r(e^{tA}) = e^{ts(A)}, \qquad \forall t \geq 0.$$

This implies that $s(A) = \omega_0(\mathbf{T})$. If $A$ is unbounded, then strict inequality $s(A) < \omega_0(\mathbf{T})$ may occur, even if the underlying Banach space is a Hilbert space; cf. Example 1.2.4 below. The important consequence of this is that *the growth of the mild solutions $T(\cdot)x$ of the abstract Cauchy problem is not controlled by the location of the spectrum of $A$.*

This pathology invites us to look for more subtle quantities to describe growth of solutions and spectrum of the generator. The first of these is motivated by the following observation. Instead of considering all mild solutions, one can consider the classical solutions only, i.e. those originating from an initial value in the domain $D(A)$. With this in mind we define the *growth bound* $\omega_1(\mathbf{T})$ as the infimum of all $\omega \in \mathbb{R}$ for which there exists a constant $M > 0$ such that

$$\|T(t)x\| \leq M e^{\omega t} \|x\|_{D(A)}, \qquad \forall x \in D(A), \, t \geq 0.$$

Thus, $\mathbf{T}$ is exponentially stable if and only if $\omega_1(\mathbf{T}) < 0$. It is obvious from the definition that $\omega_1(\mathbf{T}) \leq \omega_0(\mathbf{T})$. The following result gives more precise information.

**Theorem 1.2.3.** Let **T** be a $C_0$-semigroup on a Banach space $X$, with generator $A$.

(i) $s(A) \leq \omega_1(\mathbf{T}) \leq \omega_0(\mathbf{T})$;

(ii) $\omega_1(\mathbf{T})$ is the infimum of all $\omega \in \mathbb{R}$ with the following property: $\{\operatorname{Re} \lambda > \omega\} \subset \varrho(A)$ and
$$R(\lambda, A)x = \lim_{\tau \to \infty} \int_0^\tau e^{-\lambda t} T(t) x \, dt, \quad \forall \operatorname{Re} \lambda > \omega \text{ and } x \in X;$$

(iii) $\omega_1(\mathbf{T})$ is the infimum of all $\omega \in \mathbb{R}$ with the following property: there exists a $\lambda \in \mathbb{C}$ with $\operatorname{Re} \lambda = \omega$ such that
$$\sup_{\tau > 0} \left\| \int_0^\tau e^{-\lambda t} T(t) x \, dt \right\| < \infty, \quad \forall x \in X;$$

(iv) $\omega_1(\mathbf{T})$ is the infimum of all $\omega \in \mathbb{R}$ such that
$$\int_0^\infty e^{-\lambda t} \|T(t)x\| \, dt < \infty, \quad \forall \operatorname{Re} \lambda > \omega \text{ and } x \in D(A).$$

*Proof:* The second inequality in (i) is trivial and the first is an immediate consequence of (ii), which we prove now.

We start with the following observation: if, for some $\lambda \in \mathbb{C}$,
$$B_\lambda x := \lim_{\tau \to \infty} \int_0^\tau e^{-\lambda t} T(t) x \, dt$$

exists for all $x \in X$, then $\lambda \in \varrho(A)$ and $B_\lambda x = R(\lambda, A)x$ for all $x \in X$. This is proved as follows. For $x \in X$ and $t > 0$ we have
$$\frac{1}{t}(T(t) - I)B_\lambda x = \frac{1}{t} \int_0^\infty e^{-\lambda s}(T(t+s) - T(s))x \, ds$$
$$= \frac{1}{t}\left((e^{\lambda t} - 1)\int_0^\infty e^{-\lambda s} T(s)x \, ds - e^{\lambda t}\int_0^t e^{-\lambda s} T(s)x \, ds\right).$$

Taking the limit $t \downarrow 0$ we obtain $B_\lambda x \in D(A)$ and $AB_\lambda x = \lambda B_\lambda x - x$, i.e. $(\lambda - A)B_\lambda x = x$. Further, for $x \in D(A)$ we have
$$B_\lambda(\lambda - A)x = \lim_{\tau \to \infty} \int_0^\tau e^{-\lambda t} T(t)(\lambda - A)x \, dt$$
$$= \lim_{\tau \to \infty} (\lambda - A) \int_0^\tau e^{-\lambda t} T(t)x \, dt.$$

Hence, putting
$$F_{\tau, x} := \int_0^\tau e^{-\lambda t} T(t)x \, dt,$$

we have $\lim_{\tau\to\infty} F_{\tau,x} = B_\lambda x$ and $\lim_{\tau\to\infty}(\lambda - A)F_{\tau,x} = B_\lambda(\lambda - A)x$. From the closedness of $A$ it follows that $B_\lambda x \in D(A)$ and $(\lambda - A)B_\lambda x = B_\lambda(\lambda - A)x$.

We have proved that $B_\lambda$ defines a two-sided inverse of $\lambda - A$. Therefore, $B_\lambda$ is a closed operator and hence bounded by the closed graph theorem. It follows that $\lambda \in \varrho(A)$ and $B_\lambda = R(\lambda, A)$.

Now we can start the proof of (ii). Let $\omega_1$ denote the infimum as meant in (ii). Fix $\omega > \omega_1(\mathbf{T})$ and $\operatorname{Re}\lambda > \omega$ arbitrary. We shall prove that

$$B_\lambda x := \lim_{\tau\to\infty} \int_0^\tau e^{-\lambda t} T(t)x \, dt \qquad (1.2.3)$$

exists for all $x \in X$. Then, by what we just proved, we can conclude that $\omega_1 \leq \omega$ and hence $\omega_1 \leq \omega_1(\mathbf{T})$.

To prove (1.2.3), choose $M > 0$ such that $\|T(t)y\| \leq Me^{\omega t}\|y\|_{D(A)}$ holds for all $y \in D(A)$ and $t \geq 0$. We distinguish three cases. *Case 1*: If $x \in D(A)$, then for all $0 \leq \tau_0 \leq \tau_1$ we have

$$\left\|\int_{\tau_0}^{\tau_1} e^{-\lambda t} T(t)x \, dt\right\| \leq \int_{\tau_0}^{\tau_1} e^{-\operatorname{Re}\lambda t} \|T(t)x\| \, dt$$

$$\leq M \int_{\tau_0}^{\tau_1} e^{(\omega-\operatorname{Re}\lambda)t} \|x\|_{D(A)} \, dt$$

$$= \frac{M}{\operatorname{Re}\lambda - \omega} \left(e^{(\omega-\operatorname{Re}\lambda)\tau_0} - e^{(\omega-\operatorname{Re}\lambda)\tau_1}\right) \|x\|_{D(A)}.$$

As $\tau_i \to \infty$, $i = 0, 1$, the right hand converges to 0, and it follows that the limit in (1.2.3) exists. *Case 2*: If $x = (\lambda - A)y$ for some $y \in D(A)$, then for all $\tau > 0$ we have

$$\int_0^\tau e^{-\lambda t} T(t)x \, dt = (\lambda - A)\int_0^\tau e^{-\lambda t} T(t)y \, dt = y - e^{-\lambda\tau} T(\tau)y$$

and therefore

$$\lim_{\tau\to\infty} \int_0^\tau e^{-\lambda t} T(t)x \, dt = \lim_{\tau\to\infty} y - e^{-\lambda\tau} T(\tau)y = y. \qquad (1.2.4)$$

Here we used that $\operatorname{Re}\lambda > \omega_1(\mathbf{T})$ and $y \in D(A)$. *Case 3*: For arbitrary $x \in X$ by the resolvent identity we have $x = (\mu - \lambda)R(\mu, A)x + (\lambda - A)R(\mu, A)x$, where $\mu \in \varrho(A)$ is arbitrary but fixed. Therefore, by the two cases just considered, the limit (1.2.3) exists for all $x \in X$.

To complete the proof of (ii) we need to show that $\omega_1(\mathbf{T}) \leq \omega_1$. If $\omega \in \mathbb{R}$ is such that $\{\operatorname{Re}\lambda > \omega\} \subset \varrho(A)$ and $R(\lambda, A)x = \lim_{\tau\to\infty} \int_0^\tau e^{-\lambda t} T(t)x \, dt$ for all $x \in X$ and $\operatorname{Re}\lambda > \omega$, the identity (1.2.4) shows that $\lim_{\tau\to\infty} e^{-\lambda\tau} T(\tau)R(\lambda, A)x = 0$ for all $x \in X$ and $\operatorname{Re}\lambda > \omega$. This implies that $\omega_1(\mathbf{T}) \leq \omega$.

Next we prove (iii). Denote the infimum as meant in (iii) by $\omega_1$. For all $\omega > \omega_1(\mathbf{T})$ and $x \in X$ we have $\sup_{\tau>0}\left\|\int_0^\tau e^{-\omega t} T(t)x \, dt\right\| < \infty$ by (ii). Hence, $\omega_1 \leq \omega_1(\mathbf{T})$.

Conversely, let $\omega > \omega_1$, let $\mu \in \mathbb{C}$ with $\mathrm{Re}\,\mu = \omega$ be such that

$$\sup_{\tau > 0} \left\| \int_0^\tau e^{-\mu t} T(t) x \, dt \right\| < \infty, \quad \forall x \in X,$$

and let $\mathrm{Re}\,\lambda > \omega$ be arbitrary. Then, with $F_{\tau,x} := \int_0^\tau e^{-\mu t} T(t) x \, dt$, by partial integration we have

$$\int_0^\tau e^{-\lambda t} T(t) x \, dt = e^{-(\lambda-\mu)\tau} F_{\tau,x} + (\lambda - \mu) \int_0^\tau e^{-(\lambda-\mu)t} F_{t,x} \, dt.$$

By letting $\tau \to \infty$, it follows that

$$\lim_{\tau \to \infty} \int_0^\tau e^{-\lambda t} T(t) x \, dt = (\lambda - \mu) \int_0^\infty e^{-(\lambda-\mu)t} F_{t,x} \, dt,$$

so in particular the limit on the left hand side exists. By the observation preceding the proof of (ii), this implies that $\lambda \in \varrho(A)$ and $\lim_{\tau \to \infty} \int_0^\tau e^{-\lambda t} T(t) x \, dt = R(\lambda, A) x$ for all $x \in X$. Therefore, by (ii) $\omega_1(\mathbf{T}) \leq \omega$ and hence $\omega_1(\mathbf{T}) \leq \omega_1$.

Finally we prove (iv). For this we use the following fact. Let $F : \mathbb{R}_+ \to \mathbb{R}_+$ be an integrable function, and assume there is an integer $m$ and an interval $[0, n]$ such that $F(t + s) \leq m F(s)$ for all $s \geq 0$ and $t \in [0, n]$. Then, $\lim_{t \to \infty} F(t) = 0$. Indeed, for all $\epsilon > 0$ there exists an $a > 0$ such that $I_a := \int_a^\infty F(s) \, ds \leq m^{-1} n \epsilon$. For all $t > a + n$ there exists an $r \in [t - n, t]$ such that $F(r) \leq n^{-1} I_a$. Therefore, $F(t) = F(t - r + r) \leq m F(r) \leq m n^{-1} I_a \leq \epsilon$.

Let $\omega_1$ be the infimum in (iv). From the definition of $\omega_1(\mathbf{T})$ it is obvious that $\omega_1 \leq \omega_1(\mathbf{T})$. For the converse inequality, fix $x \in D(A)$ and let $\mathrm{Re}\,\lambda > \omega_1$. Then by the above fact applied to $F(t) = e^{-\lambda t} \|T(t) x\|$, it follows that $\|T(t) x\| \leq M_x e^{\mathrm{Re}\,\lambda t}$ for some constant $M_x > 0$ and all $t \geq 0$. By the uniform boundedness theorem, there is a constant $M > 0$ such that $\|T(t) x\| \leq M e^{\mathrm{Re}\,\lambda t} \|x\|_{D(A)}$ for all $t \geq 0$. It follows that $\omega_1(\mathbf{T}) \leq \mathrm{Re}\,\lambda$ and hence $\omega_1(\mathbf{T}) \leq \omega_1$. ////

It need not be true that $s(A) = \omega_1(\mathbf{T})$, even for $C_0$–groups on a Hilbert space. This is shown by the following example.

**Example 1.2.4.** For $n = 1, 2, 3, \ldots$, let $A_n$ be the $n \times n$ matrix acting on $\mathbb{C}^n$ defined by

$$A_n := \begin{pmatrix} 0 & 1 & 0 & 0 & \cdots \\ 0 & 0 & 1 & 0 & \cdots \\ \vdots & \vdots & \ddots & \ddots & \ddots \end{pmatrix}$$

Each matrix $A_n$ is nilpotent and therefore $\sigma(A_n) = \{0\}$. Let $X$ be the Hilbert space consisting of all sequences $x = (x_n)_{n \geq 1}$ with $x_n \in \mathbb{C}^n$ such that

$$\|x\| := \left( \sum_{n=1}^\infty \|x_n\|_{\mathbb{C}^n}^2 \right)^{\frac{1}{2}} < \infty.$$

Let **T** be the semigroup on $X$ defined coordinatewise by

$$T(t) = (e^{int} e^{tA_n})_{n \geq 1}.$$

It is easily checked that **T** is a $C_0$–semigroup on $X$ and that **T** extends to a $C_0$–group. Since $\|A_n\| = 1$ for $n \geq 2$, we have $\|e^{tA_n}\| \leq e^t$ and hence $\|T(t)\| \leq e^t$, so $\omega_0(\mathbf{T}) \leq 1$.

First, we show that $s(A) = 0$, where $A$ is the generator of **T**. To see this, we note that $A$ is defined coordinatewise by

$$A = (in + A_n)_{n \geq 1}.$$

An easy calculation shows that for all $\operatorname{Re} \lambda > 0$,

$$\lim_{n \to \infty} \|R(\lambda, A_n + in)\|_{\mathbb{C}^n} = 0.$$

It follows that the operator $(R(\lambda, A_n + in))_{n \geq 1}$ defines a bounded operator on $X$, and clearly this operator is a two-sided inverse of $\lambda - A$. Therefore $\{\operatorname{Re} \lambda > 0\} \subset \varrho(A)$ and $s(A) \leq 0$. On the other hand, $in \in \sigma(in + A_n) \subset \sigma(A)$ for all $n \geq 1$, so $s(A) = 0$.

Next, we show that $\omega_1(\mathbf{T}) = 1$. In view of $\omega_0(\mathbf{T}) \leq 1$ it suffices to show that $\omega_1(\mathbf{T}) \geq 1$. For each $n$ we put

$$x_n := n^{-\frac{1}{2}}(1, 1, ..., 1) \in \mathbb{C}^n.$$

Then, $\|x_n\|_{\mathbb{C}^n} = 1$ and

$$\|e^{tA_n} x_n\|_{\mathbb{C}^n}^2 = \frac{1}{n} \sum_{m=0}^{n-1} \left( \sum_{j=0}^{m} \frac{t^j}{j!} \right)^2$$

$$= \frac{1}{n} \sum_{m=0}^{n-1} \left( \sum_{j,k=0}^{m} \frac{t^{j+k}}{j!k!} \right)$$

$$= \frac{1}{n} \sum_{m=0}^{n-1} \sum_{i=0}^{2m} t^i \sum_{j+k=i} \frac{1}{j!k!}$$

$$= \frac{1}{n} \sum_{m=0}^{n-1} \sum_{i=0}^{2m} \frac{t^i}{i!} \sum_{j=0}^{i} \frac{i!}{j!(i-j)!}$$

$$= \frac{1}{n} \sum_{m=0}^{n-1} \sum_{i=0}^{2m} \frac{2^i t^i}{i!}$$

$$\geq \frac{1}{n} \sum_{i=0}^{2n-2} \frac{2^i t^i}{i!}.$$

For $0 < q < 1$, we define $x_q \in X$ by $x_q := (n^{\frac{1}{2}} q^n x_n)_{n \geq 1}$. It is easy to check that $x_q \in D(A)$ and

$$\|T(t)x_q\|^2 = \sum_{n=1}^{\infty} nq^{2n} \|e^{tA_n} x_n\|^2$$

$$\geq \sum_{n=1}^{\infty} nq^{2n} \left( \frac{1}{n} \sum_{i=0}^{2n-2} \frac{2^i t^i}{i!} \right)$$

$$= \sum_{i=0}^{\infty} \frac{2^i t^i}{i!} \sum_{n=\{i/2\}+1}^{\infty} q^{2n}$$

$$= \sum_{i=0}^{\infty} \frac{q^{2\{i/2\}+2}}{1-q^2} \frac{2^i t^i}{i!}$$

$$\geq \frac{q^3}{1-q^2} e^{2tq}.$$

Here $\{a\}$ denotes the least integer greater than or equal to $a$; we used that $2\{i/2\} + 2 \leq i + 3$ for all $i = 0, 1, \ldots$ Thus, $\omega_1(\mathbf{T}) \geq q$ for all $0 < q < 1$, so $\omega_1(\mathbf{T}) \geq 1$. ////

Another way of looking at the failure of the identity $s(A) = \omega_0(\mathbf{T})$ is that the location of the spectrum of $A$ alone does not contain enough information to deduce the asymptotic behaviour of $\mathbf{T}$ from it. As will become apparent later, not the numerical value of $s(A)$, but rather the growth of the resolvent along vertical lines $\{\operatorname{Re} \lambda = \omega\}$ with $\omega > s(A)$ plays an important role.

By (1.1.4), the supremum of $\|R(\lambda, A)\|$ along each vertical line $\operatorname{Re} \lambda = \omega$ with $\omega > \omega_0(\mathbf{T})$ is finite, uniformly for $\omega \geq \omega_0 > \omega_0(\mathbf{T})$. The resolvent need not be bounded on vertical lines between $s(A)$ and $\omega_0(\mathbf{T})$, however. This motivates us to define the *abscissa of uniform boundedness of the resolvent*, notation $s_0(A)$, as the infimum of all $\omega \in \mathbb{R}$ such that $\{\operatorname{Re} \lambda > \omega\} \subset \varrho(A)$ and $\sup_{\operatorname{Re} \lambda > \omega} \|R(\lambda, A)\| < \infty$. Thus,

$$s(A) \leq s_0(A) \leq \omega_0(\mathbf{T}).$$

In Section 2.2 we shall show that $s_0(A) = \omega_0(\mathbf{T})$ holds for $C_0$–semigroups on Hilbert spaces and in Sections 4.2 and 4.5 we show that $\omega_1(\mathbf{T}) \leq s_0(A)$ holds for arbitrary $C_0$–semigroups.

In Section 1.4 we shall give an example of a $C_0$–semigroup for which $s(A) = s_0(A) = \omega_1(\mathbf{T}) < \omega_0(\mathbf{T})$. In view of the identity $s_0(A) = \omega_0(\mathbf{T})$ in Hilbert spaces, Example 1.2.4 shows that also strict inequality $s(A) < s_0(A)$ can occur. Thus, neither $s_0(A) = \omega_0(\mathbf{T})$ nor $s(A) = s_0(A)$ generally holds. Similarly, neither $s(A) = \omega_1(\mathbf{T})$ nor $\omega_1(\mathbf{T}) = \omega_0(\mathbf{T})$ generally holds and strict inequality $\omega_1(\mathbf{T}) < s_0(A)$ can occur; cf. the notes at the end of the chapter. By a direct sum construction we see that all four quantities may be different.

We conclude this section with two useful rescaling techniques that will be used throughout this book.

If $A$ is the generator of a $C_0$–semigroup $\mathbf{T}$, then for all $\lambda \in \mathbb{C}$ the operator $A_\lambda := A - \lambda$ is the generator of the $C_0$–semigroup $\mathbf{T}_\lambda := \{e^{-\lambda t} T(t)\}_{t \geq 0}$. It is obvious that $\omega_0(\mathbf{T}_\lambda) = \omega_0(\mathbf{T}) - \operatorname{Re} \lambda$, $\omega_1(\mathbf{T}_\lambda) = \omega_1(\mathbf{T}) - \operatorname{Re} \lambda$, $s(A_\lambda) = s(A) - \operatorname{Re} \lambda$, and $s_0(A_\lambda) = s_0(A) - \operatorname{Re} \lambda$. Similarly, for all $\alpha > 0$ the operator $\alpha A$ is the generator of the $C_0$–semigroup $\mathbf{T}^{(\alpha)} := \{T(\alpha t)\}_{t \geq 0}$, and we have $\omega_0(\mathbf{T}^{(\alpha)}) = \alpha \omega_0(\mathbf{T})$, $\omega_1(\mathbf{T}^{(\alpha)}) = \alpha \omega_1(\mathbf{T})$, $s(\alpha A) = \alpha s(A)$, and $s_0(\alpha A) = \alpha s_0(A)$.

## 1.3. The Laplace transform and its complex inversion

The identity (1.1.5) identifies the resolvent of the generator $A$ of a $C_0$–semigroup $\mathbf{T}$ as the Laplace transform of $\mathbf{T}$ in the half-plane $\{\operatorname{Re} \lambda > \omega_0(\mathbf{T})\}$. Theorem 1.2.3 shows that the same is true in the half-plane $\{\operatorname{Re} \lambda > \omega_1(\mathbf{T})\}$. This motivates us to take a closer look at the Laplace transform of Banach space-valued functions.

Let $X$ be a Banach space. The *Laplace transform* $\mathcal{L}f$ of a function $f \in L^1_{loc}(\mathbb{R}_+, X)$ is defined by

$$\mathcal{L}f(\lambda) := \int_0^\infty e^{-\lambda t} f(t)\, dt$$

whenever this integral exists as a convergent improper integral. It is easy to prove that the integral converges for all $\operatorname{Re} \lambda > \omega$ once it converges for some $\lambda \in \mathbb{C}$ with $\operatorname{Re} \lambda = \omega$. In fact, we claim that $\mathcal{L}f(\lambda)$ converges for all $\operatorname{Re} \lambda > \omega$ if there exists a $\mu \in \mathbb{C}$, $\operatorname{Re} \mu = \omega$, such that

$$\sup_{\tau > 0} \left\| \int_0^\tau e^{-\mu t} f(t)\, dt \right\| < \infty.$$

To see this, define $F(\tau) := \int_0^\tau e^{-\mu t} f(t)\, dt$. By partial integration,

$$\int_0^\tau e^{-\lambda t} f(t)\, dt = e^{-(\lambda - \mu)\tau} F(\tau) + (\lambda - \mu) \int_0^\tau e^{-(\lambda - \mu)t} F(t)\, dt.$$

By letting $\tau \to \infty$ it follows that

$$\lim_{\tau \to \infty} \int_0^\tau e^{-\lambda t} f(t)\, dt = (\lambda - \mu) \int_0^\infty e^{-(\lambda - \mu)t} F(t)\, dt.$$

Since by assumption $F$ is bounded, the integral on the right hand side converges absolutely and the claim is proved.

Thus, the domain of convergence is either empty or contains a right half plane, and it makes sense to define the *abscissa of improper convergence* $\omega_1(f)$ as the infimum of all $\omega \in \mathbb{R}$ such that $\mathcal{L}f(\lambda)$ converges in the improper sense for all $\operatorname{Re} \lambda > \omega$. If such an $\omega$ does not exist we put $\omega_1(f) := \infty$. With this terminology, Theorem 1.2.3 can be reformulated as saying that $\omega_1(\mathbf{T})$ coincides with the abscissa of improper convergence of the Laplace transform of $\mathbf{T}$, in the sense that it is the infimum of the abscissae $\omega_1(f_x)$ of the functions $f_x(t) = T(t)x$, $x \in X$.

The Laplace transform $\mathcal{L}f$ is easily seen to be an $X$-valued holomorphic function in $\{\operatorname{Re} \lambda > \omega_1(f)\}$. It can happen that this function can be holomorphically extended to some larger open subset of $\mathbb{C}$. Whenever this is the case, this extension will be denoted by $\mathcal{L}f$ as well.

Similar to the definition of $\omega_1(f)$, one defines the *abscissa of absolute convergence* $\omega_0(f)$ of the Laplace transform of a function $f \in L^1_{loc}(\mathbb{R}_+, X)$ as the infimum of all $\omega \in \mathbb{R}$ such that $\mathcal{L}f(\lambda)$ converges absolutely for all $\operatorname{Re} \lambda > \omega$. We will see in Section 2.1 that the abscissa of absolute convergence of the Laplace transform of a $C_0$–semigroup $\mathbf{T}$ is precisely $\omega_0(\mathbf{T})$.

Often we will find ourselves in situations where we would like to derive information about $\mathbf{T}$ from information about the resolvent. For this reason, it is desirable to have inversion techniques for the Laplace transform at our disposal. We shall prove a simple complex inversion formula which is based on the inversion theorem of the Fourier transform. The *Fourier transform* of a function $f \in L^1(\mathbb{R}, X)$ is defined by

$$\hat{f}(s) := \int_{-\infty}^{\infty} e^{-is\tau} f(\tau)\, dt, \quad s \in \mathbb{R}.$$

By the Riemann-Lebesgue lemma, $\hat{f} \in C_0(\mathbb{R}, X)$. The bounded linear operator $f \mapsto \hat{f}$ from $L^1(\mathbb{R}, X)$ into $C_0(\mathbb{R}, X)$ is denoted by $\mathcal{F}$ and the notations $\hat{f}$ and $\mathcal{F}f$ will be used interchangably.

The following result is known as *Fejér's theorem*: Let $X$ be a Banach space, let $f \in L^1(\mathbb{R}, X)$ and assume that $f$ is locally of bounded variation. Then for all $\tau \in \mathbb{R}$ we have

$$\frac{1}{2\pi}(C,1) \int_{-\infty}^{\infty} e^{is\tau} \hat{f}(s)\, ds = \frac{1}{2}(f(\tau+) + f(\tau-)).$$

Here, $(C, 1)$ denotes convergence of the integral in the Cesàro mean, i.e.

$$(C,1)\int_{-\infty}^{\infty} g(s)\, ds := \lim_{t \to \infty} \frac{1}{t} \int_0^t \int_{-\tau}^{\tau} g(s)\, ds\, d\tau.$$

It is useful to have sufficient conditions that enable us to replace Cesàro integrability by integrability in the principle value sense. We have the following result in this direction.

**Lemma 1.3.1.** *Let $X$ be a Banach space and let $f \in L^1(\mathbb{R}, X)$ and $\tau \in \mathbb{R}$ be arbitrary. If $\hat{f}(\xi) = O(|\xi|^{-1})$ for $\xi \to \pm\infty$, then*

$$(C,1)\int_{-\infty}^{\infty} e^{i\xi\tau} \hat{f}(\xi)\, d\xi$$

exists if and only if
$$PV \int_{-\infty}^{\infty} e^{i\xi\tau} \hat{f}(\xi) \, d\xi$$
exists, in which case the two integrals are equal.

*Proof:* Let $\tau \in \mathbb{R}$ be fixed and define, for all $t > 0$,
$$S_t = (S_t(f))(\tau) := \int_{-t}^{t} e^{i\xi\tau} \hat{f}(\xi) \, d\xi$$
and
$$\sigma_t = (\sigma_t(f))(\tau) := \frac{1}{t} \int_0^t (S_s(f))(\tau) \, ds.$$

Noting that $\hat{f}$ is continuous, we can choose a constant $C > 0$ such that $|\hat{f}(\xi)| \le C|\xi|^{-1}$ for all $\xi \in \mathbb{R}$. Fix $\epsilon > 0$ arbitrary and choose $\lambda > 1$ such that
$$\frac{1}{\lambda - 1} \int_1^\lambda \log s \, ds \le \frac{\epsilon}{2C}.$$

By writing out the definitions of $S_t$ and $\sigma_t$ one finds the relation
$$\lambda \sigma_{\lambda t} - \sigma_t = (\lambda - 1) S_t + \frac{1}{t} \int_t^{\lambda t} \int_{(-s,-t) \cup (t,s)} e^{i\xi\tau} \hat{f}(\xi) \, d\xi \, ds.$$

Therefore,
$$\left\| \frac{\lambda \sigma_{\lambda t} - \sigma_t}{\lambda - 1} - S_t \right\| \le \frac{1}{\lambda t - t} \left| \int_t^{\lambda t} \int_{(-s,-t) \cup (t,s)} e^{i\xi\tau} \hat{f}(\xi) \, d\xi \, ds \right|$$
$$\le \frac{C}{\lambda t - t} \int_t^{\lambda t} \int_{(-s,-t) \cup (t,s)} \frac{1}{|\xi|} \, d\xi \, ds$$
$$= \frac{2C}{\lambda t - t} \int_t^{\lambda t} \log \frac{s}{t} \, ds = \frac{2C}{\lambda - 1} \int_1^\lambda \log \sigma \, d\sigma \le \epsilon.$$

This implies that $\lim_{t \to \infty} S_t$ exists whenever $\lim_{t \to \infty} \sigma_t$ exists, and that the two limits are equal. Conversely, it is trivial from the definitions that $\lim_{t \to \infty} \sigma_t$ exists whenever $\lim_{t \to \infty} S_t$ exists, and that the two limits are equal.  ////

In the following lemma, we use the obvious fact that for $\omega > \omega_0(\mathbf{T})$ and $s \in \mathbb{R}$, the resolvent $R(\omega + is, A)x$ agrees with the Fourier transform at the point $s$ of the function $t \mapsto F(t) := e^{-\omega t} T(t) x \cdot \chi_{\mathbb{R}_+}(t) \in L^1(\mathbb{R}, X)$. This follows immediately from (1.1.5), for
$$R(\omega + is, A)x = \int_0^\infty e^{-(\omega + is)t} T(t) x \, dt = \int_{-\infty}^\infty e^{-ist} F(t) \, dt = \hat{F}(s).$$

**Lemma 1.3.2.** *Let $\mathbf{T}$ be a $C_0$–semigroup on a Banach space $X$, with generator $A$, and assume that there exist $\omega_1 \in \mathbb{R}$ and $r \in \mathbb{R}$ such that the resolvent exists and is uniformly bounded in the set $\{\lambda \in \mathbb{C} : \operatorname{Re}\lambda \geq \omega_1, \operatorname{Im}\lambda \geq r\}$. Let $\omega_0 \geq \omega_1$ be arbitrary. Then for all $x \in X$ we have*
$$\lim_{s \geq r, s \to \infty} \|R(\omega + is, A)x\| = 0$$
*uniformly for $\omega \in [\omega_1, \omega_0]$.*

*Proof:* Upon replacing $\omega_0$ by some larger number, we may assume that $\omega_0 > \omega_0(\mathbf{T})$. By the Riemann-Lebesgue lemma and the observation preceding the lemma,
$$\lim_{s \to \infty} \|R(\omega_0 + is, A)x\| = 0, \quad \forall x \in X. \tag{1.3.1}$$
Let $M := \sup_{\operatorname{Re}\lambda \geq \omega_1, \operatorname{Im}\lambda \geq r} \|R(\lambda, A)\|$. By the resolvent identity (1.1.7), for all $\omega \in [\omega_1, \omega_0]$ and $s \geq r$ we have
$$R(\omega + is, A) = (I + (\omega_0 - \omega)R(\omega + is, A))R(\omega_0 + is, A)$$
and hence, for all $x \in X$,
$$\|R(\omega + is, A)x\| \leq (1 + M(\omega_0 - \omega_1))\|R(\omega_0 + is, A)x\|.$$
From this and (1.3.1), the lemma follows. ////

Of course, an analogous result holds for $s \to -\infty$ if the resolvent is uniformly bounded in $\{\lambda \in \mathbb{C} : \operatorname{Re}\lambda \geq \omega_1, \operatorname{Im}\lambda \leq -r\}$. In particular, for all $\omega_0 \geq \omega_1 > s_0(A)$ and all $x \in X$ we have
$$\lim_{s \to \pm\infty} \|R(\omega + is, A)x\| = 0$$
uniformly for $\omega \in [\omega_1, \omega_0]$. This fact will be used in the following theorem.

**Theorem 1.3.3.** *Let $\mathbf{T}$ be a $C_0$–semigroup on a Banach space $X$, with generator $A$. Then, for all $\omega > s_0(A)$ and $t > 0$ we have*
$$T(t)x = \frac{1}{2\pi i}(C,1)\int_{\operatorname{Re}\lambda = \omega} e^{\lambda t} R(\lambda, A)x \, d\lambda, \quad \forall x \in X, \tag{1.3.2}$$
*and*
$$T(t)x = \frac{1}{2\pi i} PV \int_{\operatorname{Re}\lambda = \omega} e^{\lambda t} R(\lambda, A)x \, d\lambda, \quad \forall x \in D(A). \tag{1.3.3}$$

*Proof:* For $\omega > \omega_0(\mathbf{T})$, (1.3.2) follows from Fejér's theorem applied to the function $t \mapsto e^{-\omega t}T(t)x \cdot \chi_{\mathbb{R}_+}(t)$. For general $\omega > s_0(A)$, we note that by Lemma 1.3.2 we can apply Cauchy's theorem to shift the path of integration to a vertical line to the right of $\omega_0(\mathbf{T})$.

For $x \in D(A)$ we have the identity
$$R(\lambda, A)x = \lambda^{-1}(x + R(\lambda, A)Ax).$$
Since the resolvent is uniformly bounded on the line $\operatorname{Re}\lambda = \omega$, this shows that $\|R(\lambda, A)x\| = O(|\lambda|^{-1})$ there. Hence, for $\omega > \omega_0(\mathbf{T})$ we can apply Lemma 1.3.1 to the function $t \mapsto e^{-\omega t}T(t)x \cdot \chi_{\mathbb{R}_+}(t)$. By doing so, we see that (1.3.3) follows from (1.3.2). For general $\omega > s_0(A)$, we apply Cauchy's theorem as in the first part of the proof. ////

We close this section with a result about the Laplace transform of *positive* Banach lattice-valued functions which will be useful in the next section. It is usually referred to as the *Pringsheim-Landau theorem*.

**Theorem 1.3.4.** Let $X$ be a Banach lattice and let $0 \leq f \in L^1_{loc}(\mathbb{R}_+, X)$. If $-\infty < \omega_1(f) < \infty$, then $\mathcal{L}f$ cannot be holomorphically extended to a neighbourhood of $\omega_1(f)$.

*Proof:* Upon replacing $f(t)$ by $e^{-\omega_1(f)t}f(t)$, we may assume that $\omega_1(f) = 0$. Suppose, for a contradiction, that $\mathcal{L}f$ can be holomorphically extended to a neighbourhood of $0$. Then there is an $\epsilon > 0$ such that the Taylor series of $f$ at the point $\lambda = 1$,

$$\mathcal{L}f(\lambda) = \sum_{k=0}^{\infty} \frac{(\mathcal{L}f)^{(k)}(1)}{k!}(\lambda - 1)^k,$$

has radius of convergence $1 + 2\epsilon$. In particular, the series converges absolutely in the point $\lambda = -\epsilon$. Hence, for all $x^* \in X^*$ we obtain

$$\begin{aligned}
\langle x^*, \mathcal{L}f(-\epsilon)\rangle &= \sum_{k=0}^{\infty} \frac{(-\epsilon - 1)^k}{k!} \int_0^{\infty} (-t)^k e^{-t} \langle x^*, f(t)\rangle \, dt \\
&= \int_0^{\infty} e^{-t} \sum_{k=0}^{\infty} \frac{(1+\epsilon)^k t^k}{k} \langle x^*, f(t)\rangle \, dt \\
&= \int_0^{\infty} e^{\epsilon t} \langle x^*, f(t)\rangle \, dt \\
&= \lim_{\tau \to \infty} \left\langle x^*, \int_0^{\tau} e^{\epsilon t} f(t) \, dt \right\rangle.
\end{aligned} \quad (1.3.4)$$

The interchange of integration and summation is justified by Fubini's theorem, noting that all terms are positive if $x^* \geq 0$; in general we split $x^*$ into real and imaginary part and each of them into positive and negative part.

It follows from (1.3.4) and the uniform boundedness theorem that

$$\sup_{\tau > 0} \left\| \int_0^{\tau} e^{\epsilon t} f(t) \, dt \right\| < \infty.$$

Therefore, by the discussion at the beginning of this section, $\omega_1(f) \leq -\epsilon$. This contradicts the assumption that $\omega_1(f) = 0$. ////

## 1.4. Positive semigroups

In this section we collect some elementary results about stability of positive $C_0$-semigroups that will be useful in later chapters. The main result is that $s(A) =$

$s_0(A) = \omega_1(\mathbf{T})$ always holds. At the end of the section, we present an example of a positive $C_0$–semigroup such for which we have strict inequality $s(A) < \omega_0(\mathbf{T})$. This example is important not only for the theory of positive $C_0$–semigroups, but also for the theory of general $C_0$–semigroups: whenever we associate a quantity $\omega(\mathbf{T})$ to arbitrary $C_0$–semigroups $\mathbf{T}$ that can be shown to be intermediate between $s(A)$ and $s_0(A)$ or between $s(A)$ and $\omega_1(\mathbf{T})$, it follows from this example that strict inequality $\omega(\mathbf{T}) < \omega_0(\mathbf{T})$ can occur. Several such quantities will be studied later, e.g. abscissae of scalar $p$-integrability and abscissae of polynomial boundedness of the resolvent.

A $C_0$–semigroup $\mathbf{T}$ on a Banach lattice $X$ is called *positive* if each operator $T(t)$ is positive, i.e. $T(t) \geq 0$ for all $t \geq 0$.

**Theorem 1.4.1.** *Let $\mathbf{T}$ be a positive $C_0$–semigroup on a Banach lattice $X$, with generator $A$. Then,*
$$s(A) = s_0(A) = \omega_1(\mathbf{T}).$$
*Moreover, $s(A)$ is the infimum of all $\omega \in \mathbb{R}$ such that $\{\operatorname{Re}\lambda > \omega\} \subset \varrho(A)$ and*
$$R(\lambda, A)x = \lim_{\tau \to \infty} \int_0^\tau e^{-\lambda t} T(t)x\, dt, \quad \forall \operatorname{Re}\lambda > \omega,\, x \in X. \tag{1.4.1}$$
*Furthermore, the following assertions hold.*
*(i) Either $s(A) = -\infty$ or $s(A) \in \sigma(A)$;*
*(ii) For a given $\lambda \in \varrho(A)$, we have $R(\lambda, A) \geq 0$ if and only if $\lambda > s(A)$;*
*(iii) For all $\operatorname{Re}\lambda > s(A)$ and $x \in X$, we have $|R(\lambda, A)x| \leq R(\operatorname{Re}\lambda, A)|x|$.*

*Proof:* We start by proving (i) and the identity $s(A) = \omega_1(\mathbf{T})$. The characterization of $s(A)$ follows from this and Theorem 1.2.3 (ii).

If $\omega_1(\mathbf{T}) = -\infty$, then also $s(A) \leq \omega_1(\mathbf{T}) = -\infty$ by Theorem 1.2.3 (i) and hence $s(A) = \omega_1(\mathbf{T}) = -\infty$. We may therefore assume that $\omega_1(\mathbf{T}) > -\infty$. We shall prove that $\omega_1(\mathbf{T}) \in \sigma(A)$. Assume the contrary. Then the resolvent, which is defined in $\{\operatorname{Re}\lambda > \omega_1(\mathbf{T})\}$, admits a holomorphic extension to some neighbourhood $V_\epsilon := \{\lambda \in \mathbb{C} : |\lambda - \omega_1(\mathbf{T})| < \epsilon\}$ of $\omega_1(\mathbf{T})$. Hence, for all $x \in X$, $\lambda \mapsto R(\lambda, A)x$ extends holomorphically to $V_\epsilon$. By Theorem 1.2.3 (ii), in $\{\operatorname{Re}\lambda > \omega_1(\mathbf{T})\}$ the map $\lambda \mapsto R(\lambda, A)x$ is the Laplace transform of the function $f_x(t) := T(t)x$. Therefore, for $x \geq 0$ the Pringsheim-Landau theorem implies that $\omega_1(f_x) \leq \omega_1(\mathbf{T}) - \epsilon$. By decomposing an arbitrary $x \in X$ into real and imaginary part and each of these into positive and negative part we see that $\omega_1(f_x) \leq \omega_1(\mathbf{T}) - \epsilon$ holds for all $x \in X$. But then Theorem 1.2.3 (iii) implies that $\omega_1(\mathbf{T}) \leq \omega_1(\mathbf{T}) - \epsilon$, a contradiction. This proves that $\omega_1(\mathbf{T}) \in \sigma(A)$. In particular, $s(A) \geq \omega_1(\mathbf{T})$. Since also $s(A) \leq \omega_1(\mathbf{T})$, it follows that $s(A) = \omega_1(\mathbf{T})$ and $s(A) \in \sigma(A)$.

Next we prove (ii). If $\lambda > s(A)$, then $R(\lambda, A) \geq 0$ by (1.4.1). Conversely, suppose $\lambda \in \mathbb{C}$ is such that $R(\lambda, A) \geq 0$. First we observe that $\lambda \in \mathbb{R}$. Indeed, for $0 \leq x \in X$ we let $y := R(\lambda, A)x \geq 0$ and note that
$$\overline{Ay} = \lim_{t \downarrow 0} \frac{1}{t} \overline{(T(t)y - y)} = \lim_{t \downarrow 0} \frac{1}{t}(T(t)y - y) = Ay$$

and hence the identities

$$\lambda y - Ay = x = \bar{x} = \overline{\bar{\lambda}y - Ay} = \bar{\lambda}y - Ay$$

show that $\lambda = \bar{\lambda}$. The bars denote the complex conjugates; cf. Appendix A.3. Since $R(\mu, A) \geq 0$ for all $\mu > s(A)$, for all $\mu > \max\{\lambda, s(A)\}$ the resolvent identity yields

$$R(\lambda, A) = R(\mu, A) + (\mu - \lambda) R(\lambda, A) R(\mu, A) \geq R(\mu, A) \geq 0.$$

Using Proposition 1.1.5 and the fact that $s(A) \in \sigma(A)$ whenever it is finite it follows that for all $\mu > \max\{\lambda, s(A)\}$ we have

$$(\mu - s(A))^{-1} = (\operatorname{dist}(\mu, \sigma(A)))^{-1} \leq \|R(\mu, A)\| \leq \|R(\lambda, A)\|.$$

This can only be true if $\lambda > s(A)$.

For the proof of (iii), we note that by (1.4.1) for all $x \in X$ and $\operatorname{Re}\lambda > s(A)$ we have

$$|R(\lambda, A)x| = \left| \int_0^\infty e^{-\lambda t} T(t) x \, dt \right| \leq \int_0^\infty e^{-\operatorname{Re}\lambda t} T(t) |x| \, dt = R(\operatorname{Re}\lambda, A)|x|,$$

the integrals being in the improper sense. This proves (iii).

It remains to prove that $s(A) = s_0(A)$. Let $\omega > s(A)$ be arbitrary. Assume, for a contradiction, that there is a sequence $(\lambda_n) \subset \{\operatorname{Re}\lambda > \omega\}$ such that $\lim_{n \to \infty} \|R(\lambda_n, A)\| = \infty$. By the uniform boundedness theorem, there is an $x \in X$ such that $\lim_{n \to \infty} \|R(\lambda_n, A)x\| = \infty$. Let $\lambda \geq \omega$ be an accumulation point of the sequence $(\operatorname{Re}\lambda_n)$. Then, there is a subsequence $(\lambda_{n_k})$ such that $\operatorname{Re}\lambda_{n_k} \to \lambda$ and hence by (iii),

$$\|R(\lambda, A)|x|\| = \lim_{k \to \infty} \|R(\operatorname{Re}\lambda_{n_k}, A)|x|\| \geq \limsup_{k \to \infty} \|R(\lambda_{n_k}, A)x\| = \infty.$$

This contradiction concludes the proof.  ////

The following corollary is concerned with $C_0$-*groups*. These are defined analogously to $C_0$ semigroups, the only difference being that the role of the index family $t \geq 0$ is replaced by $t \in \mathbb{R}$. The *generator* of a $C_0$-group $\mathbf{T} = \{T(t)\}_{t \in \mathbb{R}}$ is defined as the generator of the associated $C_0$-semigroup $\{T(t)\}_{t \geq 0}$.

**Corollary 1.4.2.** *If $\mathbf{T}$ is a positive $C_0$-group on a non-zero Banach lattice $X$, with generator $A$, then $\sigma(A) \neq \emptyset$.*

*Proof:* Assume for a contradiction that $\sigma(A) = \emptyset$. Then by Theorem 1.4.1 we have $R(\lambda, A) \geq 0$ for all $\lambda \in \mathbb{R}$. Since $-A$ generates the positive $C_0$-group $\{T(-t)\}_{t \in \mathbb{R}}$, the same argument shows that $R(\lambda, -A) \geq 0$ for all $\lambda \in \mathbb{R}$. But $R(\lambda, -A) = (\lambda + A)^{-1} = -R(-\lambda, A)$, and therefore $-R(-\lambda, A) \geq 0$ for all $\lambda \in \mathbb{R}$. Since also $R(-\lambda, A) \geq 0$ it follows that $R(-\lambda, A) = 0$ for all $\lambda \in \mathbb{R}$. This contradicts the injectivity of $R(-\lambda, A)$ unless $X = \{0\}$.  ////

We close this section with an example of a positive $C_0$-semigroup for which strict inequality $s(A) < \omega_0(\mathbf{T})$ holds. We need the following simple fact.

**Lemma 1.4.3.** *Let $X$ and $Y$ be Banach spaces, $Y \subset X$ with continuous inclusion. Let $A$ be a linear operator on $X$ with $D(A) \subset Y$ and let $A_Y$ denote the part of $A$ in $Y$. If $\varrho(A) \neq \emptyset$, then $\sigma(A) = \sigma(A_Y)$.*

*Proof:* Let $\mu \in \varrho(A)$. Then $R(\mu, A)Y \subset D(A) \subset Y$. It is clear that $R(\mu, A)|_Y$ defines a two-sided inverse for $\mu - A_Y$, so that $\mu \in \varrho(A_Y)$.

Conversely, let $\mu \in \varrho(A_Y)$. Define the bounded operator $Q$ on $X$ by $Qx := R(\lambda, A)x + (\lambda - \mu)R(\mu, A_Y)R(\lambda, A)x$, where $\lambda \in \varrho(A)$ is fixed but arbitrary. Then $Qx \in D(A)$ and $(\mu - A)Qx = x$ for all $x \in X$. Similarly, $Q(\lambda - A)x = x$ for all $x \in D(A)$ and hence $Q = (\mu - A)^{-1}$. ////

**Example 1.4.4.** Let $1 \leq p < q < \infty$ and let $X = L^p(1, \infty) \cap L^q(1, \infty)$. This space is a Banach space under the norm

$$\|f\| := \max\{\|f\|_p, \|f\|_q\}$$

and a Banach lattice with respect to the pointwise a.e. ordering. On $X$, we define the semigroup $\mathbf{T}$ by

$$(T(t)f)(s) = f(se^t), \quad s > 1, \, t \geq 0. \tag{1.4.2}$$

It is easy to check that this defines a $C_0$-semigroup on $X$ whose generator $A$ is given by

$$D(A) = \{f \in X : s \mapsto sf'(s) \in X\},$$
$$(Af)(s) = sf'(s), \quad s > 1, \, f \in D(A).$$

In a similar way, for $r = p$ and $r = q$ (1.4.2) defines a $C_0$-semigroup $\mathbf{T}_r$ on $L^r(1, \infty)$; its generator will be denoted by $A_r$. Note that $T_r(t)|_X = T(t)$ and $A$ is the part of $A_r$ in $X$, $r = p, q$.

For $f \in L^p(1, \infty)$ and $t \geq 0$ we have

$$\|T_p(t)f\|_p = \left(\int_1^\infty |f(se^t)|^p \, ds\right)^{\frac{1}{p}} = e^{-\frac{t}{p}} \left(\int_{e^t}^\infty |f(s)|^p \, ds\right)^{\frac{1}{p}} \leq e^{-\frac{t}{p}} \|f\|_p \tag{1.4.3}$$

with equality if the support of $f$ is contained in $(e^t, \infty)$. Thus, $s(A_p) \leq \omega_0(\mathbf{T}_p) \leq -\frac{1}{p}$. Also, using Theorem 1.4.1 and the fact that $s(A_p) < 0$, for $f \in L^p(1, \infty)$ one easily checks that

$$(R(0, A_p)f)(s) = \int_0^\infty f(se^t) \, dt = \int_s^\infty f(t) \, \frac{dt}{t}, \quad \text{a.a. } s > 1,$$

and therefore, if $p = 1$, then $|(R(0, A_p)f)(s)| \leq s^{-1}\|f\|_1$ for almost all $s > 1$. Similarly, for $1 < p < \infty$ and almost all $s > 1$ it follows that

$$|(R(0, A_p)f)(s)| \leq \left(\int_s^\infty \frac{1}{t^{p'}} \, dt\right)^{\frac{1}{p'}} \|f\|_p = (p' - 1)^{-\frac{1}{p'}} s^{\frac{1}{p'}-1} \|f\|_p,$$

where $\frac{1}{p} + \frac{1}{p'} = 1$.

In both cases it follows that $R(0, A_p)f \in L^\infty(1, \infty)$, and hence $D(A_p) \subset L^\infty(1, \infty) \cap L^p(1, \infty) \subset L^q(1, \infty)$. Since also $D(A_p) \subset L^p(1, \infty)$, it follows that $D(A_p) \subset X$. By Lemma 1.4.3, this implies that $s(A) = s(A_p) \leq -\frac{1}{p}$.

On the other hand, for all $\beta > \frac{1}{p}$ the function $f_\beta(s) := s^{-\beta}$ belongs to $D(A)$ and $(Af_\beta)(s) = sf'_\beta(s) = -\beta s^{-\beta} = -\beta f_\beta(s)$ for almost all $s > 1$. This shows that $-\beta$ is an eigenvalue of $A$, so $s(A) \geq -\beta$. Since $\beta > \frac{1}{p}$ is arbitrary it follows that $s(A) \geq -\frac{1}{p}$.

For $t > 0$ fixed we define $f_t := \chi_{(e^t, e^t+1)}$. Then by (1.4.3),

$$\|T(t)f_t\| \geq \|T_q(t)f_t\|_q = e^{-\frac{t}{q}}\|f_t\|_q = e^{-\frac{t}{q}}.$$

This implies that $\|T(t)\| \geq e^{-\frac{t}{q}}$. Since $t > 0$ was arbitrary it follows that $\omega_0(\mathbf{T}) \geq -\frac{1}{q}$.

On the other hand, by (1.4.3) for all $f \in X$ we have

$$\|T(t)f\| = \max\{\|T(t)f\|_p, \|T(t)f\|_q\} \leq \max\{e^{-\frac{t}{p}}\|f\|_p, e^{-\frac{t}{q}}\|f\|_q\}$$
$$\leq e^{-\frac{t}{q}}\max\{\|f\|_p, \|f\|_q\} = e^{-\frac{t}{q}}\|f\|,$$

so $\omega_0(\mathbf{T}) \leq -\frac{1}{q}$. Putting everything together, we obtain

$$s(A) = -\frac{1}{p} < -\frac{1}{q} = \omega_0(\mathbf{T}). \tag{1.4.4}$$

The interest of this example lies in the fact that $s(A) = \omega_0(\mathbf{T})$ always holds for positive $C_0$-semigroups in $L^p$-spaces. This will be proved in Section 3.5.

**Notes.** The abstract Cauchy problem is studied in many monographs; an encyclopaedic reference is the book [Fa]. Theorem 1.1.2 is proved in [Na]. The equivalence of (i) and (ii) is due to E. Hille [Hi] and the equivalence of (i) and (iii) is due to W. Arendt. The proof of the Hille-Yosida theorem, as well as its history, can be found in most textbooks about $C_0$-semigroups, e.g., [HP], [Pz], [Go], [Na], [Da1], [vC].

Theorem 1.2.3 was proved by F. Neubrander [Nb2]. Example 1.2.4 is due to J. Zabczyk [Zb2]. By minor modification of this example, V. Wrobel [Wr] was able to construct a $C_0$-group on a Hilbert space with $s(A) = 0$, $s_0(A) = \omega_0(\mathbf{T}) = 1$, and $\omega_n(\mathbf{T}) = 2^{-n}$, $n = 1, 2, \ldots$. The idea is to define $\mathbf{T}$ coordinatewise by $T(t) = (e^{int}e^{tA_{m(n)}})_{n \geq 1}$, where $m(n)$ is the integer part of $\log(n+1)/\log 2$ and $A_k$ are the matrices defined in Example 1.2.4, $k = 1, 2, \ldots$. This example shows that strict inequality $\omega_1(\mathbf{T}) < s_0(A)$ can occur.

The proof of Lemma 1.3.1, which is due to G.H. Hardy, is adapted from [Ka, Thm. II.2.2] where the discrete case is proved. The second statement in Theorem 1.3.3 is a special case of a theorem due to F. Neubrander [Nb1] who gives a different proof. The Pringsheim-Landau theorem for Banach lattice-valued functions appears in [GWV]; the scalar version can be found, e.g., in [Wi, Thm. II.5b].

Theorem 1.4.1 is proved in part in [GWV]; the rest appears in [Na]. The identity $s(A) = \omega_1(\mathbf{T})$ is due to F. Neubrander [Nb2]. Example 1.4.4 is due to W. Arendt [Ar2]. It was the first example of a positive semigroup on a rearrangement invariant Banach function space whose spectral bound and uniform growth bound do not coincide. Earlier, it was shown in [GWV] that the spectral bound and uniform growth bound do not coincide for the translation semigroup $\mathbf{T}$ defined by $(T(t)f)(s) = f(s+t)$ in the Banach function space $L^p(\mathbb{R}_+) \cap L^q(\mathbb{R}_+, e^{t^2} dt)$. This space is not rearrangement invariant, however. In [Ne6], Example 1.4.4 is extended in various ways to rearrangement invariant Banach function spaces over $(1, \infty)$. Roughly speaking, it turns out that $\omega_0(\mathbf{T})$ is in some sense determined by the behaviour of the so-called fundamental function (cf. Appendix A4) near 0 and $s(A)$ by the behaviour near $\infty$.

# Chapter 2

## Spectral mapping theorems

In this chapter we study conditions under which a spectral mapping theorem holds for a $C_0$-semigroup $\mathbf{T}$ and its generator $A$. If $A$ is bounded, by the Dunford functional calculus for all $t \in \mathbb{R}$ we have the spectral mapping formula
$$\sigma(\exp(tA)) = \exp(t\sigma(A)).$$
As we observed in Section 1.2, an easy consequence of this is that $s(A) = \omega_0(\mathbf{T})$. In general, however, a generator $A$ is unbounded and the spectral mapping theorem in the above form does not hold.

In Section 2.1 we start with the spectral inclusion theorem
$$\sigma(T(t)) \supset \exp(t\sigma(A)), \quad t \geq 0,$$
and the spectral mapping theorems for point spectrum and residual spectrum,
$$\sigma_p(T(t))\backslash\{0\} = \exp(t\sigma_p(A));$$
$$\sigma_r(T(t))\backslash\{0\} = \exp(t\sigma_r(A)), \quad t \geq 0,$$
valid for arbitrary $C_0$-semigroups.

In Section 2.2 we prove the spectral mapping theorems of Greiner and Gearhart. The first of these states that $1 \in \varrho(T(2\pi))$ if and only if $i\mathbb{Z} \subset \varrho(A)$ and the resolvents $R(ik, A)$ are Cesàro summable with respect to $k$. The second shows that Cesàro summability can be replaced by uniform boundedness if the underlying space is a Hilbert space.

In Section 2.3 we prove the spectral mapping theorem
$$\sigma(T(t))\backslash\{0\} = \exp(t\sigma(A)), \quad t \geq 0,$$
for eventually uniformly continuous semigroups and give necessary and sufficient conditions for a semigroup on a Hilbert space to be uniformly continuous for $t > 0$.

In Section 2.4 we prove the weak spectral mapping theorem
$$\sigma(T(t)) = \overline{\exp(t\sigma(A))}, \quad t \geq 0,$$
for $C_0$-groups of non-quasianalytic growth and show that this result actually characterizes non-quasianalyticity.

In Section 2.5 we present the Latushkin - Montgomery-Smith spectral mapping theorem: $1 \in \sigma(T(2\pi))$ if and only if $0 \in \sigma(B_p)$, where $B_p$ is the generator of the $C_0$-semigroup $\mathbf{S}_p$ on $L^p(\Gamma, X)$ defined by
$$(S_p(t)f)(e^{i\theta}) := T(t)f(e^{i(\theta-t)}), \quad t \geq 0.$$

## 2.1. The spectral mapping theorem for the point spectrum

In this section we prove the spectral inclusion theorem and the spectral mapping theorems for the point spectrum and the residual spectrum. We start with the *spectral inclusion theorem*.

**Theorem 2.1.1.** Let $\mathbf{T}$ be a $C_0$–semigroup on a Banach space $X$, with generator $A$. Then we have the spectral inclusion relation

$$\sigma(T(t)) \supset \exp(t\sigma(A)), \quad \forall t \geq 0.$$

*Proof:* By the identities (1.1.1) and (1.1.2), applied to the semigroup $\mathbf{T}_\lambda := \{e^{-\lambda t} T(t)\}_{t \geq 0}$ generated by $A - \lambda$, for all $\lambda \in \mathbb{C}$ and $t \geq 0$ we have

$$(\lambda - A) \int_0^t e^{\lambda(t-s)} T(s) x \, ds = (e^{\lambda t} - T(t)) x, \quad \forall x \in X,$$

and

$$\int_0^t e^{\lambda(t-s)} T(s)(\lambda - A) x \, ds = (e^{\lambda t} - T(t)) x, \quad \forall x \in D(A). \tag{2.1.1}$$

Suppose $e^{\lambda t} \in \varrho(T(t))$ for some $\lambda \in \mathbb{C}$ and $t \geq 0$, and denote the inverse of $e^{\lambda t} - T(t)$ by $Q_{\lambda, t}$. Since $Q_{\lambda, t}$ commutes with $T(t)$ and hence also with $A$, we have

$$(\lambda - A) \int_0^t e^{\lambda(t-s)} T(s) Q_{\lambda, t} x \, ds = x, \quad \forall x \in X,$$

and

$$\int_0^t e^{\lambda(t-s)} T(s) Q_{\lambda, t} (\lambda - A) x \, ds = x, \quad \forall x \in D(A).$$

This shows that the bounded operator $B_\lambda$ defined by

$$B_\lambda x := \int_0^t e^{\lambda(t-s)} T(s) Q_{\lambda, t} x \, ds$$

is a two-sided inverse of $\lambda - A$. It follows that $\lambda \in \varrho(A)$. ////

The converse inclusion $\exp(t\sigma(A)) \subset \sigma(T(t))\backslash\{0\}$ generally fails. For example, if $\mathbf{T}$ is the $C_0$–semigroup on the Hilbert space $H$ of Example 1.2.4, then $\omega_0(\mathbf{T}) = 1$ and hence $r(T(t)) = e^t$ for all $t$. In particular, there exist $\lambda \in \sigma(T(t))$ with $|\lambda| = e^t$. On the other hand, $s(A) = 0$ and hence $\exp(t\sigma(A)) \subset \{|\lambda| \leq 1\}$.

Next, we turn to spectral mapping theorems for certain parts of the spectrum.

The *point spectrum* $\sigma_p(A)$ of a (bounded or unbounded) linear operator $A$ on a Banach space $X$ is the set of all $\lambda \in \sigma(A)$ for which there exists a non-zero vector $x \in D(A)$ such that $Ax = \lambda x$, or equivalently, for which the operator $\lambda - A$ is not injective.

The *residual spectrum* $\sigma_r(A)$ is the set of all $\lambda \in \sigma(A)$ for which $\lambda - A$ does not have dense range. By the Hahn-Banach theorem, $\sigma_r(A) = \sigma_p(A^*)$ provided the adjoint $A^*$ of $A$ is well-defined, i.e. $A$ is densely defined.

The *approximate point spectrum* $\sigma_a(A)$ is the set of all $\lambda \in \sigma(A)$ for which there exists an *approximate eigenvector*, i.e. a sequence $(x_n)$ of norm one vectors in $X$, $x_n \in D(A)$ for all $n$, such that

$$\lim_{n\to\infty} \|Ax_n - \lambda x_n\| = 0.$$

Clearly, $\sigma_p(A) \subset \sigma_a(A)$. By Proposition 2.1.4 below, $\sigma(A) = \sigma_r(A) \cup \sigma_a(A)$.

We shall prove next that spectral mapping theorems hold for the point spectrum and the residual spectrum.

**Theorem 2.1.2.** *Let* **T** *be a* $C_0$-*semigroup on a Banach space* $X$, *with generator* $A$. *Then*

$$\sigma_p(T(t))\setminus\{0\} = \exp(t\sigma_p(A)), \quad \forall t \geq 0.$$

*Proof:* If $\lambda \in \sigma_p(A)$ and $x \in D(A)$ is an eigenvector corresponding to $\lambda$, the identity (2.1.1) shows that $T(t)x = e^{\lambda t}x$, i.e. $e^{\lambda t}$ is an eigenvalue of $T(t)$ with eigenvector $x$. This proves the inclusion $\supset$.

The inclusion $\subset$ is proved as follows. The case $t = 0$ being trivial, we fix $t > 0$. If $\lambda \in \sigma_p(T(t))\setminus\{0\}$, then $\lambda = e^{\mu t}$ for some $\mu \in \mathbb{C}$. If $x$ is an eigenvector, then $T(t)x = e^{\mu t}x$ implies that the map $s \mapsto e^{-\mu s}T(s)x$ is periodic with period $t$. Since this map is not identically zero, the uniqueness theorem for the Fourier transform implies that at least one of its Fourier coefficients is non-zero. Thus, there exists an integer $k \in \mathbb{Z}$ such that

$$x_k := \frac{1}{t}\int_0^t e^{-(2\pi i k/t)s}(e^{-\mu s}T(s)x)\,ds \neq 0.$$

We shall show that $\mu_k := \mu + 2\pi i k/t$ is an eigenvalue of $A$ with eigenvector $x_k$.

By the $t$-periodicity of $s \mapsto e^{-\mu s}T(s)x$, for all $\operatorname{Re}\nu > \omega_0(\mathbf{T})$ we have

$$R(\nu, A)x = \int_0^\infty e^{-\nu s}T(s)x\,ds$$

$$= \sum_{n=0}^\infty \int_{nt}^{(n+1)t} e^{-\nu s}T(s)x\,ds$$

$$= \sum_{n=0}^\infty \int_0^t e^{-\nu s}T(s)(e^{-\nu nt}T(nt)x)\,ds \qquad (2.1.2)$$

$$= \sum_{n=0}^\infty e^{(\mu-\nu)nt}\int_0^t e^{-\nu s}T(s)x\,ds$$

$$= \frac{1}{1 - e^{(\mu-\nu)t}}\int_0^t e^{-\nu s}T(s)x\,ds.$$

Since the integral on the right hand side is an entire function, this shows that the map $\nu \mapsto R(\nu, A)x$ admits a holomorphic continuation to $\mathbb{C}\setminus\{\mu+2\pi i n/t : n \in \mathbb{Z}\}$. Denoting this extension by $F_x(\cdot)$, by (2.1.2) and the definition of $x_k$ we have

$$\lim_{\nu \to \mu_k} (\nu - \mu_k) F_x(\nu) = x_k.$$

Also, by (2.1.2) and the $t$-periodicity of $s \mapsto e^{-\mu s}T(s)x$,

$$\lim_{\nu \to \mu_k} (\mu_k - A)\big((\nu - \mu_k)F_x(\nu)\big)$$
$$= \lim_{\nu \to \mu_k} \frac{\nu - \mu_k}{1 - e^{(\mu-\nu)t}} \left( (I - e^{-\nu t}T(t)) + (\mu_k - \nu) \int_0^t e^{-\nu s}T(s)x\,ds \right)$$
$$= \frac{1}{t}(0+0) = 0.$$

From the closedness of $A$ it follows that $x_k \in D(A)$ and $(\mu_k - A)x_k = 0$. ////

The spectral mapping theorem also holds for the residual spectrum. This follows from a duality argument, for which we need the following definitions.

If $\mathbf{T}$ is a $C_0$-semigroup on $X$, we define

$$X^\odot := \{x^* \in X^* : \lim_{t \downarrow 0} \|T^*(t)x^* - x^*\| = 0\},$$

where $T^*(t) := (T(t))^*$ is the adjoint operator. It is easy to see that $X^\odot$ is a closed $\mathbf{T}^*$-invariant subspace of $X^*$, and the restriction $\mathbf{T}^\odot$ of $\mathbf{T}^*$ to $X^\odot$ is a $C_0$-semigroup on $X^\odot$. We denote its generator by $A^\odot$.

We claim that $\sigma_p(A^*) = \sigma_p(A^\odot)$, where $A^*$ is the adjoint of the generator $A$ of $\mathbf{T}$, and $\sigma_p(T^*(t)) = \sigma_p(T^\odot(t))$, $t \geq 0$.

We start with the first of these assertions. For all $x^* \in D(A^*)$ and $x \in X$ we have

$$\langle T^*(t)x^* - x^*, x \rangle = \langle x^*, T(t)x - x \rangle$$
$$= \langle A^*x^*, \int_0^t T(s)x\,ds \rangle$$
$$= \int_0^t \langle A^*x^*, T(s)x \rangle\,ds \qquad (2.1.3)$$
$$= \int_0^t \langle T^*(s)A^*x^*, x \rangle\,ds.$$

Therefore,

$$|\langle T^*(t)x^* - x^*, x \rangle| \leq t\,\|x\|\,\|A^*x^*\| \sup_{0 \leq s \leq t} \|T(s)\|.$$

By taking the supremum over all $x \in X$ of norm $\leq 1$, it follows that

$$\lim_{t \downarrow 0} \|T^*(t)x^* - x^*\| = 0,$$

i.e. $x^* \in X^\odot$. This proves that $D(A^*) \subset X^\odot$.

Now assume that $A^*x^* = \lambda x^*$ for some $x^* \in D(A^*)$. Then $x^* \in X^\odot$ and (2.1.3) shows that

$$\langle \frac{1}{t}(T^\odot(t)x^* - x^*) - \lambda x^*, x \rangle = \frac{\lambda}{t}\int_0^t \langle T^\odot(s)x^* - x^*, x \rangle \, ds$$

and therefore,

$$\left\| \frac{1}{t}(T^\odot(t)x^* - x^*) - \lambda x^* \right\| \leq |\lambda| \, \|x^*\| \sup_{0 \leq s \leq t} \|T(s)x - x\|.$$

Letting $t \downarrow 0$, this shows that $x^* \in D(A^\odot)$ and $A^\odot x^* = \lambda x^*$, so $\lambda \in \sigma_p(A^\odot)$.

Conversely, if $\lambda \in \sigma_p(A^\odot)$ and $A^\odot x^\odot = \lambda x^\odot$ for some $x^\odot \in D(A^\odot)$, then for all $x \in D(A)$ we have

$$\langle x^\odot, Ax \rangle = \lim_{t \downarrow 0} \frac{1}{t}\langle x^\odot, T(t)x - x \rangle$$
$$= \lim_{t \downarrow 0} \frac{1}{t}\langle T^\odot(t)x^\odot - x^\odot, x \rangle = \langle A^\odot x^\odot, x \rangle = \lambda \langle x^\odot, x \rangle.$$

This shows that $x^\odot \in D(A^*)$ and $A^*x^\odot = \lambda x^\odot$, so $\lambda \in \sigma_p(A^*)$.

Next we prove that $\sigma_p(T^*(t)) = \sigma_p(T^\odot(t))$ for all $t \geq 0$. Clearly we have the inclusion $\sigma_p(T^\odot(t)) \subset \sigma_p(T^*(t))$ since $T^\odot(t)$ is a restriction of $T^*(t)$. Conversely, if $T^*(t)x^* = \lambda x^*$ for some non-zero $x^* \in X^*$, then for all $\mu \in \varrho(A^*) = \varrho(A)$ we have $R(\mu, A^*)x^* \in D(A^*) \subset X^\odot$ and $T^\odot(t)R(\mu, A^*)x^* = R(\mu, A^*)T^*(t)x^* = \lambda R(\mu, A^*)x^*$. Hence $R(\mu, A^*)x^*$ is an eigenvector of $T^\odot(t)$ with eigenvalue $\lambda$.

**Theorem 2.1.3.** *Let* **T** *be a $C_0$–semigroup on a Banach space $X$, with generator $A$. Then*

$$\sigma_r(T(t))\backslash\{0\} = \exp(t\sigma_r(A)).$$

*Proof:* By the above, $\sigma_r(T(t)) = \sigma_p(T^*(t)) = \sigma_p(T^\odot(t))$ and $\sigma_r(A) = \sigma_p(A^*) = \sigma_p(A^\odot)$. The theorem now follows from Theorem 2.1.2 applied to the $C_0$–semigroup $\mathbf{T}^\odot$.  ////

We close this section by recording for later use some propositions that give information about the approximate point spectrum.

**Proposition 2.1.4.** *Let $A$ be a closed linear operator on a Banach space $X$. Then $\sigma(A) = \sigma_r(A) \cup \sigma_a(A)$.*

*Proof:* Assume that $\lambda \in \sigma(A)\backslash\sigma_r(A)$. Then $\lambda - A$ has dense range. If $\lambda - A$ is not injective, then $\lambda \in \sigma_p(A) \subset \sigma_a(A)$ and we are done. Assume therefore that $\lambda - A$ is injective.

Assume for the moment that there exists a constant $C > 0$ such that $\|(\lambda - A)x\| \geq C\|x\|$ for all $x \in D(A)$. Then the range of $\lambda - A$ is closed. Indeed, if $y_n \to y$

with $y_n = (\lambda - A)x_n$, then $\|x_n - x_m\| \leq C^{-1}\|(\lambda - A)(x_n - x_m)\| = \|y_n - y_m\|$, so the sequence $(x_n)$ is Cauchy, with limit $x$, say. The closedness of $A$ implies that $x \in D(A)$ and $(\lambda - A)x = y$, proving that $y$ belongs to the range of $\lambda - A$. Thus, the range of $\lambda - A$ is closed. Since it is also dense, it follows that it is all of $X$. Since $\lambda - A$ is injective, the inverse $R_\lambda := (\lambda - A)^{-1}$ is well-defined as a closed linear operator on $X$ whose domain is all of $X$. Hence, $R_\lambda$ is bounded by the closed graph theorem. Thus, $\lambda - A$ is invertible, a contradiction.

It follows that a constant $C > 0$ as above does not exist. But then there is a sequence $(x_n)$ of norm one vectors, $x_n \in D(A)$ for all $n$, such that $\lim_{n\to\infty}(\lambda - A)x_n = 0$. This proves that $\lambda \in \sigma_a(A)$. ////

Of course, the union need not be disjoint.

Since the spectral mapping theorem holds for the residual spectrum, this proposition shows that the failure of the spectral mapping theorem for the entire spectrum is caused by the behaviour of the approximate point spectrum. This observation is the key to understanding the proofs of the spectral mapping theorems in Sections 2.3 and 2.5.

**Proposition 2.1.5.** *Let $A$ be a closed linear operator on a Banach space $X$. Then the topological boundary $\partial\sigma(A)$ of the spectrum $\sigma(A)$ is contained in the approximate point spectrum $\sigma_a(A)$.*

*Proof:* Let $\lambda \in \partial\sigma(A)$ be fixed and let $(\lambda_n) \subset \varrho(A)$ be a sequence such that $\lambda_n \to \lambda$. It follows from the uniform boundedness theorem and Proposition 1.1.5 that there exists an $x \in X$ such that $\lim_{n\to\infty} \|R(\lambda_n, A)x\| \to \infty$. Let $x_n := \|R(\lambda_n, A)x\|^{-1} R(\lambda_n, A)x$. Then $\|x_n\| = 1$ and

$$\lim_{n\to\infty} \|Ax_n - \lambda x_n\| = \lim_{n\to\infty} \|R(\lambda_n, A)x\|^{-1} \cdot \|(\lambda_n - \lambda)R(\lambda_n, A)x - x\| = 0.$$

////

We conclude this section with a useful observation which relates the approximate point spectra of $A$ and that of the operators $T(t)$:

**Proposition 2.1.6.** *Let $\mathbf{T}$ be a $C_0$–semigroup on a Banach space $X$, with generator $A$. An approximate eigenvector for $A$ with approximate eigenvalue $\lambda$ is also an approximate eigenvector for each operator $T(t)$, with approximate eigenvalue $e^{\lambda t}$.*

*Proof:* Let $\lambda \in \sigma_a(A)$ and choose a sequence $(x_n)$ of norm one vectors in $X$, $x_n \in D(A)$ for all $n$, such that $\lim_{n\to\infty} \|Ax_n - \lambda x_n\| \to 0$. Then, for all $t \geq 0$,

$$\|T(t)x_n - e^{\lambda t}x_n\| = \left\|e^{\lambda t}\int_0^t e^{-\lambda s}T(s)(\lambda - A)x_n\, ds\right\|$$

$$\leq t e^{2|\operatorname{Re}\lambda|t}\left(\sup_{0\leq s\leq t}\|T(s)\|\right)\|(\lambda - A)x_n\| \to 0, \quad n \to \infty,$$

and the convergence is uniform for $t$ in each interval $[0, t_0]$. ////

In particular, if $x$ is an eigenvector for $A$ with eigenvalue $\lambda$, then $x$ is also an eigenvector for each $T(t)$, with eigenvalue $e^{\lambda t}$. In a similar way one proves that if $\lambda$ is an eigenvalue of $A^*$ with eigenvector $x^*$, then $T^*(t)x^* = e^{\lambda t}$ for all $t \geq 0$.

The converse of Proposition 2.1.6 also holds: if $T(t)x = e^{\lambda t}x$ for all $t \geq 0$, then by differentiation we see that $x \in D(A)$ and $Ax = \lambda x$.

If $\mathbf{T}$ extends to a $C_0$–group, we can apply the proposition to the generators $A$ and $-A$ to obtain the same conclusions for all $t \in \mathbb{R}$.

## 2.2. The spectral mapping theorems of Greiner and Gearhart

In this section we show that the spectral mapping theorem holds if we make an additional assumption on the growth of the resolvent on vertical lines. We shall need Fejér's theorem for the circle: if $X$ is a Banach space and $f \in L^1(\Gamma, X)$ is of bounded variation, then

$$(C,1) \sum_{k=-\infty}^{\infty} \hat{f}(k) e^{ik\theta} = \frac{1}{2}(f(e^{i\theta+}) + f(e^{i\theta-})), \quad \forall \theta \in [0, 2\pi].$$

Here, $(C,1) \sum_{k=-\infty}^{\infty} := \lim_{N \to \infty} N^{-1} \sum_{n=0}^{N-1} \sum_{k=-n}^{n}$ denotes the Cesàro sum.

The following result shows that a spectral mapping theorem holds if the resolvent is Cesàro summable along the imaginary axis.

**Theorem 2.2.1.** *Let $\mathbf{T}$ be a $C_0$–semigroup on a Banach space $X$, with generator $A$. Then the following assertions are equivalent:*

(i) $1 \in \varrho(T(2\pi))$;
(ii) $i\mathbb{Z} \subset \varrho(A)$, and $(C,1) \sum_{k=-\infty}^{\infty} R(ik, A)x$ converges for all $x \in X$.
(iii) $i\mathbb{Z} \subset \varrho(A)$, and $(C,1) \sum_{k=-\infty}^{\infty} \langle x^*, R(ik, A)x \rangle$ converges for all $x \in X$ and $x^* \in X^*$.

*Proof:* Define, for each $k \in \mathbb{Z}$, the bounded operator $Q_k$ on $X$ by

$$Q_k x := \frac{1}{2\pi} R(ik, A)(I - T(2\pi)) = \frac{1}{2\pi} \int_0^{2\pi} e^{-iks} T(s)x \, ds, \quad x \in X. \quad (2.2.1)$$

Thus, $Q_k x$ is the $k$-th Fourier coefficient of the restriction to $[0, 2\pi]$ of $T(\cdot)x$. By Fejér's theorem applied to the function $f \in L^1(\Gamma, X)$ defined by $f(e^{i\theta}) := T(\theta)x$, $0 < \theta < 2\pi$, for all $x \in X$ we have

$$(C,1) \sum_{k=-\infty}^{\infty} Q_k x = \frac{1}{2}(I + T(2\pi))x. \quad (2.2.2)$$

(i)⇒(ii): By the spectral inclusion theorem we have $i\mathbb{Z} \in \varrho(A)$. Since by assumption $I - T(2\pi)$ invertible, by (2.2.1) we have

$$R(ik, A) = 2\pi (I - T(2\pi))^{-1} Q_k.$$

Therefore, (2.2.2) implies that for all $x \in X$,

$$(C,1) \sum_{k=-\infty}^{\infty} R(ik, A)x = \pi (I - T(2\pi))^{-1}(I + T(2\pi))x.$$

It is clear that (ii) implies (iii).

(iii)⇒(i): We start the proof with the following general observation: if $(x_n)$ is weakly Cauchy in $X$ and we define the functional $x^{**} : X^* \to \mathbb{C}$ by $\langle x^{**}, x^* \rangle := \lim_{n \to \infty} \langle x^*, x_n \rangle$, then $x^{**}$ is bounded, i.e. $x^{**} \in X^{**}$. Indeed, by the uniform boundedness theorem the sequence $(x_n)$ bounded, say $\sup_n \|x_n\| = M$, and for all $x^* \in X^*$ we have $|\langle x^{**}, x^* \rangle| = \lim_{n \to \infty} |\langle x^*, x_n \rangle| \leq M \|x^*\|$. Applying this to $x_n := n^{-1} \sum_{j=0}^{n-1} \sum_{k=-j}^{j} R(ik, A)x$, we see that for all $x \in X$ the map $Sx : X^* \to \mathbb{C}$ defined by

$$\langle Sx, x^* \rangle := \frac{1}{2} \langle x^*, x \rangle + \frac{1}{2\pi} (C,1) \sum_{k=-\infty}^{\infty} \langle x^*, R(ik, A)x \rangle$$

is bounded. Hence, $S$ defines a linear operator from $X$ into $X^{**}$. Moreover, $\|S\| \leq \sup_n \|n^{-1} \sum_{j=0}^{n-1} \sum_{k=-j}^{j} R(ik, A)\|$, and the latter is finite by the uniform boundedness theorem. It follows that $S$ is bounded as an operator from $X$ into $X^{**}$. We claim that $S$ actually maps $X$ into itself.

Using (2.2.1) and (2.2.2), for all $x \in X$ and $x^* \in X^*$ we have

$$\langle (I - T(2\pi))^{**} Sx, x^* \rangle = \langle S(I - T(2\pi))x, x^* \rangle = \langle x^*, x \rangle. \quad (2.2.3)$$

The second identity shows that $S(I - T(2\pi))$ maps $X$ into itself. We claim that $I - T(2\pi)$ has dense range in $X$. For if this were not the case, the Hahn-Banach theorem implies the existence of a non-zero $x_0^* \in X^*$ such that $(I - T(2\pi))^* x_0^* = 0$. Choose $x_0 \in X$ such that $\langle x_0^*, x_0 \rangle \neq 0$. Then by (2.2.3),

$$\langle x_0^*, x_0 \rangle = \langle Sx_0, (I - T(2\pi))^* x_0^* \rangle = \langle Sx_0, 0 \rangle = 0,$$

a contradiction.

Thus, (2.2.3) implies that $Sy \in X$ for all $y$ in the dense subspace $(I - T(2\pi))X$ of $X$. It follows that $S$ maps $X$ into itself and the claim is proved. But then (2.2.3) shows that $S$ is a two-sided inverse of $I - T(2\pi)$.  ////

A rescaling argument leads to the following characterization of uniform exponential stability in terms of Cesàro convergence of the resolvent.

**Corollary 2.2.2.** *Let* **T** *be a $C_0$-semigroup on a Banach space $X$, with generator $A$. Then the following assertions are equivalent:*

(i) **T** *is uniformly exponentially stable;*
(ii) $\{\operatorname{Re}\lambda \geq 0\} \subset \varrho(A)$, *and for all $x \in X$ and $\operatorname{Re}\lambda \geq 0$ we have $(C,1)\sum_{k=-\infty}^{\infty} R(\lambda+ik, A)x$ converges in $X$.*
(iii) $\{\operatorname{Re}\lambda \geq 0\} \subset \varrho(A)$, *and for all $x \in X$, $x^* \in X^*$, and $\operatorname{Re}\lambda \geq 0$ and we have $(C,1)\sum_{k=-\infty}^{\infty} \langle x^*, R(\lambda+ik, A)x\rangle$ converges.*

**Corollary 2.2.3.** *Let* **T** *be a $C_0$-semigroup on a Banach space $X$, with generator $A$. Assume*

(i) $i\mathbb{Z} \subset \varrho(A)$ *and* $\sup_{k\in\mathbb{Z}} \|R(ik, A)\| < \infty$;
(ii) *There exists an $\omega > \omega_0(\mathbf{T})$ such that*

$$\sum_{k=-\infty}^{\infty} \|R(\omega+ik, A)x\|^2 < \infty, \quad \forall x \in X$$

*and*

$$\sum_{k=-\infty}^{\infty} \|R(\omega+ik, A^*)x^*\|^2 < \infty, \quad \forall x^* \in X^*.$$

*Then $1 \in \varrho(T(2\pi))$.*

*Proof:* By the resolvent identity

$$R(ik, A) = (I + \omega R(ik, A))R(\omega+ik, A)$$

and the assumptions (i) and (ii), we have

$$\sum_{k=-\infty}^{\infty} \|R(ik, A)x\|^2 < \infty, \quad \forall x \in X.$$

Therefore, the sum

$$\sum_{k=-\infty}^{\infty} \langle R(\omega+ik, A^*)x^*, R(ik, A)x\rangle$$

is absolutely convergent by Hölder's inequality. Also, by Corollary 2.2.2, the Cesàro sum $(C,1)\sum_{k=-\infty}^{\infty} R(\omega+ik, A)x$ is convergent in $X$. It follows that for all $x \in X$ and $x^* \in X^*$ we have

$$(C,1)\sum_{k=-\infty}^{\infty} \langle x^*, R(ik, A)x\rangle = (C,1)\sum_{k=-\infty}^{\infty} \langle x^*, R(\omega+ik, A)x\rangle$$

$$+ \omega \sum_{k=-\infty}^{\infty} \langle R(\omega+ik, A^*)x^*, R(ik, A)x\rangle,$$

showing that the Cesàro sum $(C,1)\sum_{k=-\infty}^{\infty} \langle x^*, R(ik, A)x\rangle$ converges. Thus, (iii) of Theorem 2.2.1 is satisfied.    ////

A continuous version of this theorem will be proved in Section 4.6.

Before we can prove the our next result we need one more fact from the theory of adjoint semigroups: the adjoint $\mathbf{T}^*$ of a $C_0$-semigroup $\mathbf{T}$ on a reflexive space $X$ is strongly continuous, i.e. $X^\odot = X^*$. The easiest way to see this is to note that the adjoint $A^*$ of a densely defined closed operator $A$ is always weak*-densely defined in $X^*$. If $X$ is reflexive, it follows that $A^*$ is weakly densely defined and hence densely defined by the Hahn-Banach theorem. Since further $R(\lambda, A^*) = R(\lambda, A)^*$ for all $\lambda \in \varrho(A) = \varrho(A^*)$, the resolvent of $A^*$ satisfies the estimates of the Hille-Yosida theorem. Therefore, by that theorem, $A^*$ is the generator of a $C_0$-semigroup, and it is easy to verify that this semigroup is precisely $\mathbf{T}^*$.

In Hilbert spaces, the Cesàro summability of Theorem 2.2.1 can be replaced by mere boundedness. This is the content of the following result, usually referred to as Gearhart's theorem.

**Theorem 2.2.4.** *Let $\mathbf{T}$ be a $C_0$-semigroup on a Hilbert space $H$, with generator $A$. Then the following assertions are equivalent:*

(i) $1 \in \varrho(T(2\pi))$;
(ii) $i\mathbb{Z} \subset \varrho(A)$ and $\sup_{k \in \mathbb{Z}} \|R(ik, A)\| < \infty$.

*Proof:* (i)$\Rightarrow$(ii): By the spectral inclusion theorem we have $i\mathbb{Z} \subset \varrho(A)$. Since by (2.2.1),

$$R(ik, A)x = (I - T(2\pi))^{-1} \int_0^{2\pi} e^{-iks} T(s) x \, ds, \quad \forall x \in X,$$

it is evident that $\sup_{k \in \mathbb{Z}} \|R(ik, A)\| < \infty$.

(ii)$\Rightarrow$(i): Fix $\omega > \omega_0(\mathbf{T})$. Since

$$R(\omega + ik, A)x = (I - e^{-2\pi\omega} T(2\pi))^{-1} \int_0^{2\pi} e^{-iks}(e^{-\omega s} T(s) x) \, ds,$$

the Hilbert space-valued Plancherel theorem for the circle implies that for all $x \in H$,

$$\sum_{k=-\infty}^{\infty} \|R(\omega + ik, A)x\|^2 < \infty.$$

By the same argument applied to the $C_0$-semigroup $\mathbf{T}^*$, the analogous inequality holds for the adjoint of the resolvent. We can now apply Corollary 2.2.3. ////

**Corollary 2.2.5.** *Let $\mathbf{T}$ be a $C_0$-semigroup on a Hilbert space $H$. If the resolvent of the generator $A$ exists and is uniformly bounded in the right half plane $\{\operatorname{Re} z > 0\}$, then $\mathbf{T}$ is uniformly exponentially stable.*

Indeed, we obtain that the spectrum of $T(2\pi)$ is contained in the open unit disc and therefore $r(T(2\pi)) = e^{2\pi\omega_0(\mathbf{T})} < 1$. This corollary admits a very short proof based on the Hilbert space-valued Plancherel theorem and Paley-Wiener theorem, which will be given in Section 3.1. The corollary breaks down for $C_0$-semigroups **T** in Banach spaces, even if **T** is a positive semigroup on a uniformly convex Banach lattice. Taking into account the identity $s(A) = s_0(A)$ for positive semigroups, a counterexample is obtained by rescaling the semigroup of Example 1.4.4.

## 2.3. Eventually uniformly continuous semigroups

A $C_0$-semigroup **T** on a Banach space $X$ is said to be *eventually uniformly continuous* if there exists a $t_0 \geq 0$ such that the map $t \mapsto T(t)$ is continuous with respect to the uniform operator topology for $t > t_0$. Examples of eventually uniformly continuous semigroups are eventually compact semigroups and holomorphic semigroups.

In this section we shall prove that the spectral mapping theorem

$$\sigma(T(t))\setminus\{0\} = \exp(t\sigma(A)), \quad t \geq 0,$$

holds for eventually uniformly continuous semigroups. The proof depends on an embedding construction that extends **T** to a $C_0$-semigroup $\hat{\mathbf{T}}$ on a space $\hat{X}$ containing $X$ isometrically. This space is constructed in such a way that the spectrum of the generator $A$ of **T** coincides with the spectrum of the generator $\hat{A}$ of $\hat{\mathbf{T}}$ and the approximate point spectrum of $A$ coincides with the point spectrum of $\hat{A}$. In this way, the spectral mapping theorem becomes a consequence of the spectral mapping theorem for the point spectrum.

We define the space $l_0^\infty(X)$ as the set of all bounded sequences $(x_n) \subset X$ such that

$$\lim_{t \downarrow 0} \left( \sup_n \|T(t)x_n - x_n\| = 0 \right).$$

It is trivial to check that $l_0^\infty(X)$ is a closed subspace of $l^\infty(X)$, the space of all bounded sequences in $X$. By the very definition of $l_0^\infty(X)$, **T** extends to a $C_0$-semigroup $\mathbf{T}_0^\infty$ on $l_0^\infty(X)$ by coordinatewise action; its generator will be denoted by $A_0^\infty$. Note that

$$D(A_0^\infty) = \{(x_n) \in l_0^\infty(X) : x_n \in D(A) \text{ for all } n \text{ and } (Ax_n) \in l_0^\infty(X)\}.$$

We claim that $c_0(X) \subset l_0^\infty(X)$, where $c_0(X)$ is the space of all sequences in $X$ that converge to 0. Indeed, let $(x_n) \in c_0(X)$. Put $M := \limsup_{t \downarrow 0} \|T(t)\|$, let $\epsilon > 0$ be arbitrary and choose $n_0 \in \mathbb{N}$ such that $\|x_n\| \leq \epsilon/(M+1)$ for all $n \geq n_0$. Then, by decomposing $(x_n)$ as the sum of $(x_0, x_1, ..., x_{n_0-1}, 0, 0, ...)$ and $(0, ..., 0, x_{n_0}, x_{n_0+1}, ...)$, we have

$$\limsup_{t \downarrow 0} \left( \sup_n \|T(t)x_n - x_n\| \right) \leq \limsup_{t \downarrow 0} \left( \sup_{n \geq n_0} \|T(t)x_n - x_n\| \right) \leq \epsilon.$$

This proves the claim. We now define $\hat{X}$ as the quotient space

$$\hat{X} = l_0^\infty(X)/c_0(X).$$

Since $c_0(X)$ is $\mathbf{T}_0^\infty$-invariant, the quotient semigroup $\hat{\mathbf{T}}$ is well-defined. Being the quotient of a $C_0$–semigroup, $\hat{\mathbf{T}}$ is a $C_0$–semigroup on $\hat{X}$. We denote by $\hat{A}$ its generator. By general facts concerning quotient mappings, its domain is given by $D(\hat{A}) = D(A_0^\infty) + c_0(X)$.

Note that $X$ embeds isometrically into $\hat{X}$ in the natural way: the embedding is given by $x \mapsto (x,x,...)+c_0(X)$. Obviously, as a subspace of $\hat{X}$, $X$ is $\hat{\mathbf{T}}$-invariant.

The following proposition describes the spectrum of $\hat{A}$:

**Proposition 2.3.1.** *In the above situation, $\sigma(\hat{A}) = \sigma(A)$.*

*Proof:* Let $\lambda \in \sigma(A)$.

First assume $\lambda \in \sigma_a(A)$. Then there is an approximate eigenvector $(x_n)$ such that $\lim_{n\to\infty} \|Ax_n - \lambda x_n\| = 0$. In particular, the sequence $(Ax_n)$ is bounded. We have

$$\lim_{t\downarrow 0}\left(\sup_n \|T(t)x_n - x_n\|\right) = \lim_{t\downarrow 0}\left(\sup_n \left\|\int_0^t T(s)Ax_n\, ds\right\|\right)$$
$$\leq \lim_{t\downarrow 0} Mt \sup_n \|Ax_n\| = 0,$$

where $M := \limsup_{t\downarrow 0} \|T(t)\|$. Hence, $(x_n) \in l_0^\infty(X)$. Also, since $(Ax_n) = \lambda(x_n) + (y_n)$, where $(y_n) := ((A-\lambda)x_n) \in c_0(X)$, we see that $(Ax_n) \in l_0^\infty(X)$ as well. Hence, $(x_n) \in D(A_0^\infty)$. By passing to the quotient $\hat{X} = l_0^\infty(X)/c_0(X)$, we have $\hat{x} := (x_n) + c_0(X) \in D(\hat{A})$ and $\hat{A}\hat{x} = \lambda \hat{x}$. Since $\hat{x} \neq 0$ this shows that $\lambda \in \sigma_p(\hat{A})$.

If $\lambda \in \sigma_r(A)$, there exists an $x_0 \in X$ and an $\epsilon > 0$ such that $\|x_0 - (\lambda - A)y\| \geq \epsilon$ for all $y \in D(A)$. Let $y = (y_n) \in D(A_0^\infty)$ be arbitrary and let $x = (x_0, x_0, ...) \in l_0^\infty(X)$. Then, $\|x - (\lambda - A_0^\infty)y\|_{l_0^\infty(X)} \geq \epsilon$ and by passing to the quotient $\hat{X}$, we see that $\|\hat{x} - (\lambda - \hat{A})\hat{y}\|_{\hat{X}} \geq \epsilon$ for all $\hat{y} \in D(\hat{A})$. Therefore, $\lambda \in \sigma_r(\hat{A})$.

Since by Proposition 2.1.4 we have $\sigma(A) = \sigma_a(A) \cup \sigma_r(A)$, it follows that $\sigma(A) \subset \sigma(\hat{A})$.

Conversely, assume that $\lambda - A$ is invertible. We claim that $\lambda - \hat{A}$ is injective. Indeed, if there is a non-zero $\hat{x} \in D(\hat{A})$ such that $\hat{A}\hat{x} = \lambda \hat{x}$, then the definition of $\hat{X}$ implies that there is a bounded sequence $(x_n) \subset X$ with $\limsup_{n\to\infty} \|x_n\| > 0$ such that $x_n \in D(A)$ for all $n$ and $\lim_n \|Ax_n - \lambda x_n\| = 0$. By passing to a subsequence if necessary we may assume that $\inf_n \|x_n\| \geq \epsilon$ for some $\epsilon > 0$. But then the sequence $(y_n) := (\|x_n\|^{-1} x_n)$ is an approximate eigenvector for $A$. This contradicts the assumption that $\lambda \in \varrho(A)$ and the claim is proved.

Let $\hat{x} \in \hat{X}$ be arbitrary, say $\hat{x} = (x_n) + c_0(X)$. Then, $\hat{y} := (R(\lambda, A)x_n) + c_0(X) \in D(\hat{A})$ and $(\lambda - \hat{A})\hat{y} = (x_n) + c_0(X) = \hat{x}$. This shows that $\lambda - \hat{A}$ is also surjective. The closed graph theorem implies that the inverse $(\lambda - \hat{A})^{-1}$ is bounded, and hence $\lambda \in \varrho(\hat{A})$. ////

**Theorem 2.3.2.** *Let* **T** *be an eventually uniformly continuous semigroup on a Banach space* $X$, *with generator* $A$. *Then the spectral mapping theorem holds, i.e.*

$$\sigma(T(t))\setminus\{0\} = \exp(t\sigma(A)), \quad \forall t \geq 0.$$

*Proof:* Noting that $\sigma(T(t)) = \sigma_r(T(t)) \cup \sigma_a(T(t))$ for all $t \geq 0$, in view of the spectral inclusion theorem and the spectral mapping theorem for the residual spectrum it suffices to prove that there exists an $\alpha \in \sigma(A)$ such that $e^{\alpha\tau} = \lambda$ whenever $\tau > 0$ and $0 \neq \lambda \in \sigma_a(T(\tau))$.

To this end, let $(x_n) \in X$ be an approximate eigenvector of $T(\tau)$ with approximate eigenvalue $\lambda$. Let **T** be uniformly continuous for $t > t_0$ and fix $k \in \mathbb{N}$ such that $k\tau > t_0$. Since $\lim_n \|T(\tau)x_n - \lambda x_n\| = 0$, we also have $\lim_n \|T(k\tau)x_n - \lambda^k x_n\| = 0$. By passing to a subsequence of $(x_n)$ if necessary we may assume that $\|T(k\tau)x_n\| \geq \epsilon$ for all $n$ and some $\epsilon > 0$ (we could take $\epsilon = |\lambda|^k/2$) and we can define $y_n := \|T(k\tau)x_n\|^{-1}T(k\tau)x_n$. We claim that $(y_n) \in l_0^\infty(X)$. Indeed, the uniform continuity of **T** for $t > t_0$ implies that

$$\limsup_{h\downarrow 0}\left(\sup_n \|T(h)y_n - y_n\|\right)$$

$$\leq \limsup_{h\downarrow 0}\left(\sup_n \|T(k\tau)x_n\|^{-1}\|T(k\tau+h) - T(k\tau)\|\,\|x_n\|\right)$$

$$\leq \limsup_{h\downarrow 0} \epsilon^{-1}\|T(k\tau+h) - T(k\tau)\| = 0$$

proving the claim. Since

$$\limsup_{n\to\infty} \|T(\tau)y_n - \lambda y_n\| \leq \limsup_{n\to\infty} \|T(k\tau)x_n\|^{-1}\|T(k\tau)\|\,\|T(\tau)x_n - \lambda x_n\| = 0,$$

it follows that $\hat{T}(\tau)\hat{y} = \lambda\hat{y}$, where $\hat{y} = (y_n) + c_0(X)$. We have proved that $\lambda \in \sigma_p(\hat{T}(\tau))$. By the spectral mapping theorem for the point spectrum, there exists an $\alpha \in \sigma_p(\hat{A})$ such that $e^{\alpha\tau} = \lambda$. But then $\alpha \in \sigma(A)$ by Proposition 2.3.1. ////

As the proof shows, we do not need really need Proposition 2.3.1: it suffices to know that $\sigma_p(\hat{A}) \subset \sigma_a(A)$ and this is a trivial consequence of the definition of $\hat{A}$.

The next result shows that the uniform continuity condition can be somewhat relaxed, at the price that we only get a spectral mapping theorem for the peripheral spectrum. Recall that for a bounded operator $T$, the peripheral spectrum was defined as $\sigma(T) \cap \{\lambda \in \mathbb{C} : |\lambda| = r(T)\}$; for a(n unbounded) generator $A$ we defined the peripheral spectrum as $\sigma(A) \cap \{\lambda \in \mathbb{C} : \mathrm{Re}\,\lambda = s(A)\}$.

A $C_0$-semigroup **T** is said to be *uniformly continuous at infinity* if

$$\lim_{t\to\infty}\left(\limsup_{h\downarrow 0} e^{-\omega_0(\mathbf{T})t}\|T(t+h) - T(t)\|\right) = 0.$$

This condition is fulfilled, e.g., if there exist operator families $\mathbf{U}_0 = \{U_0(t)\}_{t\geq 0}$ and $\mathbf{U}_1 = \{U_1(t)\}_{t\geq 0}$ such that $T(t) = U_0(t) + U_1(t)$, $\mathbf{U}_0$ is eventually uniformly continuous, and $\lim_{t\to\infty} e^{-\omega_0(\mathbf{T})t}\|U_1(t)\| = 0$.

**Theorem 2.3.3.** Let **T** be a $C_0$-semigroup which is uniformly continuous at infinity. Then the spectral mapping theorem holds for the peripheral spectrum of **T**, i.e. for all $t \geq 0$ we have

$$(\sigma(T(t))\backslash\{0\}) \cap \{\lambda \in \mathbb{C} : |\lambda| = r(T(t))\} = \exp(t\sigma(A)) \cap \{\lambda \in \mathbb{C} : |\lambda| = r(T(t))\}.$$

In particular, $s(A) = \omega_0(\mathbf{T})$.

*Proof:* As in Theorem 2.3.2, it suffices to prove that $\lambda \in \sigma_p(\hat{T}(\tau))$ whenever $\tau > 0$ and $0 \neq \lambda \in \sigma_a(T(\tau)) \cap \{\lambda \in \mathbb{C} : |\lambda| = r(T(\tau))\}$. If $r(T(\tau)) = 0$, there is nothing to prove. Therefore, by rescaling we may assume that $\omega_0(\mathbf{T}) = 0$. Then $r(T(\tau)) = 1$. Let $\lambda \in \sigma_a(T(\tau)) \cap \{\lambda \in \mathbb{C} : |\lambda| = 1\}$ and choose an approximate eigenvector $(x_n)$ such that $\lim_n \|T(\tau)x_n - \lambda x_n\| = 0$. We claim that $(x_n) \in l_0^\infty(X)$. To this end, let $\epsilon > 0$ be arbitrary and choose $k \in \mathbb{N}$ and $\delta > 0$ such that $\|T(k\tau + h) - T(k\tau)\| \leq \epsilon$ for all $0 \leq h \leq \delta$. This is possible by the definition of uniform continuity at infinity. Since $\lim_n \|T(k\tau)x_n - \lambda^k x_n\| = 0$, we can choose $n_0 \in \mathbb{N}$ such that $\|T(k\tau)x_n - \lambda^k x_n\| \leq \epsilon/(M+1)$ for all $n \geq n_0$, where $M := \sup_{0 \leq h \leq \delta} \|T(h)\|$. Then, noting that $|\lambda| = 1$, for all $n \geq n_0$ and $0 \leq h \leq \delta$ we have

$$\begin{aligned} \|T(h)x_n - x_n\| &= \|(T(h) - I)\lambda^k x_n\| \\ &\leq \|(T(h) - I)(T(k\tau)x_n - \lambda^k x_n)\| + \|(T(h) - I)T(k\tau)x_n\| \\ &\leq \epsilon + \epsilon = 2\epsilon. \end{aligned}$$

Together with the strong continuity on the first $n_0$ vectors, this yields

$$\lim_{h \downarrow 0} \left( \sup_n \|T(h)x_n - x_n\| \right) \leq 2\epsilon.$$

Since $\epsilon > 0$ was arbitrary, the claim is proved.

Since $\|x_n\| = 1$ for all $n$, its class $\hat{x} = (x_n) + c_0(X)$ in $\hat{X}$ is non-zero. Since $(x_n)$ is an approximate eigenvector for $T(\tau)$, in $\hat{X}$ we have $\hat{T}(\tau)\hat{x} = \lambda \hat{x}$, i.e. $\lambda$ is an eigenvalue of $\hat{T}(\tau)$. ////

We continue with a sufficient condition for uniform continuity at infinity for semigroups on Hilbert space.

**Lemma 2.3.4.** Let $A$ be a closed operator on a Banach space $X$ and let $\Omega \subset \varrho(A)$ be such that

$$\sup_{\lambda \in \Omega} \|R(\lambda, A)\| \leq M.$$

Let $\delta := (2M)^{-1}$ and define $\Omega_\delta := \{\lambda \in \mathbb{C} : \text{dist}(\lambda, \Omega) \leq \delta\}$. Then $\Omega_\delta \subset \varrho(A)$ and

$$\sup_{\lambda \in \Omega_\delta} \|R(\lambda, A)\| \leq 2M.$$

*Proof:* For all $\lambda \in \Omega$ and $z \in \mathbb{C}$ with dist $(\lambda, z) \leq \delta$, the series

$$R_z := \sum_{n=0}^{\infty} (\lambda - z)^n R(\lambda, A)^{n+1}$$

converges absolutely. Therefore the function $z \mapsto R_z$ is a holomorphic operator-valued function in $\Omega_\delta$, and by the resolvent identity it coincides with $R(z, A)$ whenever $z \in \varrho(A)$. Therefore, by Proposition 1.1.6, $\Omega_\delta \subset \varrho(A)$ and $R_z = R(z, A)$ for all $z \in \Omega_\delta$. ////

In particular, if the resolvent is uniformly bounded in the open right half plane $\{\operatorname{Re} \lambda > 0\}$, then $s_0(A) < 0$. This will be used frequently in later sections.

**Theorem 2.3.5.** *Let $\mathbf{T}$ be a $C_0$-semigroup on a Hilbert space $H$, with generator $A$. Let $r_0 > 0$ be such that the resolvent of $A$ exists and is uniformly bounded in the set $\{\lambda \in \mathbb{C} : \operatorname{Re} \lambda \geq 0, |\operatorname{Im} \lambda| \geq r_0\}$. Then, there are constants $\epsilon > 0$ and $M > 0$ such that for all $t > 0$,*

$$\limsup_{h \downarrow 0} \|T(t+h) - T(t)\| \leq Me^{-\epsilon t} \limsup_{s \to \pm\infty} \|R(is, A)\|.$$

*Proof:* Choose $\epsilon > 0$ such that the resolvent of $A$ exists and is uniformly bounded in $\{\lambda \in \mathbb{C} : \operatorname{Re} \lambda \geq -\epsilon, |\operatorname{Im} \lambda| \geq r_0\}$. Such $\epsilon$ exists by virtue of Lemma 2.3.4. Let $r \geq r_0$ be arbitrary, fix $\omega_0 > \omega_0(\mathbf{T})$, and let $\Gamma = \Gamma_1 \cup \ldots \cup \Gamma_5$ be the upwards oriented path defined by

$$\Gamma_1 = \{\lambda = -\epsilon + i\eta : -\infty < \eta \leq -r\};$$
$$\Gamma_2 = \{\lambda = \xi - ir : -\epsilon \leq \xi \leq \omega_0\};$$
$$\Gamma_3 = \{\lambda = \omega_0 + i\eta : -r \leq \eta \leq r\};$$
$$\Gamma_4 = \{\lambda = \xi + ir : -\epsilon \leq \xi \leq \omega_0\};$$
$$\Gamma_5 = \{\lambda = -\epsilon + i\eta : r \leq \eta < \infty\}.$$

For each $x \in H$ the map $t \mapsto e^{-\omega_0 t} T(t)x$ defines an element of $L^2(\mathbb{R}_+, H)$ of norm $\leq C\|x\|$, where $C$ is a constant independent of $x$. By the Plancherel theorem, the map $s \mapsto R(\omega_0 + is, A)x$ is in $L^2(\mathbb{R}, H)$, of norm $\leq C\sqrt{2\pi}\|x\|$. By the resolvent identity,

$$R(-\epsilon + is, A) = (I + (\omega_0 + \epsilon)R(-\epsilon + is, A))R(\omega_0 + is, A), \quad |s| \geq r_0, \quad (2.3.1)$$

and therefore there is a constant $c > 0$, independent of $r \geq r_0$, such that

$$\int_r^\infty \|R(-\epsilon \pm is, A)x\|^2 \, ds \leq c^2 \|x\|^2, \quad \forall x \in H. \quad (2.3.2)$$

Since the adjoint semigroup $\mathbf{T}^*$ is strongly continuous (cf. the remarks preceding Theorem 2.2.4), we can apply the same reasoning to $\mathbf{T}^*$, and by replacing $c$ by some larger constant if necessary we also have

$$\int_r^\infty \|R(-\epsilon \pm is, A^*)y\|^2 \, ds \leq c^2 \|y\|^2, \quad \forall y \in H. \tag{2.3.3}$$

Put
$$K_r := \sup_{|s| \geq r} \|R(is, A)\|$$

and note that by (2.3.1) (with $\omega_0$ replaced by 0) there is a constant $N > 0$ such that
$$\sup_{|s| \geq r} \|R(-\epsilon + is, A)\| \leq NK_r.$$

By a double partial integration, Lemma 1.3.2, and the fact that $\frac{d}{d\lambda} R(\lambda, A) = -R(\lambda, A)^2$, for all $x, y \in H$ and $t > 0$ we have

$$\left| \int_{\Gamma_1} e^{\lambda t} \langle R(\lambda, A)x, y \rangle \, d\lambda \right|$$
$$\leq e^{-\epsilon t} \left( \frac{1}{t} |\langle R(-\epsilon - ir, A)x, y \rangle| + \frac{1}{t^2} |\langle R(-\epsilon - ir, A)^2 x, y \rangle| \right)$$
$$+ e^{-\epsilon t} \left| \int_r^\infty \frac{e^{-ist}}{t^2} \langle R(-\epsilon - is, A)^3 x, y \rangle \, ds \right|$$
$$\leq e^{-\epsilon t} \Big\{ \left( \frac{NK_r}{t} + \frac{N^2 K_r^2}{t^2} \right) \|x\| \|y\|$$
$$+ \frac{NK_r}{t^2} \left( \int_r^\infty \|R(-\epsilon - is, A)x\|^2 \, ds \right)^{\frac{1}{2}} \left( \int_r^\infty \|R(-\epsilon - is, A^*)y\|^2 \, ds \right)^{\frac{1}{2}} \Big\}$$
$$\leq e^{-\epsilon t} \left( \frac{NK_r}{t} + \frac{N^2 K_r^2}{t^2} + \frac{c^2 NK_r}{t^2} \right) \|x\| \|y\|,$$

the first integral being in the improper sense. Here, we estimated
$$|\langle R(-\epsilon - is, A)^3 x, y \rangle| \leq \|R(-\epsilon - is, A)^2 x\| \, \|R(-\epsilon - is, A^*)y\|$$
$$\leq NK_r \|R(-\epsilon - is, A)x\| \, \|R(-\epsilon - is, A^*)y\|$$

and used Hölder's inequality along with (2.3.2) and (2.3.3). An analogous estimate holds for the integral over $\Gamma_5$.

For $t > 0$ we define the bounded operators $T_r(t)$ by
$$T_r(t)x := \frac{1}{2\pi i} \int_{\Gamma_2 \cup \Gamma_3 \cup \Gamma_4} e^{\lambda t} R(\lambda, A) x \, d\lambda, \quad x \in H.$$

By the complex Laplace inversion formula (Theorem 1.3.3) and Cauchy's theorem (which can be applied by virtue of Lemma 1.3.2), for all $x \in D(A)$ and $y \in H$ we have

$$\langle T(t)x, y \rangle = \frac{1}{2\pi i} \left\langle PV \int_\Gamma e^{\lambda t} R(\lambda, A) x \, d\lambda, y \right\rangle$$

$$= \langle T_r(t)x, y \rangle + \frac{1}{2\pi i} \int_{\Gamma_1 \cup \Gamma_5} e^{\lambda t} \langle R(\lambda, A) x, y \rangle \, d\lambda.$$

Therefore, for all $x \in D(A)$, $y \in H$, and $t > 0$ we have

$$|\langle T(t)x - T_r(t)x, y \rangle| \leq \frac{1}{\pi} e^{-\epsilon t} \left( \frac{NK_r}{t} + \frac{N^2 K_r^2}{t^2} + \frac{c^2 NK_r}{t^2} \right) \|x\| \|y\|.$$

Since $D(A)$ is dense, it follows that

$$\|T(t) - T_r(t)\| \leq \frac{1}{\pi} e^{-\epsilon t} \left( \frac{NK_r}{t} + \frac{N^2 K_r^2}{t^2} + \frac{c^2 NK_r}{t^2} \right), \quad \forall t > 0.$$

Next, we observe that

$$\|T(t+h) - T(t)\| \leq \|T(t+h) - T_r(t+h)\| + \|T_r(t+h) - T_r(t)\| + \|T_r(t) - T(t)\|.$$

Since the map $t \mapsto T_r(t)$ is uniformly continuous for $t > 0$, it follows that

$$\limsup_{h \downarrow 0} \|T(t+h) - T(t)\| \leq \frac{2}{\pi} e^{-\epsilon t} \left( \frac{NK}{t} + \frac{N^2 K^2}{t^2} + \frac{c^2 NK}{t^2} \right), \quad \forall t > 0, \quad (2.3.4)$$

where

$$K := \limsup_{r \geq r_0, \, r \to \infty} K_r.$$

If $K = 0$ we are done. Otherwise, we choose $N_0 > 0$ such that

$$\sup_{0 < t < 1} \sup_{0 < h < 1} \|T(t+h) - T(t)\| \leq \frac{2}{\pi} e^{-\epsilon} N_0 K. \quad (2.3.5)$$

By (2.3.4) with $t \geq 1$ and (2.3.5),

$$\limsup_{h \downarrow 0} \|T(t+h) - T(t)\| \leq M e^{-\epsilon t} K \quad \forall t > 0,$$

where $M = 2\pi^{-1} \max\{N_0, N + N^2 K + c^2 N\}$. ////

As a corollary, we obtain the following characterization of semigroups in Hilbert space which are uniformly continuous for $t > 0$ due to P. You.

**Corollary 2.3.6.** *Let $\mathbf{T}$ be a $C_0$-semigroup on a Hilbert space $H$, with generator $A$. Then the following assertions are equivalent:*

  (i) *$\mathbf{T}$ is uniformly continuous for $t > 0$;*
  (ii) *For all $\omega > \omega_0(\mathbf{T})$ we have $\lim_{s \to \pm \infty} \|R(\omega + is, A)\| = 0$.*
  (iii) *There exist $\omega \in \mathbb{R}$ and $r \geq 0$ such that $\lim_{|s| \geq r,\ s \to \pm \infty} \|R(\omega + is, A)\| = 0$.*

*Proof:* If $\mathbf{T}$ is uniformly continuous for $t > 0$, then the Riemann-Lebesgue lemma applied to the function $t \mapsto e^{-\omega t} T(t) \cdot \chi_{\mathbb{R}_+}(t)$, where $\omega > \omega_0(\mathbf{T})$, shows that (i) implies (ii). Trivially, (ii) implies (iii). Assume that (iii) holds for some $r \geq 0$ and $\omega \in \mathbb{R}$. By rescaling, we may assume that $\omega = 0$. If $\omega_0(\mathbf{T}) < 0$, we can apply Theorem 2.3.5 and we are done. If $\omega_0(\mathbf{T}) \geq 0$ we have to be slightly more careful, due to the fact that there may be spectrum in the strip $\{0 \leq \operatorname{Re} \lambda \leq \omega_0(\mathbf{T})\}$. Lemma 2.3.4 shows that for each $|s| \geq r$, the resolvent set of $A$ contains a disc with centre $is$ and radius $\frac{1}{2}\|R(is, A)\|^{-1}$; moreover, on that disc the resolvent is bounded by $2\|R(is, A)\|$. Since $\lim_{|s| \to \infty} \|R(is, A)\| = 0$, there exists an $r_0 \geq r$ such that $\frac{1}{2}\|R(\pm is, A)\|^{-1} > \omega_0(\mathbf{T}) + \epsilon$ for all $|s| \geq r_0$. But then the resolvent is uniformly bounded on the set $\{\lambda \in \mathbb{C} : 0 \leq \operatorname{Re} \lambda \leq \omega_0(\mathbf{T}) + \epsilon,\ |\operatorname{Im} \lambda| \geq r_0\}$. Since the resolvent is also uniformly bounded in the half-plane $\{\operatorname{Re} \lambda \geq \omega_0(\mathbf{T}) + \epsilon\}$, we are in a position to apply Theorem 2.3.5 and (i) follows. ////

## 2.4. Groups of non-quasianalytic growth

In this section we shall study conditions under which the weak spectral mapping theorem

$$\sigma(T(t)) = \overline{\exp(t\sigma(A))}$$

holds. The main result is that this is the case for $C_0$-groups satisfying a non-quasi-analytic growth condition.

In order to state the results, we need the following terminology. A measurable, locally bounded function $\omega : \mathbb{R} \to \mathbb{R}$ is called a *weight* if

$$1 \leq \omega(t) \quad \text{and} \quad \omega(t+s) \leq \omega(t)\omega(s), \quad \forall t, s \in \mathbb{R}.$$

It is an easy consequence of the definition that every weight is exponentially bounded. A weight $\omega$ is said to be *non-quasianalytic* if

$$\int_{-\infty}^{\infty} \frac{\log \omega(t)}{1 + t^2} \, dt < \infty.$$

**Lemma 2.4.1.** *If $\omega$ is a non-quasianalytic weight on $\mathbb{R}$, then for all $t \in \mathbb{R} \setminus \{0\}$ we have*

$$\lim_{n \to \infty} \omega(nt)^{\frac{1}{n}} = 1.$$

*Proof:* Fix $t \neq 0$. Since $\omega(-t)$ also defines a non-quasianalytic weight, we may assume that $t > 0$. Since $\omega \geq 1$, it follows that $\liminf_{n \to \infty} \omega(nt)^{\frac{1}{n}} \geq 1$. In order to prove that $\limsup_{n \to \infty} \omega(nt)^{\frac{1}{n}} \leq 1$ we define $\rho(s) := \log \omega(s)$, $s \in \mathbb{R}$. Then $\rho \geq 0$ and $\rho$ is subadditive. We shall show that $\limsup_{n \to \infty} \frac{1}{n}\rho(nt) \leq 0$. If not, then there exists a sequence $n_k \to \infty$ and an $\epsilon > 0$ such that $\rho(n_k t) \geq n_k \epsilon$ for all $k = 0, 1, 2, \ldots$. By passing to a subsequence, we may assume that $1 \leq n_0 < n_1 < \ldots$ and
$$\lim_{k \to \infty} \frac{n_k}{n_{k+1}} = 0.$$

For each $k$, we define
$$H_k := \{0 \leq m \leq n_{k+1} : \rho(mt) < \frac{1}{2}n_{k+1}\epsilon\}$$

and claim that $\#H_k \leq \frac{1}{2}n_{k+1}$. Indeed, letting
$$G_k := \{0 \leq m \leq n_{k+1} : n_{k+1} - m \in H_k\},$$

for any $m \in G_k \cap H_k$ we have $\rho(mt) < \frac{1}{2}n_{k+1}\epsilon$ and $\rho((n_{k+1} - m)t) < \frac{1}{2}n_{k+1}\epsilon$. By the subadditivity of $\rho$, it follows that $\rho(n_{k+1}t) \leq \rho(mt) + \rho((n_{k+1} - m)t) < n_{k+1}\epsilon$. Since this contradicts our assumptions, it follows that $G_k \cap H_k = \emptyset$. If, however, $\#H_k = \#G_k > \frac{1}{2}n_{k+1}$, then necessarily $G_k \cap H_k \neq \emptyset$. This proves the claim.

Therefore, $\#I_k \geq \frac{1}{2}n_{k+1} - n_k - 1$, where
$$I_k := \{n_k \leq m \leq n_{k+1} - 1 : \rho(mt) \geq \frac{1}{2}n_{k+1}\epsilon\}.$$

Note that the sets $I_k$ are pairwise disjoint. Let $M := \sup_{0 \leq s \leq t} \omega(s)$. Then $M \geq 1$ and the submultiplicativity of $\omega$ implies that for all $m \in \mathbb{N}$ and $s \in [(m-1)t, mt]$ we have $\omega(s) \geq M^{-1}\omega(mt)$. Therefore, with $\beta := \log M^{-1} \int_{-\infty}^{\infty} (1 + s^2)^{-1} ds$,

$$\int_{-\infty}^{\infty} \frac{\log \omega(s)}{1 + s^2} ds \geq \beta + \sum_{k=0}^{\infty} \sum_{m \in I_k} \int_{(m-1)t}^{mt} \frac{\log \omega(s)}{1 + s^2} ds$$

$$\geq \beta + \sum_{k=0}^{\infty} \sum_{m \in I_k} \frac{\log(\omega(mt))}{1 + m^2 t^2}$$

$$\geq \beta + \sum_{k=0}^{\infty} (\frac{1}{2}n_{k+1} - n_k - 1) \left( \frac{\frac{1}{2}n_{k+1}\epsilon}{1 + (n_{k+1} - 1)^2 t^2} \right) = \infty.$$

////

Examples of non-quasianalytic weights are the functions $\omega(t) = e^{|t|^\alpha}$, $0 \leq \alpha < 1$.

We are going to show next that weak spectral mapping theorem holds for $C_0$-groups $\mathbf{T}$ satisfying $\|T(t)\| \leq \omega(t)$ for all $t \in \mathbb{R}$, where $\omega$ is a non-quasianalytic weight on $\mathbb{R}$. For the proof we need some facts about the Beurling algebras $L_\omega(\mathbb{R})$.

Let $\omega$ be a weight on $\mathbb{R}$. The *Beurling algebra* $L_\omega(\mathbb{R})$ is the set of all $f \in L^1(\mathbb{R})$ such that

$$\|f\|_\omega := \int_{-\infty}^\infty |f(t)|\omega(t)\,dt < \infty.$$

With respect to convolution, $L_\omega(\mathbb{R})$ is Banach algebra and a subalgebra of the commutative Banach algebra $L^1(\mathbb{R})$.

For the proof of the next lemma we need the following theorem of Paley and Wiener: A function $0 \leq \psi \in L^2(\mathbb{R})$ satisfies the condition

$$\int_{-\infty}^\infty \frac{|\log \psi(s)|}{1+s^2}\,ds < \infty$$

if and only if there exists a function $\psi_0 \in L^2(\mathbb{R})$ such that $|\psi_0| = \psi$ a.e. and the Fourier-Plancherel transform of $\psi_0$ vanishes a.e. on the negative semi-axis. A proof of this theorem can be found in the book [PW].

**Lemma 2.4.2.** *Let $\omega$ be a non-quasianalytic weight on $\mathbb{R}$. Then for each $0 < \epsilon < \frac{1}{2}$ there exists a continuous function $\phi \in L_\omega(\mathbb{R})$ and a constant $M > 0$ such that:*

*(i) $0 \leq \hat\phi \leq 1$, $\hat\phi \equiv 1$ on $[-\frac{1}{2}+\epsilon, \frac{1}{2}-\epsilon]$ and $\hat\phi \equiv 0$ outside $(-\frac{1}{2}+\frac{1}{2}\epsilon, \frac{1}{2}-\frac{1}{2}\epsilon)$;*
*(ii) $|\phi(-t)|\omega(t) \leq M(1+t^2)^{-1}$ for almost all $t \in \mathbb{R}$.*

*Moreover, the set of all $f \in L_\omega(\mathbb{R})$ whose Fourier transform has compact support is a dense subspace of $L_\omega(\mathbb{R})$.*

*Proof:* We start with the proof of (i) and (ii). By replacing $\omega$ by the non-quasianalytic weight $\tilde\omega(t) := \omega(t)\omega(-t) \geq \omega(t)$, we may assume that $\omega(t) = \omega(-t)$ for all $t \in \mathbb{R}$.

The proof is divided into several steps.

*Step 1.* First we observe the following property of the function $\psi(t) := (1+t^2)^{-1}(\omega(t))^{-1}$: $\psi(t) = \psi(-t)$ and

$$0 \leq (\psi * \psi)(t) = \int_{-\infty}^\infty \frac{1}{(1+s^2)(1+(t-s)^2)\omega(s)\omega(t-s)}\,ds$$

$$\leq \frac{1}{\omega(t)} \int_{-\infty}^\infty \frac{1}{(1+s^2)(1+(t-s)^2)}\,ds$$

$$= \frac{2\pi}{(t^2+4)\omega(t)} \leq 2\pi\psi(t).$$

The second identity follows by a simple contour integration.

*Step 2.* In this step we fix an arbitrary $\delta > 0$ and construct a function $\phi_0 \in L_\omega(\mathbb{R})$ with the following properties:

(a) The Fourier transform $\hat\phi_0$ is non-negative;
(b) $\hat\phi_0$ vanishes outside the interval $(-\delta, \delta)$ and satisfies $\hat\phi_0(0) \neq 0$;
(c) There is a constant $C > 0$ such that $|\phi_0(t)| \leq C\psi(t)$ for almost all $t \in \mathbb{R}$.

In fact, it suffices to construct a function $\phi_1 \in L_\omega(\mathbb{R})$ that satisfies only (b) and (c). Indeed, once such a function is found, we define $\phi_2(t) := \overline{\phi_1(-t)}$ and $\phi_0 := \phi_1 * \phi_2$. Since the convolution of two functions in $L_\omega(\mathbb{R})$ is in $L_\omega(\mathbb{R})$, we have $\phi_0 \in L_\omega(\mathbb{R})$. Moreover, by Step 1 $\phi_0$ satisfies (c), and we have $\hat{\phi}_0 = |\hat{\phi}_1|^2$, so (a) and (b) hold for $\phi_0$.

The function $\phi_1$ is constructed as follows. Since $\omega$ is non-quasianalytic, we can apply the Paley-Wiener theorem to $\psi$ and obtain a function $\psi_0$ such that $|\psi_0| = \psi$ a.e. and whose Fourier-Plancherel transform vanishes a.e. on the negative semi-axis. Since $\psi \in L^1(\mathbb{R})$, so is $\psi_0$ and hence its Fourier-Plancherel transform coincides with the Fourier transform and is continuous. Upon replacing $\psi_0(s)$ by $e^{i\lambda s}\psi_0(s)$ for some $\lambda > 0$, we may assume that $\sup\{t \in \mathbb{R} : \hat{\psi}_0(s) = 0$ for all $s \leq t\} = 0$. Choose $0 < \delta_0 \leq \delta$ such that $\hat{\psi}_0(\delta_0) \neq 0$. Put

$$\psi_0^+(t) := e^{-i\delta_0 t}\psi_0(t), \quad \psi_0^-(t) := e^{i\delta_0 t}\overline{\psi_0(t)}$$

and define $\phi_1 \in L^1(\mathbb{R})$ by $\phi_1 := \psi_0^+ * \psi_0^-$. We claim that $\phi_1$ is the function we are looking for. Since

$$\hat{\phi}_1(s) = \hat{\psi}_0(s + \delta_0)\overline{\hat{\psi}_0(-s + \delta_0)},$$

the Fourier transform of $\phi_1$ vanishes outside $(-\delta_0, \delta_0)$, hence outside $(-\delta, \delta)$, and $\hat{\phi}_1(0) = |\hat{\psi}_0(\delta_0)|^2 \neq 0$, so $\phi_1$ satisfies (b). By Step 1,

$$|\phi_1| \leq \psi * \psi \leq 2\pi\psi.$$

This shows that $\phi_1$ satisfies (c). Since $\psi \in L_\omega(\mathbb{R})$, this also shows that $\phi_1 \in L_\omega(\mathbb{R})$.

*Step 3.* Let $\phi_0$ be the function of Step 2 satisfying (a), (b) and (c) with $\delta := \epsilon/4$. Upon multiplying $\phi_0$ with a positive real number, we may assume that

$$\int_{-\infty}^{\infty} \hat{\phi}_0(s)\, ds = 1. \tag{2.4.1}$$

Let $\chi$ denote the characteristic function of the interval $(-\frac{1}{2} + \frac{3}{4}\epsilon, \frac{1}{2} - \frac{3}{4}\epsilon)$. Then $\hat{\chi} \in C_0(\mathbb{R}) \cap L^2(\mathbb{R})$, $|\hat{\chi}|$ is bounded, say by a constant $K$, and

$$\phi(t) := \frac{1}{2\pi}\phi_0(t) \cdot \hat{\chi}(-t)$$

defines a function $\phi \in L_\omega(\mathbb{R})$. For almost all $t \in \mathbb{R}$ we have

$$|\phi(-t)| \leq \frac{K}{2\pi}|\phi_0(-t)| \leq \frac{KC}{2\pi}\psi(-t) = \frac{KC}{2\pi}\psi(t)$$

and therefore (ii) is satisfied.

Clearly, $\hat{\phi} = \hat{\phi}_0 * \chi$. Since $0 \leq \chi \leq 1$ and $\hat{\phi}_0 \geq 0$, from the normalization (2.4.1) we have $0 \leq \hat{\phi} \leq 1$. If $|s| \leq \frac{1}{2} - \epsilon$, then

$$\hat{\phi}(s) = \int_{-\infty}^{\infty} \hat{\phi}_0(t)\chi(s-t)\,dt = \int_{-\frac{\epsilon}{4}}^{\frac{\epsilon}{4}} \hat{\phi}_0(t)\chi(s-t)\,dt$$

$$= \int_{-\frac{\epsilon}{4}}^{\frac{\epsilon}{4}} \hat{\phi}_0(t)\,dt = \int_{-\infty}^{\infty} \hat{\phi}_0(t)\,dt = 1.$$

On the other hand, if $\hat{\phi}(s) \neq 0$, then there is a $t \in \operatorname{supp} \hat{\phi}_0 = (-\frac{1}{4}\epsilon, \frac{1}{4}\epsilon)$ such that $s - t \in \operatorname{supp} \chi = (-\frac{1}{2} + \frac{3}{4}\epsilon, \frac{1}{2} - \frac{3}{4}\epsilon)$ and hence $s \in (-\frac{1}{2} + \frac{1}{2}\epsilon, \frac{1}{2} - \frac{1}{2}\epsilon)$. Consequently, $\hat{\phi}$ vanishes outside the interval $(-\frac{1}{2} + \frac{1}{2}\epsilon, \frac{1}{2} - \frac{1}{2}\epsilon)$. This proves that $\phi$ also satisfies (i).

It remains to prove the final assertion. Let $Y$ be the subspace of all functions in $L_\omega(\mathbb{R})$ whose Fourier transform has compact support. If $Y$ is not dense, there is a function $\phi \in (L_\omega(\mathbb{R}))^* = L_\omega^\infty(\mathbb{R})$, the space of all functions $\psi$ such that $\operatorname{ess\,sup}_{t \in \mathbb{R}} |\psi(t)|/\omega(t) < \infty$, such that

$$\langle \phi, f \rangle = \int_{-\infty}^{\infty} f(t)\phi(t)\,dt = 0, \quad \forall f \in Y.$$

By what we have already proved the space $Y$ is non-empty, and clearly $Y$ is invariant under translations and multiplication by the functions $t \mapsto e^{i\omega t}$, $\omega \in \mathbb{R}$. Hence, if $f \in Y$, then

$$\int_{-\infty}^{\infty} e^{-i\omega t} f(t-s)\phi(t)\,dt = 0, \quad \forall s \in \mathbb{R}, \omega \in \mathbb{R}.$$

Thus, for all $s \in \mathbb{R}$ the Fourier transform of the function $t \mapsto f(t-s)\phi(t)$ vanishes. Consequently, for all $s \in \mathbb{R}$ the function $t \mapsto f(t-s)\phi(t)$ is zero a.e., which implies that $\phi = 0$ a.e.    ////

**Lemma 2.4.3.** *Let $\omega$ be a non-quasianalytic weight on $\mathbb{R}$ and let $\mathbf{T}$ be a $C_0$-group on a Banach space $X$, with generator $A$, and assume that*

$$\|T(t)\| \leq \omega(t), \quad t \in \mathbb{R}.$$

*Then the following assertions are true.*
  (i) $\sigma(A) \subset i\mathbb{R}$;
  (ii) *For all $f \in L_\omega(\mathbb{R})$ and $x \in X$, the integral*

$$\hat{f}(\mathbf{T})x := \int_{-\infty}^{\infty} f(t)T(t)x\,dt$$

converges absolutely. The map $f \mapsto \hat{f}(\mathbf{T})$ is a continuous algebra homomorphism of $L_\omega(\mathbb{R})$ into $\mathcal{L}(X)$. If the Fourier transform $\hat{f}$ of $f$ belongs to $L^1(\mathbb{R})$, then

$$\hat{f}(\mathbf{T})x = \frac{1}{2\pi} \lim_{\delta \downarrow 0} \int_{-\infty}^{\infty} \hat{f}(-t) \left( R(\delta + it, A) - R(-\delta + it, A) \right) x \, dt. \quad (2.4.2)$$

In particular, if $\hat{f}$ is compactly supported and vanishes in a neighbourhood of $i\sigma(A)$, then $\hat{f}(\mathbf{T}) = 0$.

(iii) If $X \ne \{0\}$, then $\sigma(A) \ne \emptyset$.

*Proof:* The convergence of the integral is a trivial consequence of the definition of $L_\omega(\mathbb{R})$ and the growth assumption on $\mathbf{T}$. Also, one easily checks that $f \mapsto \hat{f}(\mathbf{T})$ is a continuous algebra homomorphism.

By Lemma 2.4.1, for all $t \in \mathbb{R}$ we have

$$r(T(t)) = \lim_{n \to \infty} \|T(nt)\|^{\frac{1}{n}} \le \limsup_{n \to \infty} \omega(nt)^{\frac{1}{n}} = 1.$$

Therefore, by Proposition 1.2.2 the uniform growth bounds of the $C_0$–semigroups $\{T(t)\}_{t \ge 0}$ and $\{T(-t)\}_{t \ge 0}$ are equal to zero. This implies that the spectral bounds of both $A$ and $-A$ are less than or equal to zero, which proves (i). Also, this implies that for all $\delta > 0$ and $x \in X$ we have

$$R(\delta - it, A)x = \int_0^\infty e^{-(\delta - is)t} T(t)x \, dt$$

and

$$R(-\delta - it, A)x = -R(\delta + it, -A) = -\int_0^\infty e^{-(\delta + is)t} T(-t)x \, dt,$$

both integrals being absolutely convergent.

If $\hat{f} \in L^1(\mathbb{R})$, by the formula for the inverse Fourier transform we have

$$f(t) = \frac{1}{2\pi} \int_{-\infty}^\infty \hat{f}(s) e^{its} \, ds, \quad \text{a.a. } t \in \mathbb{R}.$$

Using these facts, by the dominated convergence theorem and Fubini's theorem it follows that

$$\hat{f}(\mathbf{T})x = \lim_{\delta \downarrow 0} \int_{-\infty}^\infty e^{-\delta|t|} f(t) T(t) x \, dt$$

$$= \frac{1}{2\pi} \lim_{\delta \downarrow 0} \int_{-\infty}^\infty e^{-\delta|t|} \left( \int_{-\infty}^\infty e^{ist} \hat{f}(s) \, ds \right) T(t) x \, dt$$

$$= \frac{1}{2\pi} \lim_{\delta \downarrow 0} \int_{-\infty}^\infty \hat{f}(s) \left( \int_{-\infty}^\infty e^{-\delta|t|} e^{ist} T(t) x \, dt \right) ds$$

$$= \frac{1}{2\pi} \lim_{\delta \downarrow 0} \int_{-\infty}^\infty \hat{f}(s) \left( R(\delta - is, A) - R(-\delta - is, A) \right) x \, ds.$$

This proves (2.4.2).

If $\hat{f}$ is compactly supported and vanishes on a neighbourhood of $i\sigma(A)$, then $\hat{f}(\mathbf{T})x = 0$ for all $x \in X$ by (2.4.2) and the dominated convergence theorem.

Finally, assume $\sigma(A) = \emptyset$. Then (ii) implies that $\hat{f}(\mathbf{T}) = 0$ for all $f \in L_\omega(\mathbb{R})$ whose Fourier transform $\hat{f}$ has compact support. By Lemma 2.4.2, these functions are dense in $L_\omega(\mathbb{R})$ and thus $\hat{f}(\mathbf{T}) = 0$ for all $f \in L_\omega(\mathbb{R})$. In particular, by defining $f_0(t) := e^{-t}$ for $t \geq 0$ and $f_0(t) := 0$ for $t < 0$ we have $f_0 \in L_\omega(\mathbb{R})$ and $R(1,A) = \hat{f}_0(\mathbf{T}) = 0$. This implies $X = \overline{R(1,A)X} = \{0\}$. ////

Let us note that for *uniformly bounded* $C_0$-groups, Lemma 2.4.3 is entirely elementary. Indeed, all we need is that the set of all functions in $L^1(\mathbb{R})$ whose Fourier transform is compactly supported is dense; this follows by approximating a given $f$ in $L^1(\mathbb{R})$ by $f * K_\lambda$, where $K_\lambda$ is the Fejér kernel (cf. Section 5.2); these functions have compactly supported Fourier transforms. This observation is important because many of the results in Chapter 5 depend on this particular case of the lemma.

Before proving the first main result of this section, we recall some classical facts concerning spectral projections.

Let $A$ be the generator of a $C_0$-semigroup $\mathbf{T}$ on $X$ and suppose that $\sigma(A)$ is the disjoint union of two non-empty closed sets $\Omega_0 \cup \Omega_1$ with $\Omega_0$ compact. Let $\Gamma$ be a counterclockwise oriented smooth simple curve around $\Omega_0$ not enclosing any point of $\Omega_1$. In this situation we can define a bounded operator $P$ by

$$Px := \frac{1}{2\pi i} \int_\Gamma R(z,A)x\, dz, \quad x \in X.$$

The operator $P$ is a projection, and we have a direct sum decomposition $X = X_0 \oplus X_1$ by putting $X_0 := PX$ and $X_1 := (I-P)X$. Since $P$ commutes with $\mathbf{T}$, both $X_0$ and $X_1$ are $\mathbf{T}$-invariant. Denoting the generators of the restricted semigroups by $A_0$ and $A_1$ respectively, we have $\sigma(A_0) = \Omega_0$ and $\sigma(A_1) = \Omega_1$. The operator $P$ is called the *spectral projection* corresponding to $\Omega_0$.

**Theorem 2.4.4.** *Let $\omega$ be a non-quasianalytic weight on $\mathbb{R}$ and let $\mathbf{T}$ be a $C_0$-group on a Banach space $X$, with generator $A$, such that*

$$\|T(t)\| \leq \omega(t), \quad \forall t \in \mathbb{R}.$$

*Then,*

$$\sigma(T(t)) = \overline{\exp(t\,\sigma(A))}, \quad t \in \mathbb{R}.$$

*Proof:* If $X = \{0\}$, there is nothing to prove. Therefore, we assume that $X \neq \{0\}$. The proof of the theorem is divided into several steps.

*Step 1.* In view of the spectral inclusion theorem and a rescaling argument it suffices to show that $-1 \notin \sigma(T(2\pi))$ if $-1 \notin \overline{\exp(2\pi\sigma(A))}$. To this end, for $0 < \epsilon < \frac{1}{2}$ let

$$S(\epsilon) := \bigcup_{k \in \mathbb{Z}} i[k + \frac{1}{2} - \epsilon, k + \frac{1}{2} + \epsilon].$$

Then
$$\{-1\} = \bigcap_{0<\epsilon<\frac{1}{2}} \exp(2\pi\, S(\epsilon)).$$

Since, by assumption, $-1 \notin \overline{\exp(2\pi\, \sigma(A))}$, we can find some $0 < \epsilon < \frac{1}{4}$ such that
$$S(\epsilon) \cap \sigma(A) = \emptyset.$$

For $k \in \mathbb{Z}$ define
$$\sigma_k(A) := i[k - \tfrac{1}{2}, k + \tfrac{1}{2}] \cap \sigma(A).$$

Then
$$\sigma_k(A) \subset i(k - \tfrac{1}{2} + \epsilon, k + \tfrac{1}{2} - \epsilon).$$

Since $\sigma(A) \subset i\mathbb{R}$ we find that $\sigma(A)$ is decomposed into disjoint, closed subsets $\sigma_k(A)$, i.e. $\sigma(A) = \bigcup_{k \in \mathbb{Z}} \sigma_k(A)$.

*Step 2.* For $k \in \mathbb{Z}$ let $P_k$ be the spectral projection corresponding to $\sigma_k(A)$, which is given by
$$P_k = \frac{1}{2\pi i} \int_{\Gamma_k} R(z, A)\, dz,$$

where $\Gamma_k$ is the counterclockwise oriented circle with centre $ik$ and radius $\frac{1}{2} - \epsilon$. By Lemma 2.4.2 we can find a continuous function $\phi \in L_\omega(\mathbb{R})$ and a constant $M$ such that (i) and (ii) of that lemma hold. We claim that

$$P_k x = \int_{-\infty}^{\infty} e^{-ikt} \phi(-t) T(t) x\, ds, \quad x \in X. \tag{2.4.3}$$

Indeed, by Cauchy's theorem we may replace the path of integration $\Gamma_k$ in by the rectangle spanned by the points $\delta + i(k - \tfrac{1}{2} + \epsilon)$, $\delta + i(k + \tfrac{1}{2} - \epsilon)$, $-\delta + i(k + \tfrac{1}{2} - \epsilon)$, and $-\delta + i(k - \tfrac{1}{2} + \epsilon)$. By letting $\delta \downarrow 0$ and noting that $i(k + \tfrac{1}{2} - \epsilon) \in \varrho(A)$ and $i(k - \tfrac{1}{2} + \epsilon) \in \varrho(A)$, the integrals over the two horizontal parts tend to 0, and hence

$$P_k x = \frac{1}{2\pi} \lim_{\delta \downarrow 0} \int_{k-\frac{1}{2}+\epsilon}^{k+\frac{1}{2}-\epsilon} (R(\delta + it, A) - R(-\delta + it, A)) x\, dt$$
$$= \frac{1}{2\pi} \lim_{\delta \downarrow 0} \int_{k-\frac{1}{2}+\epsilon}^{k+\frac{1}{2}-\epsilon} \hat\phi(t - k)(R(\delta + it, A) - R(-\delta + it, A)) x\, dt, \quad x \in X.$$

Since $i(k + \tfrac{1}{2} - \epsilon, k + \tfrac{1}{2} - \tfrac{1}{2}\epsilon) \subset \varrho(A)$ and $i(k - \tfrac{1}{2} + \tfrac{1}{2}\epsilon, k - \tfrac{1}{2} + \epsilon) \subset \varrho(A)$, it follows that

$$\frac{1}{2\pi} \lim_{\delta \downarrow 0} \int_{k+\frac{1}{2}-\epsilon}^{k+\frac{1}{2}-\frac{1}{2}\epsilon} \hat\phi(t-k)(R(\delta+it,A)-R(-\delta+it,A))x\,dt$$
$$= \frac{1}{2\pi} \lim_{\delta \downarrow 0} \int_{k-\frac{1}{2}+\frac{1}{2}\epsilon}^{k-\frac{1}{2}+\epsilon} \hat\phi(t-k)(R(\delta+it,A)-R(-\delta+it,A))x\,dt = 0, \quad \forall x \in X.$$

Also, since $\hat{\phi}(\cdot - k)$ is identically zero outside the interval $(k - \frac{1}{2} + \frac{1}{2}\epsilon, k + \frac{1}{2} - \frac{1}{2}\epsilon)$, it follows from Lemma 2.4.3 applied to the function $t \mapsto e^{-ikt}\hat{\phi}(-t)$ that

$$P_k x = \frac{1}{2\pi} \lim_{\delta \downarrow 0} \int_{k-\frac{1}{2}+\frac{1}{2}\epsilon}^{k+\frac{1}{2}-\frac{1}{2}\epsilon} \hat{\phi}(t-k)(R(\delta+it, A) - R(-\delta+it, A))x \, dt$$

$$= \frac{1}{2\pi} \lim_{\delta \downarrow 0} \int_{-\infty}^{\infty} \hat{\phi}(t-k)(R(\delta+it, A) - R(-\delta+it, A))x \, dt$$

$$= \int_{-\infty}^{\infty} e^{ikt}\phi(-t)T(t)x \, dt, \quad \forall x \in X.$$

*Step 3.* In this step we prove that $\cup_{k \in \mathbb{Z}} P_k X$ is dense in $X$.

Define $Y$ as the the closure of $\cup_{k \in \mathbb{Z}} P_k X$. Then $Y$ is a **T**-invariant closed subspace of $X$; let $\mathbf{T}_{X/Y}$ be the quotient group on $X/Y$ induced by **T** and let $A_{X/Y}$ be its generator. By Proposition 1.1.7, applied to the semigroups $\{T(t)\}_{t \geq 0}$ and $\{T(-t)\}_{t \geq 0}$, it follows

$$\sigma(A_{X/Y}) \subset \sigma(A) = \cup_k \sigma_k(A), \qquad (2.4.4)$$

where $A_{X/Y}$ is the generator of the quotient group $\mathbf{T}_{X/Y}$ on $X/Y$.

On the other hand, we have $X = X_k^0 \oplus X_k^1$ with $X_k^0 := P_k X$ and $X_k^1 := (I - P_k)X$, both being **T**-invariant. Let $\mathbf{T}_k^i$ be the respective restrictions and let $A_k^i$ be their generators, $i = 0, 1$. Since $P_k$ is spectral projection corresponding to $\sigma_k(A)$, we have $\sigma(A_k^0) = \sigma_k(A)$ and $\sigma(A_k^1) \cap \sigma_k(A) = \emptyset$. But $X_k^0 = P_k X$ is contained in $Y$, and therefore $X/Y$ is a quotient of $X/X_k^0 = X_k^1$. Applying Proposition 1.1.7 once more, it follows that $\sigma(A_{X/Y}) \subset \sigma(A_k^1)$ and hence

$$\sigma(A_{X/Y}) \cap \sigma_k(A) = \emptyset.$$

This being true for all $k$, together with (2.4.4) it follows that $\sigma(A_{X/Y}) = \emptyset$. Since $\|T_{X/Y}\| \leq \|T(t)\| \leq \omega(t)$ for all $t \in \mathbb{R}$, Lemma 2.4.3 implies that $X/Y = \{0\}$, i.e. $Y = X$.

*Step 4.* In this step we prove that

$$2\pi \sum_{m \in \mathbb{Z}} \phi(-2\pi m)T(2\pi m) = I,$$

the convergence being in the operator norm.

By the estimate of $\phi$ we have

$$\sum_{m \in \mathbb{Z}} \|\phi(-s - 2\pi m)T(s + 2\pi m)\| \leq \sum_{m \in \mathbb{Z}} |\phi(-s - 2\pi m)|\omega(s + 2\pi m)$$

$$\leq M \sum_{m \in \mathbb{Z}} (1 + (s + 2\pi m)^2)^{-1}. \qquad (2.4.5)$$

This shows that the series $\sum_{m\in\mathbb{Z}} \phi(-s-2\pi m)T(s+2\pi m)$ converges absolutely with respect to the operator norm of $\mathcal{L}(X)$. Hence, we can rewrite (2.4.3) as follows:

$$P_k = \int_{-\infty}^{\infty} e^{-iks}\phi(-s)T(s)\,ds$$

$$= \sum_{m\in\mathbb{Z}} \int_{2\pi m}^{2\pi(m+1)} e^{-iks}\phi(-s)T(s)\,ds$$

$$= \int_0^{2\pi} e^{-iks}\left(\sum_{m\in\mathbb{Z}} \phi(-s-2\pi m)T(s+2\pi m)\right) ds.$$

Observe that for all $x \in X$ the $2\pi$-periodic continuous function

$$\xi_x(s) := 2\pi \sum_{m\in\mathbb{Z}} \phi(-s-2\pi m)T(s+2\pi m)x, \quad s\in\mathbb{R},$$

is continuous for every $x \in X$ and that $P_k x$ is the $k$-th Fourier coefficient of $\xi_x$. Since $P_k P_l = 0$ if $k \neq l$, for all $x \in \mathrm{lin}\{P_k E : k \in \mathbb{Z}\}$ we have $x = \sum_{k\in\mathbb{Z}} P_k x$. Hence by Fejér's theorem,

$$x = \sum_{k\in\mathbb{Z}} P_k x = (C,1)\sum_{k\in\mathbb{Z}} P_k x = (C,1)\sum_{k\in\mathbb{Z}} \hat{\xi}_x(k) = (C,1)\sum_{k\in\mathbb{Z}} \hat{\xi}_x(k)\cdot e^{-ik0}$$
$$= \xi_x(0) = 2\pi \sum_{m\in\mathbb{Z}} \phi(-2\pi m)T(2\pi m)x. \tag{2.4.6}$$

Since the linear span of the spaces $P_k X$ is dense in $X$ by Step 3, and since the operators $\phi(-2\pi m)T(2\pi m)$ are absolutely summable by (2.4.5), it follows from (2.4.6) that the series $2\pi \sum_{m\in\mathbb{Z}} \phi(-2\pi m)T(2\pi m)$ converges to the identity in the operator norm.

*Step 5.* To finish the proof we now assume $-1 \in \sigma(T(2\pi))$. Since the growth bounds of $\{T(t)\}_{t\geq 0}$ and $\{T(-t)\}_{t\geq 0}$ are both zero, we have $\sigma(T(t)) \subset \Gamma$ for all $t \geq 0$. Thus, $-1$ is an approximate eigenvalue of $\sigma(T(2\pi))$, i.e. there exists sequence $(x_n) \subset X$ of norm one vectors such that

$$\|T(2\pi)x_n + x_n\| \to 0 \quad \text{as } n\to\infty.$$

Let $\epsilon > 0$ be arbitrary and choose $N_0$ so large that

$$\left\|I - 2\pi \sum_{m=-N}^{N} \phi(-2\pi m)T(2\pi m)\right\| \leq \epsilon, \quad \forall N \geq N_0.$$

Then, using that $\|x_n\| = 1$, for all $n$ we have

$$\left|1 - 2\pi \sum_{m=-N}^{N} \phi(-2\pi m)(-1)^m\right|$$

$$\leq \epsilon + \left\|2\pi \sum_{m=-N}^{N} \phi(-2\pi m)T(2\pi m)x_n - 2\pi \sum_{m=-N}^{N} \phi(-2\pi m)(-1)^m x_n\right\|$$

$$\leq \epsilon + 2\pi \sum_{m=-N}^{N} |\phi(-2\pi m)| \, \|T(2\pi m)x_n - (-1)^m x_n\|.$$

By letting $n \to \infty$, the right hand side tends to zero. Therefore,

$$\left|1 - 2\pi \sum_{m=-N}^{N} \phi(-2\pi m)(-1)^m\right| \leq \epsilon, \quad \forall N \geq N_0.$$

This proves that
$$2\pi \sum_{m \in \mathbb{Z}} \phi(-2\pi m)(-1)^m = 1. \qquad (2.4.7)$$

We derive a contradiction from this as follows. Let

$$\psi(t) := \frac{1}{2\pi}(1 + e^{-2\pi i t})^{-1} \hat{\phi}(t).$$

Since $\text{supp}\,\hat{\phi} \subset (-\frac{1}{2}+\frac{1}{2}\epsilon, \frac{1}{2}-\frac{1}{2}\epsilon)$, $\psi$ is a continuous, compactly supported function. Moreover,
$$\phi(-s) = \hat{\psi}(s) + \hat{\psi}(s + 2\pi), \quad s \in \mathbb{R}.$$

In particular,
$$\phi(-2\pi m) = \hat{\psi}(2\pi m) + \hat{\psi}(2\pi m + 2\pi), \quad m \in \mathbb{Z}.$$

Hence for all $N \in \mathbb{N}$ we have

$$\sum_{m=-N}^{N} \phi(-2\pi m)(-1)^m = \sum_{m=-N}^{N} (-1)^m (\hat{\psi}(2\pi m) + \hat{\psi}(2\pi m + 2\pi))$$
$$= (-1)^N \hat{\psi}(2\pi N + 2\pi) + (-1)^{-N} \hat{\psi}(-2\pi N),$$

which tends to zero as $N \to \infty$ by the Riemann-Lebesgue lemma. This contradicts (2.4.7). ////

Although $\exp(t\sigma(A))$ is dense in $\sigma(T(t))$, it can be a very small subset of $\sigma(T(t))$. This is shown by the following simple example.

Let $X = c_0$, the space of all complex scalar sequences with the supremum norm, and let **T** be defined by

$$T(t)x_n = e^{int}x_n, \quad n = 0, 1, 2, ...,$$

where $x_n = (0, ..., 0, 1, 0, ...)$, is the $n$-th unit vector of $c_0$. The generator $A$ of **T** is easily checked to be given by

$$D(A) = \{(y_n) \in c_0 : \lim_{n\to\infty} ny_n = 0\},$$
$$A(y_n)_{n\in\mathbb{N}} = (iny_n)_{n\in\mathbb{N}}.$$

Therefore, $\sigma(A) = \{in : n \in \mathbb{N}\}$ and $\exp(t\sigma(A)) = \{e^{int} : n \in \mathbb{N}\}$.

On the other hand, the definition of **T** shows that $e^{int} \in \sigma(T(t))$ for all $n \in \mathbb{N}$ and $t \geq 0$. Since the spectrum of a bounded operator is always a closed set it follows that for $t/2\pi$ irrational we have

$$\Gamma = \overline{\{e^{int} : n \in \mathbb{N}\}} \subset \sigma(T(t)).$$

Since $T(t)$ is an invertible isometry, we also have $\sigma(T(t)) \subset \Gamma$, and hence $\sigma(T(t)) = \Gamma$ for these $t$.

As a corollary of Theorem 2.4.4, we have the following analogue of Gearhart's theorem.

**Corollary 2.4.5.** Let **T** be a $C_0$-group on a Banach space $X$, with generator $A$, and let $\omega$ be a non-quasianalytic weight such that $\|T(t)\| \leq \omega(t)$ for all $t \in \mathbb{R}$. Then the following assertions are equivalent:

(i) $1 \in \varrho(T(2\pi))$;
(ii) $i\mathbb{Z} \subset \varrho(A)$ and $\sup_{k\in\mathbb{Z}} \|R(ik, A)\| < \infty$.

*Proof:* (i)⇒(ii): This is proved as in Theorem 2.2.4. (ii)⇒(i): By Lemma 2.3.4, there is an $\epsilon > 0$ such that an $\epsilon$-neighbourhood of $i\mathbb{Z}$ is contained in the resolvent set of $A$. But then $1 \notin \overline{\exp(2\pi\sigma(A))}$. /////

The non-quasianalytic growth condition is the best possible. In fact, for a weight $\omega$ that fails to be non-quasianalytic we are going to show that the restriction of the translation group **T** on the weighted space $L^p(\mathbb{R}, (\omega(t))^{-p}dt)$ to a certain closed invariant subspace satisfies $\|T(t)\| \leq \omega(t)$ and $\sigma(A) = \emptyset$. Thus, the weak spectral mapping theorem fails for this restriction of **T**.

The idea of the construction is very elegant. It is possible to define a weighted Hardy space $H^p(\mathbb{R}, (\omega(t))^{-p}dt)$ over the upper half-plane, which is a closed subspace of $L^p(\mathbb{R}, (\omega(t))^{-p}dt)$. The crucial fact is that **T** leaves $H^p(\mathbb{R}, (\omega(t))^{-p}dt)$

invariant and can be extended holomorphically to a $C_0$-family $\{T(z)\}_{\operatorname{Im} z \geq 0}$ on $H^p(\mathbb{R}, (\omega(t))^{-p} dt)$. Along the vertical semi-axis, we then show that

$$\lim_{y \to \infty} \frac{\log \|T(iy)\|}{y} = -\infty.$$

Therefore, the spectrum of the generator $iA$ of the $C_0$-semigroup $\{T(iy)\}_{y \geq 0}$ is empty. Hence also $\sigma(A)$ is empty. On the other hand, the definition of the space $L^p(\mathbb{R}, (\omega(t))^{-p} dt)$ immediately implies that translation over $t$ defines a bounded operator of norm $\leq \omega(t)$ on $L^p(\mathbb{R}, (\omega(t))^{-p} dt)$, and this estimate is inherited by passing to the closed subspace $H^p(\mathbb{R}, (\omega(t))^{-p} dt)$.

We note that passing to a subspace of $L^p(\mathbb{R}, (\omega(t))^{-p} dt)$ is necessary: since $L^p(\mathbb{R}, (\omega(t))^{-p} dt)$ is a Banach lattice and translation on this space defines a positive $C_0$-group, the spectrum of its generator is non-empty by Corollary 1.4.2.

In order to work out the details of the approach outlined above, we need some preliminary lemmas. Let $\rho : \mathbb{R} \to \mathbb{R}_+$ be a non-negative measurable function such that

$$\rho(t) \leq M(1 + |t|), \quad t \in \mathbb{R}. \tag{2.4.8}$$

For $x \in \mathbb{R}$ and $y > 0$ we define

$$u(x + iy) = u_\rho(x + iy) := \frac{y}{\pi} \int_{-\infty}^{\infty} \left( \frac{1}{(t-x)^2 + y^2} - \frac{1}{t^2 + 1} \right) \rho(t) \, dt.$$

By (2.4.8), this integral is absolutely convergent.

**Lemma 2.4.6.** *The function $u$ is harmonic in $\{z = x + iy : y > 0\}$, and for almost all $x \in \mathbb{R}$ we have*

$$\lim_{z \to x} u(z) = \rho(x)$$

*non-tangentially. If $\rho$ is continuous, the limit exists for all $x \in \mathbb{R}$.*

*Proof:* Fix $1 \leq r < R$ arbitrary and define the function $\rho_R : \mathbb{R} \to \mathbb{R}$ by

$$\rho_R(t) = \begin{cases} \rho(-R), & t < -R; \\ \rho(t), & -R \leq t \leq R; \\ \rho(R), & t > R. \end{cases}$$

Then $\rho_R$ is a bounded measurable function, and therefore its Poisson integral

$$(P\rho_R)(x + iy) := \frac{y}{\pi} \int_{-\infty}^{\infty} \frac{1}{(t-x)^2 + y^2} \rho_R(t) \, dt$$

is a harmonic function in the open upper half-plane, and $\lim_{z \to x}(P\rho_R)(z) = \rho_R(x)$ non-tangentially for almost all $x \in \mathbb{R}$. If $\rho$ is continuous, so is $\rho_R$ and $P\rho_R$ is continuous up to the boundary. Since the function $x + iy \mapsto y$ is harmonic as well, it follows that $u_R := u_{\rho_R}$ is harmonic. Moreover, $\lim_{z \to x}(u_R)(z) = \rho_R(x)$ non-tangentially for almost all $x \in \mathbb{R}$; if $\rho$ is continuous this holds for all $x \in \mathbb{R}$.

We claim that $\lim_{R \to \infty} u_R(x + iy) = u(x + iy)$, uniformly on the semi-disc $\{x^2 + y^2 \leq r^2, y > 0\}$. Once this is proved, the lemma follows easily. In order to establish the claim, we note that $\rho_R = \rho$ on $[-R, R]$ and therefore, using that $|x| \leq r$ and $R \geq 1$,

$$|u(x+iy) - u_R(x+iy)|$$
$$= \frac{y}{\pi}\left|\int_{|t|\geq R}\left(\frac{1}{(t-x)^2+y^2} - \frac{1}{t^2+1}\right)(\rho(t) - \rho_R(t))\,dt\right|$$
$$\leq \frac{2My}{\pi}\int_{|t|\geq R}\frac{t^2+1-(t^2-2tx+x^2+y^2)}{((t-x)^2+y^2)(t^2+1)}(1+|t|)\,dt$$
$$\leq \frac{4My}{\pi}\int_{|t|\geq R}\frac{1+2r|t|+r^2}{((t-x)^2+y^2)(t^2+1)}|t|\,dt$$
$$\leq \frac{12My}{\pi}(1+r^2)\int_{|t|\geq R}\frac{1}{(t-x)^2+y^2}\,dt,$$

where in the last estimate we used that $(1+2r|t|+r^2)|t| \leq t^2 + 3r^2 t^2 \leq 3(1+r^2)(t^2+1)$ for all $|t| \geq 1$ and $r \geq 1$. From this, the claim is an immediate consequence.
////

We shall need a number of estimates for $u$. In the following lemmas, $M$ is the constant of (2.4.8).

**Lemma 2.4.7.** *For all $x \in \mathbb{R}$ and $y > 0$ we have*

$$|u(x+iy)| \leq M\left(2|x| + y + 4 + \frac{y}{\pi}\log(x^2+y^2)\right).$$

*Proof:* First we estimate $|u(iy)|$. If $y \geq 1$ we have

$$|u(iy)| \leq \frac{2My}{\pi}\int_0^\infty \left(\frac{1}{t^2+1} - \frac{1}{t^2+y^2}\right)(1+t)\,dt$$
$$= M(y - 1 + \frac{y}{\pi}\log y^2).$$

Similarly, if $0 < y \leq 1$, then

$$|u(iy)| \leq M(1 - y - \frac{y}{\pi}\log y^2). \qquad (2.4.9)$$

This proves the lemma for $x = 0$.

Next, we estimate $|u(x+iy) - u(iy)|$. First we let $x > 0$. The function $t \mapsto ((t-x)^2 + y^2)^{-1} - (t^2+y^2)^{-1}$ is positive for $t > \frac{x}{2}$ and negative for $t < \frac{x}{2}$.

Also, $|t|$ equals $t$ for $t \geq 0$ and equals $-t$ for $t \leq 0$. Accordingly, we split the integral into $\int_{\frac{x}{2}}^{\infty} + \int_0^{\frac{x}{2}} + \int_{-\infty}^0$ and calculate each of them explicitly. This gives

$$|u(x+iy) - u(iy)| \leq M \frac{y}{\pi} \int_{-\infty}^{\infty} \left| \frac{1}{(t-x)^2 + y^2} - \frac{1}{t^2+y^2} \right| (1+|t|)\, dt$$

$$= M \left( x + \frac{y}{\pi} \log(x^2+y^2) - \frac{y}{\pi} \log y^2 + \frac{4+2x}{\pi} \arctan \frac{x}{2y} - \frac{2x}{\pi} \arctan \frac{x}{y} \right)$$

$$\leq M \left( 2x + 2 + \frac{y}{\pi} \log(x^2+y^2) - \frac{y}{\pi} \log y^2 \right).$$

In the last estimate we used that $|\arctan s_0 - \arctan s_1| \leq \frac{\pi}{2}$ for all $s_0, s_1 \geq 0$.
Hence, if $x > 0$ and $y \geq 1$,

$$|u(x+iy)| \leq |u(x+iy) - u(iy)| + |u(iy)| \leq M \left( 2x + y + 1 + \frac{y}{\pi} \log(x^2+y^2) \right)$$

and if $x > 0$ and $0 < y < 1$,

$$|u(x+iy)| \leq M \left( 2x + 3 - y + \frac{y}{\pi} \log(x^2+y^2) - \frac{2y}{\pi} \log y^2 \right)$$

$$\leq M \left( 2x + 3 + \frac{4}{\pi e} + \frac{y}{\pi} \log(x^2+y^2) \right),$$

using that $|y \log y| \leq -e^{-1}$ for all $0 < y < 1$. Combining these estimates, the lemma follows for $x > 0$.
By a similar proof, the same estimate holds for $x < 0$. ////

From now an, we assume in addition that $\rho$ is subadditive. Note that for non-negative, locally bounded, subadditive functions the estimate (2.4.8) is automatically satisfied.

**Lemma 2.4.8.** *For all $x \in \mathbb{R}$ and $y > 0$ we have*

$$u(x+iy) - u(iy) \leq \rho(x) + M(1 + \min\{|x|, y\}).$$

*Proof:* Fix $y > 0$ and $x \geq 0$ arbitrary. Noting that

$$\frac{y}{\pi} \int_{-\frac{x}{2}}^{\infty} \left( \frac{1}{t^2+y^2} - \frac{1}{(t+x)^2+y^2} \right) dt \leq \frac{y}{\pi} \int_{-\infty}^{\infty} \frac{1}{t^2+y^2}\, dt = 1,$$

we have

$$u(x+iy) - u(iy) = \frac{y}{\pi} \int_{-\frac{x}{2}}^{\infty} \left( \frac{1}{t^2+y^2} - \frac{1}{(t+x)^2+y^2} \right) \rho(t+x)\, dt$$

$$+ \frac{y}{\pi} \int_{-\infty}^{\frac{x}{2}} \left( \frac{1}{(t-x)^2+y^2} - \frac{1}{t^2+y^2} \right) \rho(t)\, dt$$

$$\leq \rho(x) + \frac{y}{\pi} \int_{-\frac{x}{2}}^{\infty} \left( \frac{1}{t^2+y^2} - \frac{1}{(t+x)^2+y^2} \right) \rho(t)\, dt$$

$$+ \frac{y}{\pi} \int_{-\frac{x}{2}}^{\frac{x}{2}} \left( \frac{1}{(t-x)^2+y^2} - \frac{1}{t^2+y^2} \right) \rho(t)\, dt$$

$$= \rho(x) + \frac{y}{\pi} \int_{\frac{x}{2}}^{\infty} \left( \frac{1}{t^2+y^2} - \frac{1}{(t+x)^2+y^2} \right) \rho(t)\, dt$$

$$+ \frac{y}{\pi} \int_{-\frac{x}{2}}^{\frac{x}{2}} \left( \frac{1}{(t-x)^2+y^2} - \frac{1}{(t+x)^2+y^2} \right) \rho(t)\, dt$$

$$\leq \rho(x) + \frac{My}{\pi} \int_{\frac{x}{2}}^{\infty} \left( \frac{1}{t^2+y^2} - \frac{1}{(t+x)^2+y^2} \right) (1+t)\, dt$$

$$+ \frac{My}{\pi} \int_{0}^{\frac{x}{2}} \left( \frac{1}{(t-x)^2+y^2} - \frac{1}{(t+x)^2+y^2} \right) (1+t)\, dt$$

$$= \rho(x) + \frac{My}{\pi} \left( \frac{x}{y} \left( \frac{\pi}{2} - \arctan\frac{x}{2y} - \arctan\frac{3x}{2y} \right) \right.$$

$$\left. + \frac{2}{y} \left( \arctan\frac{x}{y} - \arctan\frac{x}{2y} \right) + \frac{1}{2}\log\left(\frac{9x^2}{4} + y^2\right) - \frac{1}{2}\log(x^2+y^2) \right)$$

$$\leq \rho(x) + \frac{Mx}{\pi} \left( \frac{\pi}{2} - \arctan\frac{x}{2y} \right) + M.$$

In the last estimate we used the fact that

$$\frac{1}{2}\log\left(\frac{9x^2}{4} + y^2\right) - \frac{1}{2}\log(x^2+y^2) - \frac{x}{y}\arctan\frac{3x}{2y}$$

$$= \int_{x}^{\frac{3x}{2}} \frac{t}{t^2+y^2}\, dt - \int_{0}^{\frac{3x}{2}} \frac{x}{t^2+y^2}\, dt$$

$$\leq \int_{x}^{\frac{3x}{2}} \frac{\frac{3}{2}x}{t^2+y^2}\, dt - \int_{0}^{\frac{3x}{2}} \frac{x}{t^2+y^2}\, dt$$

$$\leq \frac{1}{2} \int_{0}^{\frac{x}{2}} \frac{x}{(t+x)^2+y^2}\, dt - \int_{0}^{x} \frac{x}{t^2+y^2}\, dt$$

$$\leq \frac{1}{2} \int_{0}^{\frac{x}{2}} \frac{x}{t^2+y^2}\, dt - \int_{0}^{x} \frac{x}{t^2+y^2}\, dt \leq 0.$$

Now, $\dfrac{Mx}{\pi}(\dfrac{\pi}{2} - \arctan \dfrac{x}{2y}) \leq Mx$ and also

$$\dfrac{Mx}{\pi}(\dfrac{\pi}{2} - \arctan \dfrac{x}{2y}) = \dfrac{M}{\pi}\int_{\frac{1}{2}}^{\infty} \dfrac{x^2 y}{t^2 x^2 + y^2}\, dt \leq \dfrac{My}{\pi}\int_{\frac{1}{2}}^{\infty} \dfrac{1}{t^2}\, dt = \dfrac{2My}{\pi}.$$

By combining these estimates, the lemma follows for $x \geq 0$. Analogously, for $x \leq 0$ one obtains $u(x+iy) - u(iy) \leq \rho(x) + M(1 + \min\{|x|, y\})$. ////

**Lemma 2.4.9.** For all $x \in \mathbb{R}$ and $y > 0$ and all $\xi \in \mathbb{R}$ we have

$$\limsup_{z \to \xi} u(z + x) - u(z) \leq 6M + \rho(x)$$

and

$$\limsup_{z \to \xi} u(z + iy) - u(z) \leq 7M(1 + y) + u(iy).$$

*Proof:* Define

$$\tilde{\rho}(t) := \int_{-\frac{1}{2}}^{\frac{1}{2}} \rho(t+s)\, ds.$$

Since $-\rho(-s) \leq \rho(t+s) - \rho(t) \leq \rho(s)$ by the subadditivity of $\rho$, from (2.4.8) it follows that

$$|\tilde{\rho}(t) - \rho(t)| = \left|\int_{-\frac{1}{2}}^{\frac{1}{2}} \rho(t+s) - \rho(t)\, ds\right| \leq \sup_{-\frac{1}{2} \leq s \leq \frac{1}{2}} \rho(s) \leq \dfrac{3}{2}M, \quad \forall t \in \mathbb{R}.$$

The function $\tilde{\rho}$ is subadditive and continuous and satisfies

$$\tilde{\rho}(t) \leq M \int_{-\frac{1}{2}}^{\frac{1}{2}} 1 + |t + s|\, ds \leq M\left(1 + |t| + \int_{-\frac{1}{2}}^{\frac{1}{2}} |s|\, ds\right) = M(\dfrac{5}{4} + |t|).$$

It follows that Lemma 2.4.6 applies to $\tilde{\rho}$; we put $\tilde{u} := u_{\tilde{\rho}}$.

Fix $x \in \mathbb{R}$ and $y > 0$ arbitrary. Then,

$$|\tilde{u}(x+iy) - u(x+iy)| \leq \dfrac{3My}{2\pi}\int_{-\infty}^{\infty} \dfrac{1}{(t-x)^2 + y^2} + \dfrac{1}{t^2 + 1}\, dt = \dfrac{3}{2}M(1+y).$$

Moreover, since $\tilde{\rho}$ is continuous, for all $\xi \in \mathbb{R}$ we have $\lim_{z \to \xi} \tilde{u}(z) = \tilde{\rho}(\xi)$ and $\lim_{z \to \xi} \tilde{u}(z + x) = \tilde{\rho}(\xi + x)$ by Lemma 2.4.6. It follows that

$$\limsup_{z \to \xi} u(z+x) - u(z) \leq \limsup_{z \to \xi} 3M(1 + \operatorname{Im} z) + \tilde{u}(z+x) - \tilde{u}(z)$$
$$= 3M + \tilde{\rho}(\xi + x) - \tilde{\rho}(\xi)$$
$$\leq 6M + \rho(\xi + x) - \rho(\xi)$$
$$\leq 6M + \rho(x)$$

using the subadditivity of $\rho$. Similarly, using Lemma 2.4.8 we have

$$\limsup_{z \to \xi} u(z+iy) - u(z) \leq \frac{3}{2}M(1+y) + \frac{3}{2}M + \tilde{u}(\xi + iy) - \tilde{\rho}(\xi)$$
$$\leq 3M(1+y) + 3M + u(\xi + iy) - \rho(\xi)$$
$$\leq 4M(1+y) + 3M + u(iy).$$

////

**Lemma 2.4.10.** *For all $0 < y_0 \leq y_1$ we have*

$$u(iy_1) - u(iy_0) \leq 4M.$$

*Proof:* First we observe that for all $0 < y \leq 1$ by (2.4.9) we have $|u(iy)| \leq 2M$. Therefore, in order to prove the lemma it suffices to show that

$$u(iy_1) - u(iy_0) \leq 2 \sup_{0 < y \leq 1} |u(iy)|.$$

It is an immediate consequence of the definition of $u$ that $u(iy) \leq 0$ for all $y \geq 1$. Hence it is enough to prove that $y \mapsto u(iy)$ is non-increasing for $y \geq 1$.

To see this, we differentiate under the integral sign in the definition of $u$ and obtain

$$\frac{du}{dy}(iy) = \frac{1}{\pi}\int_{-\infty}^{\infty}\left(\frac{t^2 - y^2}{(t^2+y^2)^2} - \frac{1}{t^2+1}\right)\rho(t)\,dt.$$

If $y \geq 1$, then the right hand side is less than or equal to zero. ////

In the proof of Lemma 2.4.9 we had to introduce the function $\tilde{\rho}$ in order to obtain the limes superior for *all* $\xi \in \mathbb{R}$ (rather than for almost all $\xi \in \mathbb{R}$). This enables us to apply the following version of the Phragmen-Lindelöf theorem: Let $\gamma \geq \frac{1}{2}$ and let $\Omega := \{z \in \mathbb{C} : -\frac{\pi}{2\gamma} < \arg z < \frac{\pi}{2\gamma}\}$. If $f$ is a holomorphic function in $\Omega$ such that

(i) $\limsup_{z \to \zeta} |f(z)| \leq M$ for some constant $M > 0$ and all $\zeta \in \partial\Omega$,

(ii) $\limsup_{|z| \to \infty} \dfrac{\log(\log^+ |f(z)|)}{\log |z|} < \gamma,$

then $\sup_{z \in \Omega} |f(z)| \leq M$.

**Lemma 2.4.11.** *For all $x \in \mathbb{R}$ and $y > 0$ we have*

$$\sup_{\text{Im} z > 0}(u(z + (x+iy)) - u(z)) \leq 13M(1+y) + \rho(x) + u(iy).$$

*Proof:* We denote the supremum on the left hand side by $L_{x+iy}$. For $x \in \mathbb{R}$ and $y > 0$, define the holomorphic function $H_{x+iy}(\cdot)$ in the upper half-plane by

$$H_{x+iy}(z) := \frac{G(z)}{G(z + (x+iy))},$$

where $G(z) := e^{-u(z)-iv(z)}$ and $v$ is the harmonic conjugate of $u$. Then,

$$|H_{x+iy}(z)| = \exp(u(z+(x+iy)) - u(z)).$$

By Lemma 2.4.9, for all $x \in \mathbb{R}$, $y > 0$, and $\xi \in \mathbb{R}$ we have

$$\limsup_{z \to \xi} |H_x(z)| \leq \exp(6M + \rho(x)) \tag{2.4.10}$$

and

$$\limsup_{z \to \xi} |H_{iy}(z)| \leq \exp(7M(1+y) + u(iy)). \tag{2.4.11}$$

On the other hand, by Lemma 2.4.8 for all $x \in X$ and $\eta > 0$ we have $u(x + i\eta) - u(i\eta) \leq \rho(x) + M(1+|x|)$ and therefore

$$|H_x(i\eta)| \leq \exp(\rho(x) + M(1+|x|)). \tag{2.4.12}$$

Similarly, by Lemma 2.4.10 for all $y > 0$ and $\eta > 0$ we have $u(iy+i\eta) - u(iy) \leq 4M$ and hence

$$H_{iy}(i\eta) \leq \exp(4M). \tag{2.4.13}$$

By Lemma 2.4.7,

$$\limsup_{\text{Im } z > 0, |z| \to \infty} \frac{\log(\log^+ |H_{x+iy}(z)|)}{\log |z|} < 2 \tag{2.4.14}$$

for all $x \in \mathbb{R}$ and $y > 0$. By (2.4.10)-(2.4.14), the Phragmen-Lindelöf theorem applied to functions $H_x$ and $H_{iy}$ and $\gamma = 2$ implies that both $|H_x(\cdot)|$ and $|H_{iy}(\cdot)|$ are uniformly bounded in the sectors $\{0 < \arg z < \frac{\pi}{2}\}$ and $\{\frac{\pi}{2} < \arg z < \pi\}$. It follows that both functions are uniformly bounded in $\{\text{Im } z > 0\}$. Applying the Phragmen-Lindelöf theorem once more, now with $\gamma = 1$, it follows from (2.4.10) and (2.4.11) that

$$\sup_{\text{Im } z > 0} |H_x(z)| \leq \exp(6M + \rho(x))$$

and

$$\sup_{\text{Im } z > 0} |H_{iy}(z)| \leq \exp(7M(1+y) + u(iy)). \tag{2.4.15}$$

These inequalities imply that $L_x \leq 6M + \rho(x)$ for all $x \in X$ and $L_{iy} \leq 7M(1+y) + u(iy)$ for all $y > 0$. Therefore, the lemma follows from the obvious fact that $L_{x+iy} \leq L_x + L_{iy}$. ////

Now let $\omega$ be a weight on $\mathbb{R}$. Since weights are exponentially bounded, the function $\rho := \log \omega(t)$ satisfies (2.4.8) for some constant $M$. Also, the submultiplicativity of $\omega$ implies that $\rho$ is subadditive.

Fix $1 \leq p < \infty$ and define $H_\omega^p$ to be the space of all holomorphic functions in the upper half-plane such that $f \cdot G \in H^p$, the usual Hardy space over the upper half-plane; $G$ is the function from the proof of Lemma 2.4.11. The space $H^p_\omega$ is easily seen be a Banach space, and for each $f \in H^p_\omega$ the non-tangential limits

$$f(x) := \lim_{z \to x} f(z)$$

exists for almost all $x$. Indeed, the almost everywhere non-tangential limits exist for all functions in $H^p$ and by Lemma 2.4.6 the non-tangential limit of $|G|$ exists almost everywhere (and equals $\omega$). In terms of this boundary function the norm of $II^p_\omega$ is given by

$$\|f\|_{H^p_\omega} = \left(\int_{-\infty}^{\infty} |f(x)|^p (\omega(x))^{-p}\, dx\right)^{\frac{1}{p}}.$$

Thus, $H^p_\omega$ can be isometrically identified with a closed subspace of the weighted space $L^p_\omega := L^p(\mathbb{R}, (\omega(x))^{-p}\, dx)$. On the latter space, we define the translation group $\mathbf{T}$ by

$$(T(t)f)(x) = f(x+t), \quad x \in \mathbb{R},\, t \in \mathbb{R}.$$

**Lemma 2.4.12.** *The family $\mathbf{T} = \{T(t)\}_{t \in \mathbb{R}}$ is a $C_0$-group on $L^p_\omega$. Moreover,*

$$\|T(t)\|_{L^p_\omega} \leq \omega(t), \quad t \in \mathbb{R}.$$

*Proof:* The estimate for $\|T(t)\|$ is an immediate consequence of the estimate

$$\|T(t)f\|^p_{L^p_\omega} = \int_{-\infty}^{\infty} |T(t)f(x)|(\omega(x))^{-p}\, dx$$
$$= \int_{-\infty}^{\infty} |f(x+t)|(\omega(x))^{-p}\, dx$$
$$= \int_{-\infty}^{\infty} |f(x)|^p (\omega(x-t))^{-p}\, dx$$
$$\leq (\omega(t))^p \int_{-\infty}^{\infty} |f(x)|(\omega(x))^{-p}\, dx$$
$$= (\omega(t))^p \|f\|^p_{L^p_\omega}.$$

By Lusin's theorem, the compactly supported continuous functions are dense in $L^p_\omega$. For each such function $f$, it is evident that

$$\lim_{t \downarrow 0} \|T(t)f - f\|_{L^p_\omega} = 0.$$

Since $\sup_{0 \leq t \leq 1} \|T(t)\| < \infty$, the strong continuity of $\mathbf{T}$ follows from this. ////

Next, we are going to show that **T** leaves $H_\omega^p$ invariant and that the restriction to this space can be holomorphically extended to the closed upper half plane.

For a function $f \in H_\omega^p$ and $x \in \mathbb{R}$ and $y \geq 0$, we define the holomorphic function $T(x+iy)f$ by

$$(T(x+iy)f)(z) := f(z+(x+iy)), \quad \text{Im } z > 0.$$

**Theorem 2.4.13.** *For all $x \in \mathbb{R}$ and $y \geq 0$, the operator $T(x+iy)$ defines a bounded linear operator on $H_\omega^p$. The family $\{T(z)\}_{\text{Im } z \geq 0}$ is strongly continuous on $H_\omega^p$ and*

$$\|T(x)\|_{H_\omega^p} \leq \omega(x), \quad x \in \mathbb{R}.$$

*If $\omega$ fails to be non-quasianalytic, then the generator $A$ of the $C_0$–group $\mathbf{T} = \{T(x)\}_{x \in \mathbb{R}}$ has empty spectrum.*

*Proof:* Fix $f \in H_\omega^p$. First assume $y = 0$. It is evident that for all $x \in \mathbb{R}$, $T(x)$ maps $H_\omega^p$ into itself. Since $H_\omega^p$ is isometrically isomorphic to a closed subspace of $L_\omega^p$, the estimate $\|T(x)\| \leq \omega(x)$ is a consequence of Lemma 2.4.12.

For $x \in \mathbb{R}$, $y > 0$, and Im $z > 0$ we have

$$|(T(x+iy)f)(z)G(z)| = |f(z+(x+iy))G(z)|$$
$$= |f(z+(x+iy))G(z+(x+iy))| \exp(u(z+(x+iy)) - u(z)).$$

Therefore, By Lemma 2.4.11 and the fact that the $p$-norm along horizontal lines $z = y$ of $H^p$-functions is non-increasing with $y$, we see that $(T(x+iy)f)G \in H^p$ and

$$\|T(x+iy)f\|_{H_\omega^p} \leq e^{13M(1+y)+\rho(x)+u(iy)} \|f\|_{H_\omega^p}. \tag{2.4.16}$$

This proves that $T(x+iy)$ defines a bounded operator on $H_\omega^p$.

For the proof of strong continuity it suffices to check that $\lim_{y \downarrow 0} \|T(iy)f - f\| = 0$ for all $f \in H_\omega^p$; indeed, together with the strong continuity along the horizontal axis it then follows that $\lim_{\text{Im } z \geq 0, z \to 0} \|T(z)f - f\| = 0$ for all $f \in H_\omega^p$.

Let $f \in H_\omega^p$ be fixed. For all Im $z > 0$ and $y > 0$ we have

$$(f(z+iy) - f(z))G(z) = (f(z+iy)G(z+iy) - f(z)G(z))H_{iy}(z) \\ - f(z)G(z)(1 - H_{iy}(z)), \tag{2.4.17}$$

where we use the notation of Lemma 2.4.11. By (2.4.15) and (2.4.9), for $0 < y < 1$ we have

$$\sup_{\text{Im } z > 0} |H_{iy}(z)| \leq \exp(14M + |u(iy)|) \leq \exp(16M).$$

Hence by raising (2.4.17) to the $p$-th power and integrating along horizontal lines Im $z =$ constant we obtain

$$\limsup_{y \downarrow 0} \|f(\cdot+iy) - f\|_{H_\omega^p} \leq \exp(16M) \limsup_{y \downarrow 0} \|(fG)(\cdot+iy) - (fG)(\cdot)\|_{H^p}. \tag{2.4.18}$$

Note that the integral arising from the second term in the right hand side of (2.4.17) tends to 0 by the dominated convergence theorem since $\lim_{y \downarrow 0} H_{iy}(z) = 1$ for all $z$. The right hand side of (2.4.18) tends to 0 by an elementary result on $H^p$-functions. This concludes the proof of strong continuity.

Now assume that $\omega$ fails to be non-quasianalytic, i.e.
$$\int_{-\infty}^{\infty} \frac{\log \omega(t)}{1+t^2} \, dt = \infty.$$

Let $N > 0$ be arbitrary and choose $\tau > 0$ so large that
$$\int_{-\tau}^{\tau} \frac{\log \omega(t)}{1+t^2} \, dt \geq N.$$

Choose $y \geq 1$ so large that $(t^2 + y^2)^{-1} \leq \frac{1}{2}(t^2 + 1)^{-1}$ for all $t \in [-\tau, \tau]$. Then,
$$\begin{aligned}
\frac{u(iy)}{y} &= \frac{1}{\pi} \int_{-\infty}^{\infty} \left( \frac{1}{t^2 + y^2} - \frac{1}{t^2 + 1} \right) \log \omega(t) \, dt \\
&\leq \frac{1}{\pi} \int_{-\tau}^{\tau} \left( \frac{1}{t^2 + y^2} - \frac{1}{t^2 + 1} \right) \log \omega(t) \, dt \\
&\leq -\frac{1}{2\pi} \int_{-\tau}^{\tau} \frac{1}{t^2 + 1} \log \omega(t) \, dt \leq -\frac{N}{2\pi}.
\end{aligned}$$

Therefore, by (2.4.16),
$$\lim_{y \to \infty} \frac{\log \|T(iy)\|}{y} \leq \lim_{y \to \infty} \frac{13M(1+y) + \rho(0) + u(iy)}{y} = -\infty.$$

It follows that the uniform growth bound of the $C_0$-semigroup $\{T(iy)\}_{y \geq 0}$ is $-\infty$. But then its generator, which is $iA$, has empty spectrum by the Hille-Yosida theorem. Therefore, also $\sigma(A) = \emptyset$. ////

Since the spectrum of a bounded operator is always non-empty, the weak spectral mapping theorem necessarily fails if the spectrum of the generator is empty. Therefore, combining Theorems 2.4.4 and 2.4.13 we obtain:

**Corollary 2.4.14.** *Let $\omega$ be a weight on $\mathbb{R}$. Then the following assertions are equivalent:*
(i) *$\omega$ is non-quasianalytic;*
(ii) *For all Banach spaces $X$ and all $C_0$-groups $\mathbf{T}$ on $X$ satisfying*
$$\|T(t)\| \leq \omega(t), \quad t \in \mathbb{R},$$
*we have $\sigma(A) \neq \emptyset$;*
(iii) *For all Banach spaces $X$ and all $C_0$-groups $\mathbf{T}$ on $X$ satisfying*
$$\|T(t)\| \leq \omega(t), \quad t \in \mathbb{R},$$
*the weak spectral mapping theorem holds.*

Since $H^2_\omega$ is a Hilbert space, the word 'Banach space' in the corollary may be replaced by 'Hilbert space'.

Another interesting corollary is the following.

**Corollary 2.4.15.** *Let $\omega$ be a weight on $\mathbb{R}$ which is not non-quasianalytic. Then there exists a lower semicontinuous weight $\tilde{\omega}$ with $1 \leq \tilde{\omega} \leq \omega$ which also fails to be non-quasianalytic.*

*Proof:* Let **T** be the group on $H^p_\omega$ of Theorem 2.4.13 and define the weight $\tilde{\omega}(t) := \max\{1, \|T(t)\|\}$. Clearly, $1 \leq \tilde{\omega} \leq \omega$ and $\tilde{\omega}$ is submultiplicative. If $\tilde{\omega}$ were non-quasianalytic, then in view of $\|T(t)\| \leq \tilde{\omega}$ the weak spectral mapping theorem would hold for **T** and hence $\sigma(A) \neq \emptyset$, a contradiction.  ////

## 2.5. Latushkin - Montgomery-Smith theory

In this section we prove the surprising fact that if one tensors an arbitrary $C_0$-semigroup on a Banach space $X$ with the rotation semigroup on $L^p(\Gamma)$ or $C(\Gamma)$, the spectral mapping theorem holds for the resulting semigroup on $L^p(\Gamma, X)$ resp. $C(\Gamma, X)$. This important result, due to Y. Latushkin and S. Montgomery-Smith, is the basis of L. Weis's stability theorem for positive semigroups on $L^p$-spaces in Section 3.5.

The idea behind the proof is that in $L^p(\Gamma, X)$ and $C(\Gamma, X)$ one has more space to construct approximate eigenvectors than in $X$ itself; as we have seen in Section 2.1, the failure of the spectral mapping theorem for **T** is caused by the lack of approximate eigenvectors in $X$.

We start with the precise definitions. If **T** is a $C_0$-semigroup on a Banach space $X$, then for $1 \leq p < \infty$ we define the semigroup $\mathbf{S}_p$ on $L^p(\Gamma, X)$ by

$$(S_p(t)f)(e^{i\theta}) := T(t)f(e^{i(\theta - t)}), \quad f \in L^p(\Gamma, X), \ \theta \in [0, 2\pi], \ t \geq 0. \qquad (2.5.1)$$

It is easily checked that $\mathbf{S}_p$ is a $C_0$-semigroup on $L^p(\Gamma, X)$; we denote its generator by $B_p$. Similarly, we can use (2.5.1) to define a $C_0$-semigroup $\mathbf{S}_\infty$ on $C(\Gamma, X)$; the generator of this semigroup will be denoted by $B_\infty$.

**Lemma 2.5.1.** *Let **T** be a $C_0$-semigroup on a Banach space $X$, let $1 \leq p \leq \infty$, and let $\mathbf{S}_p$ be defined as above. If $1 \in \sigma_a(T(2\pi))$, then $0 \in \sigma_a(B_p)$.*

*Proof:* Since $1 \in \sigma_a(T(2\pi))$ we can choose vectors $x_n \in X$ such that $\|x_n\| = 1$ and $\|T(2\pi)x_n - x_n\| \leq n^{-1}$, $n = 1, 2, \ldots$ Note that $\|T(2\pi)x_n\| \geq 1 - n^{-1}$.

Let $a : [0, 2\pi] \to [0, 1]$ be a smooth function with bounded derivative such that $a(\theta) = 0$ for $\theta \in [0, \frac{2}{3}\pi]$ and $a(\theta) = 1$ for $\theta \in [\frac{4}{3}\pi, 2\pi]$. Define the functions $a_n : \Gamma \to X$ by

$$a_n(e^{i\theta}) := (1 - a(\theta))T(2\pi + \theta)x_n + a(\theta)T(\theta)x_n, \quad 0 < \theta < 2\pi.$$

Note $a_n(e^{i(0+)}) = a_n(e^{i(2\pi-)}) = T(2\pi)x_n$, so that this function is well-defined as a continuous function on $\Gamma$. In particular it belongs to $L^p(\Gamma, X)$. Also,

$$(S_p(t)a_n)(e^{i\theta}) = (1 - a((\theta - t) \bmod 2\pi))T(2\pi + \theta)x_n + a((\theta - t) \bmod 2\pi)T(\theta)x_n.$$

Since $a_n$ is continuously differentiable with bounded derivative, it follows that $a_n \in D(B_p)$ and

$$(B_p a_n)(e^{i\theta}) = a'_n(e^{i\theta})T(\theta)(T(2\pi)x_n - x_n), \quad 0 < \theta < 2\pi.$$

Let $M := \sup_{t \in [0,2\pi]} \|T(t)\|$ and $N := \sup_{\theta \in [0,2\pi]} |a'(\theta)|$.
First we consider the case $1 \le p < \infty$. We have

$$\|B_p a_n\|_{L^p(\Gamma, X)} \le (2\pi)^{\frac{1}{p}} \frac{MN}{n}.$$

On the other hand, noting that for all $0 \le t \le 2\pi$ we have $\|T(2\pi)x_n\| \le M\|T(t)x_n\|$,

$$\|a_n\|^p_{L^p(\Gamma, X)} \ge \int_{\frac{4}{3}\pi}^{2\pi} \|T(\theta)x_n\|^p \, d\theta$$

$$\ge \frac{2\pi}{3} M^{-p} \|T(2\pi)x_n\|^p$$

$$\ge \frac{2\pi}{3} M^{-p} \left(1 - \frac{1}{n}\right)^p.$$

Combining these two estimates, for $n \ge 2$ it follows that

$$\|B_p a_n\|_{L^p(\Gamma, X)} \le (2\pi)^{\frac{1}{p}} \frac{MN}{n} \le \frac{3^{\frac{1}{p}} M^2 N}{n-1} \|a_n\|_{L^p(\Gamma, X)}.$$

This shows that $0 \in \sigma_a(B_p)$.
For the case $p = \infty$, we estimate

$$\|B_\infty a_n\|_{C(\Gamma, X)} \le \frac{MN}{n}$$

and

$$\|a_n\|_{C(\Gamma, X)} \ge \|a_n(0)\| = \|T(2\pi)x_n\| \ge 1 - \frac{1}{n}.$$

Hence

$$\|B_\infty a_n\|_{C(\Gamma, X)} \le \frac{MN}{n-1} \|a_n\|_{C(\Gamma, X)}, \quad n \ge 2,$$

and $0 \in \sigma_a(B_\infty)$.  ////

In the next theorem, we use the fact that $\sigma(B_p)$ is invariant under translation by $i$. To see this, consider the operator $L_k$ defined by $(L_k f)(e^{i\theta}) := e^{ik\theta} f(e^{i\theta})$. Then one easily checks that $B_p L_k = L_k(B - ik)$ for all $k \in \mathbb{Z}$ and $f \in D(B)$. Since $L_k$ is invertible for all $k \in \mathbb{Z}$, it follows that $B_p$ is invertible if and only if $B - ik$ is invertible.

**Theorem 2.5.2.** *Let $\mathbf{T}$ be a $C_0$-semigroup on a Banach space $X$, let $1 \le p \le \infty$, and let $\mathbf{S}_p$ be defined as above. Then the following assertions are equivalent:*

(i) $0 \in \varrho(B_p)$;
(ii) $1 \in \varrho(S_p(2\pi))$;
(iii) $1 \in \varrho(T(2\pi))$.

*Proof:* The implication (ii)$\Rightarrow$(i) follows from the spectral inclusion theorem 2.1.1. Since
$$(S_p(2\pi)f)(e^{i\theta}) = T(2\pi)(f(e^{i\theta}))$$
it follows that $I - S_p(2\pi)$ is invertible in $L^p(\Gamma, X)$ if $I - T(2\pi)$ is invertible in $X$. This proves (iii)$\Rightarrow$(ii).

(i)$\Rightarrow$(iii): Let $1 \in \sigma(T(2\pi)) = \sigma_a(T(2\pi)) \cup \sigma_r(T(2\pi))$. If $1 \in \sigma_a(T(2\pi))$, then $0 \in \sigma_a(B_p)$ by Lemma 2.5.1. If $1 \in \sigma_r(T(2\pi))$, we claim that $1 \in \sigma_r(S_p(2\pi))$. Once this is proved, it follows that $ik \in \sigma(B_p)$ for some $k \in \mathbb{Z}$ by the spectral mapping theorem for the residual spectrum. Since $\sigma(B_p)$ is invariant under translation by $i$, it follows that $0 \in \sigma(B_p)$ and the theorem is proved. To prove the claim, assume that $I - T(2\pi)$ does not have dense range. Then there exists a non-zero $x_0^* \in X^*$ such that
$$\langle x_0^*, (I - T(2\pi))x \rangle = 0, \quad \forall x \in X.$$
Let $g := 1 \otimes x_0^* \in L^q(\Gamma, X^*)$, $\frac{1}{p} + \frac{1}{q} = 1$. Identifying $L^q(\Gamma, X^*)$ in the natural way with a closed subspace of the dual $(L^p(\Gamma, X))^*$, for all $f \in L^p(\Gamma, X)$ we have
$$\langle g, (I - S_p(2\pi))f \rangle = \int_\Gamma \langle x_0^*, (I - S_p(2\pi))f(e^{i\theta}) \rangle \, d\theta$$
$$= \int_\Gamma \langle x_0^*, (I - T(2\pi))f(e^{i\theta}) \rangle \, d\theta = 0.$$
Hence, $1 \in \sigma_r(S_p(2\pi))$. ////

**Corollary 2.5.3.** *In the above situation,*
$$s(B_p) = \omega_0(\mathbf{S}_p) = \omega_0(\mathbf{T}).$$

*Proof:* Clearly, $s(B_p) \le \omega_0(\mathbf{S}_p) \le \omega_0(\mathbf{T})$, so it remains to prove that $\omega_0(\mathbf{T}) \le s(B_p)$. If $s(B_p) = -\infty$, then a rescaling argument and Theorem 2.5.2 imply that $\sigma(T(2\pi)) = \{0\}$. Hence, $r(T(2\pi)) = 0$ and $\omega_0(\mathbf{T}) = -\infty$ by Proposition 1.2.2. If $s(B_p) > -\infty$, by rescaling we may assume that $s(B_p) = 0$. Theorem 2.5.2 then implies that $\sigma(T(2\pi)) \subset \{|\lambda| \le 1\}$. Therefore, $r(T(2\pi)) \le 1$ and $\omega_0(\mathbf{T}) \le 0$. ////

As another corollary of Theorem 2.5.2, we have the following spectral mapping theorem. By $e_k$ we denote the function $e_k(e^{i\theta}) := e^{ik\theta}$, $\theta \in [0, 2\pi]$.

**Theorem 2.5.4.** *Let $\mathbf{T}$ be a $C_0$–semigroup on a Banach space $X$, with generator $A$, and let $E(\Gamma, X)$ be one of the spaces $L^p(\Gamma, X)$, $1 \le p < \infty$, or $C(\Gamma, X)$. Then the following assertions are equivalent:*

(i) $1 \in \varrho(T(2\pi))$;

(ii) $i\mathbb{Z} \subset \varrho(A)$, *and there exists a constant $M > 0$ such that for all finite sequences $x_{-n}, ..., x_n \in X$ we have*

$$\left\| \sum_{k=-n}^{n} e_k \otimes R(ik, A) x_k \right\|_{E(\Gamma, X)} \le M \left\| \sum_{k=-n}^{n} e_k \otimes x_k \right\|_{E(\Gamma, X)}. \quad (2.5.2)$$

*Proof:* To start the proof, let us assume that $i\mathbb{Z} \subset \varrho(A)$. Let $x_{-n}, ..., x_n \in X$ and define $f, g \in E_p(\Gamma, X)$ by

$$f := \sum_{k=-n}^{n} e_k \otimes x_k, \quad g := -\sum_{k=-n}^{n} e_k \otimes R(ik, A) x_k. \quad (2.5.3)$$

Here, $E_p(\Gamma, X) = L^p(\Gamma, X)$ if $1 \le p < \infty$ and $E_p(\Gamma, X) = C(\Gamma, X)$ if $p = \infty$. Then, $g \in D(B_p)$ and

$$\begin{aligned} B_p g &= \frac{d}{dt}\Big|_{t=0} S_p(t) g \\ &= -\sum_{k=-n}^{n} (e_k \otimes AR(ik, A) x_k - ik e_k \otimes R(ik, A) x_k) = f. \end{aligned} \quad (2.5.4)$$

The set of all $f$ as in (2.5.3) is dense in $E_p(\Gamma, X)$. We claim that the set of all $g$ as in (2.5.3) is dense in $D(B_p)$ with respect to the graph norm. To see this, fix $\lambda > \omega_0(\mathbf{T}) = \omega_0(\mathbf{S}_p)$, let $h \in D(B_p)$ and let $\epsilon > 0$. Choose $y_{-n}, ..., y_n \in X$ such that

$$\left\| (\lambda - B_p) h - \sum_{k=-n}^{n} e_k \otimes y_k \right\| \le \epsilon.$$

Consider the function

$$g := -\sum_{k=-n}^{n} e_k \otimes R(ik, A)(-I + \lambda R(\lambda + ik, A)) y_k.$$

By the above we have $g \in D(B_p)$ and, using the resolvent identity,

$$(\lambda - B_p) g = \sum_{k=-n}^{n} e_k \otimes (\lambda R(\lambda + ik, A) + (I - \lambda R(\lambda + ik, A))) y_k = \sum_{k=-n}^{n} e_k \otimes y_k.$$

This shows that $\|(\lambda - B_p)(h - g)\| \leq \epsilon$. Since $\lambda \in \varrho(B_p)$, the norm $\|f\| := \|(\lambda - B_p)f\|$ is equivalent to the graph norm on $D(B_p)$ (apply the open mapping theorem). This proves that $h$ can be approximated in the norm of $D(B_p)$ by functions $g$ as in (2.5.3).

Now we can prove (ii)$\Rightarrow$(i): The assumption (2.5.2) implies that
$$\|B_p g\|_{E_p(\Gamma,X)} = \|f\|_{E_p(\Gamma,X)} \geq M^{-1}\|g\|_{E_p(\Gamma,X)}$$
for all functions $f$ and $g$ as in (2.5.3). By the claim these are dense in $E_p(\Gamma, X)$ and $D(B_p)$, respectively, and therefore it follows that
$$\|B_p g\|_{E_p(\Gamma,X)} \geq M^{-1}\|g\|_{E_p(\Gamma,X)}$$
holds for *all* $g \in D(B_p)$. But this means that $0 \notin \sigma_a(B_p)$, and then Lemma 2.5.1 implies that $1 \notin \sigma_a(T(2\pi))$. We also have $1 \notin \sigma_r(T(2\pi))$ since otherwise the spectral mapping theorem for the residual spectrum would imply the existence of a $k \in \mathbb{Z}$ such that $ik \in \sigma_r(A)$. Therefore, $1 \notin \sigma(T(2\pi))$.

Conversely, assume (i) and let $1 \leq p \leq \infty$. By the spectral inclusion theorem, $i\mathbb{Z} \subset \varrho(A)$. Moreover, $0 \in \varrho(B_p)$ by Theorem 2.5.2. Therefore, by (2.5.4), for all $f$ and $g$ as in (2.5.3) we have
$$\|g\|_{E_p(\Gamma,X)} = \|B_p^{-1} f\|_{E_p(\Gamma,X)} \leq \|B_p^{-1}\| \|f\|_{E_p(\Gamma,X)}.$$
Hence (ii) follows with $M = \|B_p^{-1}\|$. ////

If $X$ is a Hilbert space, then Parseval's identity implies
$$\left\|\sum_{k=-n}^{n} y_k \otimes e_k\right\|_{L^2(\Gamma,X)} = \left(2\pi \sum_{k=-n}^{n} \|y_k\|^2\right)^{\frac{1}{2}}$$
for all finite sequences $y_{-n}, ..., y_n \in X$. Since
$$\left(\sum_{k=-n}^{n} \|R(ik,A)x_k\|^2\right)^{\frac{1}{2}} \leq M \left(\sum_{k=-n}^{n} \|x_k\|^2\right)^{\frac{1}{2}}$$
holds for all finite sequences $x_{-n}, ..., x_n \in X$ if and only if
$$\sup_{k \in \mathbb{Z}} \|R(ik,A)\| \leq M,$$
it follows that Gearhart's theorem 2.2.4 can be viewed as a special case of Theorem 2.5.4.

In Section 3.3, we shall need a version of Lemma 2.5.1 for translations on the half-line. If $\mathbf{T}$ is a $C_0$-semigroup on a Banach space $X$ and $1 \leq p < \infty$, we define a $C_0$-semigroup $\mathbf{S}_p$ on $L^p(\mathbb{R}_+, X)$ by
$$(S_p(t))f(s) = \begin{cases} T(t)f(s-t), & s-t \geq 0; \\ 0, & \text{else.} \end{cases}$$
We denote by $B_p$ its infinitesimal generator. Similarly, by using the right translation semigroup on $C_{00}(\mathbb{R}_+)$, the subspace of all $f \in C_0(\mathbb{R}_+)$ such that $f(0) = 0$, we obtain a $C_0$-semigroup $\mathbf{S}_\infty$ on the corresponding $X$-valued space $C_{00}(\mathbb{R}_+, X)$. Its generator will be denoted by $B_\infty$.

**Lemma 2.5.5.** Let $1 \leq p \leq \infty$. If $e^{i\theta} \in \sigma_a(T(1))$ for some $\theta \in [0, 2\pi]$, then $0 \in \sigma_a(B_p)$.

*Proof:* First assume $1 \leq p < \infty$. Since $e^{i\theta} \in \sigma_a(T(1))$, for each $n = 1, 2, \ldots$ one can find a norm one vector $x_n \in X$ such that

$$\|T(k)x_n - e^{ik\theta}x_n\| \leq \frac{1}{2}, \quad k = 1, \ldots, n.$$

In particular, $\frac{1}{2} \leq \|T(k)x_n\| \leq \frac{3}{2}$ for all $n$ and $k = 1, \ldots, n$. Using the local boundedness of **T**, it is easy to see that there are constants $0 < \alpha \leq \beta < \infty$ such that

$$\alpha \leq \|T(t)x_n\| \leq \beta, \quad t \in [0, n]; \quad n = 1, 2, \ldots$$

For each $n$, let $a_n : \mathbb{R}_+ \to [0, 1]$ be a continuously differentiable function such that $a_n = 0$ on $[0, \frac{1}{8}] \cup [n - \frac{1}{8}, \infty)$, $a_n = 1$ on $[\frac{1}{4}, n - \frac{1}{4}]$, and $a'_n(t) \leq 10$ for all $t \geq 0$ and $n = 1, 2, \ldots$. Let $g_n := c_n^{-1} a_n(t) T(t) x_n$, where

$$c_n := \left( \int_0^n \|T(t)x_n\|^p \, dt \right)^{\frac{1}{p}}.$$

Then, using that $a_n = 1$ on $[\frac{1}{4}, n - \frac{1}{4}]$, we have

$$\frac{\alpha}{2^{\frac{1}{p}} \beta} \leq \frac{\alpha(n - \frac{1}{2})^{\frac{1}{p}}}{\beta n^{\frac{1}{p}}} \leq \|g_n\|_{L^p(\mathbb{R}_+, X)} \leq 1; \quad n = 1, 2, \ldots$$

Also, by direct calculation one checks that

$$B_p g_n(t) = -c_n^{-1} a'_n(t) T(t) x_n, \quad t \geq 0.$$

Since $|a'_n| \leq 10$ on $[\frac{1}{8}, \frac{1}{4}] \cup [n - \frac{1}{4}, n - \frac{1}{8}]$ and 0 elsewhere, we have

$$\|B_E g_n\|_{L^p(\mathbb{R}_+, X)} \leq \frac{10\beta \cdot 8^{-\frac{1}{p}} + 10\beta \cdot 8^{-\frac{1}{p}}}{\alpha n^{\frac{1}{p}}}; \quad n = 1, 2, \ldots$$

Therefore $\lim_{n \to \infty} \|B_p g_n\|_{L^p(\mathbb{R}_+, X)} = 0$, showing that $\left( \|g_n\|_{L^p(\mathbb{R}_+, X)}^{-1} \cdot g_n \right)_{n \geq 1}$ is an approximate eigenvector for $B_p$ with approximate eigenvalue 0.

Next, we show how to modify this argument for $p = \infty$. In this case, we choose a $C^1$-function $a_n$ that vanishes on $[0, \frac{1}{4}] \cup [n - \frac{1}{4}, \infty)$, and further satisfies $a_n(\frac{1}{2}n) = 1$, $\|a_n\|_\infty = 1$, and $\|a'_n\|_\infty \leq 5n^{-1}$. Then $g_n := a_n(t) T(t) x_n$, $n = 1, 2, \ldots$, defines an approximate eigenvector for $B_\infty$ with approximate eigenvalue 0. ////

This lemma implies that the uniform growth bounds of **T** and $\mathbf{S}_p$ and the spectral bound of $B_p$ coincide:

**Corollary 2.5.6.** For all $1 \leq p \leq \infty$ we have $s(B_p) = \omega_0(\mathbf{S}_p) = \omega_0(\mathbf{T})$.

*Proof:* Clearly, $s(B_p) \leq \omega_0(\mathbf{S}_p) \leq \omega_0(\mathbf{T})$. It remains to prove that $\omega_0(\mathbf{T}) \leq s(B_p)$. If $\omega_0(\mathbf{T}) = -\infty$ there is noting to prove. Therefore by rescaling we may assume that $\omega_0(\mathbf{T}) = 0$, and it suffices to show that this implies that $s(B_p) \geq 0$. Since $r(T(1)) = e^{\omega_0(\mathbf{T})} = 1$, there is a $\theta \in [0, 2\pi]$ such that $e^{i\theta} \in \sigma(T(1))$, and in fact, $e^{i\theta} \in \sigma_a(T(1))$ since the boundary spectrum is always contained in the approximate point spectrum by Lemma 2.1.5. By Lemma 2.5.5, this implies that $0 \in \sigma(B_p)$.
////

**Notes.** The proofs of Theorems 2.1.1 and 2.1.2 are taken from Pazy's book. In our definition of residual spectrum, we follow [Na].

A detailed treatment of adjoint semigroups is given in [Ne1].

The equivalence (i)⇔(ii) of Theorem 2.2.1 is due to G. Greiner [Gr]; see also [Na], where our proof is taken from. For contraction semigroups, Theorem 2.2.4 is due to L. Gearhart [Ge]; the general case was proved independently by I.W. Herbst [He], J.S. Howland [Hw], and J. Prüss [Pr2]. The approach to Gearhart's theorem via Theorem 2.2.1 is due to G. Greiner [Gr]; our proof via 2.2.1 (iii) is new.

The spectral mapping theorem for eventually uniformly continuous semigroups is due to E. Hille and R.S. Phillips [HP]. The concept of uniform continuity at infinity was introduced by J. Mazon and J. Martinez [MM], who proved Theorem 2.3.3. Corollary 2.3.5 is due to P. You [Yo]. A simple proof was given by K.-J. Engel and O. ElMennaoui [EE]. Theorem 2.3.4 is an improvement of Theorem 2.1 of [MM]. The proof combines the ideas from [MM] and [EE].

The proof of Lemma 2.4.2 is taken from [Ly]. The formula for $\hat{f}(\mathbf{T})$ in Lemma 2.4.3 is due to D.E. Evans [Ev], who used it to show that the spectrum of the generator of a uniformly bounded $C_0$-group $\mathbf{T}$ and the so-called Arveson spectrum of $\mathbf{T}$ can be indentified. This result will be proved in Section 5.4.

Part (iii) of Lemma 2.4.3 was obtained independently in [NH] and [Vu3]. Theorem 2.4.4 was proved by G. Greiner for polynomially bounded groups; see Theorem 7.4 in [Na, A-III]. The extension to $C_0$-groups of non-quasianalytic growth is due to E. Marschall [Ml]. It was obtained independently by Yu. I. Lyubich and Vũ Quôc Phóng [LV2] and Sen-Zhong Huang and R. Nagel [NH]. We followed the proof of [NH]. In [LV2] the proof is only sketched and some essential details are omitted. Corollary 2.4.5 is in [LV2]. A generalization to non-quasianalytic group representations is given in [HS3].

Theorem 2.4.13 is due to Sen-Zhong Huang [HS1]; we follow his proof. That the spectrum of the generator of a $C_0$-groups may be empty has been known for a long time; an example is the boundary group $\{I(it)\}_{t \in \mathbb{R}}$ of the fractional integration semigroup $\{I^z\}_{\operatorname{Re} z > 0}$ on $L^p(\mathbb{R})$ [HP, p. 665]. Further examples are given by J. Zabczyk (Example 1.2.4) and M. Wolff [Wo]. These groups, however, have exponential growth.

Lemma 2.5.1, Theorem 2.5.2 and its corollaries are taken from [LM]. The analogues of these results for translation along the real line also hold; for separable $X$ this was proved in [LM] and the general case is due to F. Räbiger and R. Schnaubelt [RS]. Some special cases were obtained earlier by R. Rau [Ra]. Lemma 2.5.5 and its corollary and are slight modifications of the corresponding results the real line proved in [RS] and are taken from [Ne4].

# Chapter 3

## Uniform exponential stability

In this chapter we shall study the uniform growth bound $\omega_0(\mathbf{T})$ in more detail. Our main concern is finding necessary and sufficient conditions for uniform exponential stability.

In Section 3.1 we prove the Datko-Pazy theorem which asserts that $\mathbf{T}$ is uniformly exponentially stable if and only if there exists a $1 \leq p < \infty$ such that

$$\int_0^\infty \|T(t)x\|^p \, dt < \infty, \quad \forall x \in X.$$

Actually, we shall prove the more general result that it is necessary and sufficient that all orbits lie in certain vector-valued function spaces over $\mathbb{R}_+$.

In Section 3.2 as a special case we derive Rolewicz's theorem which states that $\mathbf{T}$ is uniformly exponentially stable if and only if

$$\int_0^\infty \phi(\|T(t)x\|) \, dt < \infty, \quad \forall x \in X,$$

for some strictly positive, non-decreasing function $\phi$.

In Section 3.3 we prove that $\mathbf{T}$ is uniformly exponentially stable if and only if convolution with $\mathbf{T}$,

$$(\mathbf{T} * f)(t) := \int_0^t T(s)f(t-s) \, ds,$$

maps $L^p(\mathbb{R}_+, X)$ or $C_0(\mathbb{R}_+, X)$ into itself.

This result is applied in Section 3.4, where we prove that $\mathbf{T}$ is uniformly exponentially stable if and only if

$$\sup_{s>0} \left\| \int_0^s T(t)g(t) \, dt \right\| < \infty, \quad \forall g \in AP(\mathbb{R}_+, X),$$

where $AP(\mathbb{R}_+, X)$ is the space of $X$-valued almost periodic functions.

In Sections 3.5 and 3.6 we consider the problem of finding sufficient conditions for the equality $s(A) = \omega_0(\mathbf{T})$ in the absence of a spectral mapping theorem. In Section 3.5 we prove the theorem of Weis that this equality holds for positive $C_0$-semigroups on $L^p$-spaces and in Section 3.6 we give a sufficient condition for $s(A) = \omega_0(\mathbf{T})$ in terms of the essential spectrum of $\mathbf{T}$.

## 3.1. The theorem of Datko and Pazy

We start with the following general situation.

**Lemma 3.1.1.** *Let $X, Y$, and $Z$ be Banach spaces and let $S : X \times Y \to Z$ be a separately continuous bilinear map. Then there exists a constant $K > 0$ such that*

$$\|S(x,y)\| \leq K\|x\|\,\|y\|, \qquad \forall x \in X,\, y \in Y.$$

*Proof:* For $x \in X$ define $S_x : Y \to Z$, $S_x y := S(x,y)$. Then each $S_x$ is bounded by the continuity in the $Y$-variable. Using the continuity in the $X$-variable, it is easy to see that the map $x \mapsto S_x$ is closed, and hence bounded by the closed graph theorem. Denoting the norm of this map by $K$, for all $x \in X$ and $y \in Y$ we have $\|S(x,y)\| \leq \|S_x\|\,\|y\| \leq K\|x\|\,\|y\|$. ////

We specialize this to the case where $Y = X^*$ and $Z$ is a Banach function space. For more details about Banach function spaces we refer to Appendix A.4.

**Lemma 3.1.2.** *Let $E$ be a Banach function space over a positive $\sigma$-finite measure space $(\Omega, \mu)$. Let $X$ be a Banach space and assume that $S : \Omega \to \mathcal{L}(X)$ is a mapping such that for all $x \in X$ and $x^* \in X^*$, the function*

$$S_{x,x^*}(\omega) := \langle x^*, S(\omega)x \rangle, \quad \omega \in \Omega,$$

*defines an element of $E$. Then, there exists a constant $K > 0$ such that*

$$\|S_{x,x^*}\|_E \leq K\|x\|\,\|x^*\|, \qquad \forall x \in X,\, x^* \in X^*.$$

*Proof:* Consider the map $\mathcal{S} : X \times X^* \to E$ defined by $\mathcal{S}(x, x^*) := S_{x,x^*}$. We claim that this bilinear map is separately continuous. Indeed, fix $x^* \in X^*$. We will show that $S_{x^*} : X \to E$ defined by $S_{x^*}x := S_{x,x^*}$ is closed. To this end, let $x_n \to x$ in $X$ and $S_{x^*}x_n \to f$ in $E$. Since Cauchy sequences in Banach function spaces have pointwise a.e. convergent subsequences, and since $(S_{x^*}x_n)(\omega) \to (S_{x^*}x)(\omega)$ for all $\omega$, it follows that $S_{x^*}x = f$, proving closedness. Therefore, each operator $S_{x^*}$ is bounded. Similarly, each operator $S_x : X^* \to E$, $S_x x^* := S_{x,x^*}$, is bounded. Therefore, $\mathcal{S}$ is separately continuous and we can apply Lemma 3.1.1. ////

We are going to apply this lemma to $C_0$-semigroups whose orbits define elements of a given Banach function space over $\mathbb{R}_+$. The key observation is the following simple fact. It states that the scalar orbits of an operator with spectral radius one can have arbitrarily long initial parts staying away from zero.

**Lemma 3.1.3.** *Suppose $T$ is a bounded operator on a Banach space $X$ whose spectral radius satisfies $r(T) \geq 1$. Then for all $N \in \mathbb{N}$ and $0 < \epsilon < 1$, there exist norm one vectors $x \in X$ and $x^* \in X^*$ such that*

$$|\langle x^*, T^n x \rangle| \geq 1 - \epsilon, \quad n = 0, ..., N.$$

*Proof:* Fix $N \in \mathbb{N}$ and $0 < \epsilon < 1$ and fix $\lambda \in \sigma(T)$ with $|\lambda| = r(T)$. Then $\lambda \in \partial\sigma(T)$, the boundary of the spectrum of $T$, so $\lambda$ is an approximate eigenvalue and $|\lambda| \geq 1$. Let $(y_n)$ be an approximate eigenvector corresponding to $\lambda$. Since for each $k \in \mathbb{N}$,
$$\lim_{n\to\infty} \|T^k y_n - \lambda^k y_n\| = 0,$$
we may choose $n_0$ so large that $\|T^k y_{n_0} - \lambda^k y_{n_0}\| \leq \frac{\epsilon}{2}$, $k = 0, ..., N$. Put $x = y_{n_0}$ and let $x^* \in X^*$ be any norm one vector such that $|\langle x^*, x \rangle| \geq 1 - \frac{\epsilon}{2}$. Then for $k = 0, ..., N$ we have
$$|\langle x^*, T^k x \rangle| \geq |\lambda|^k |\langle x^*, x \rangle| - \frac{\epsilon}{2} \geq 1 - \epsilon.$$
////

We remind the reader of the convention that unless otherwise stated all Banach spaces are complex. For real Banach spaces, Lemma 3.1.3 is wrong. A simple counterexample is rotation over $\alpha$ in $X = \mathbb{R}^2$, with $\alpha/(2\pi)$ irrational. For this operator, Lemma 3.1.3 fails for every choice of $\epsilon$. We leave the easy proof to the reader.

For a Banach function space $E(\mathbb{N})$ over $\mathbb{N}$ we define
$$\varphi_{E(\mathbb{N})}(n) = \|\chi_{\{0,...,n-1\}}\|_{E(\mathbb{N})}, \qquad n = 1, 2, ...,$$

**Theorem 3.1.4.** *Let $T$ be a bounded operator on a Banach space $X$. Let $E = E(\mathbb{N})$ be a Banach function space over $\mathbb{N}$ with $\lim_{n\to\infty} \varphi_E(n) = \infty$. If, for some sequence $k_n \to \infty$ and all $x \in X$ and $x^* \in X^*$, the function $n \mapsto \langle x^*, T^{k_n} x \rangle$ belongs to $E$, then $r(T) < 1$.*

*Proof:* For each $x \in X$ and $x^* \in X^*$, define $f_{x,x^*}(n) := \langle x^*, T^{k_n} x \rangle$. Then $\|f_{x,x^*}\|_E \leq K\|x\| \|x^*\|$, where $K$ is the constant of Lemma 3.1.2.

Suppose for contradiction that $r(T) \geq 1$. Fix $0 < \epsilon < 1$ and let $N \in \mathbb{N}$ be arbitrary. Set $\Omega_N = \{0, ..., N-1\}$ and let $x$ and $x^*$ norm one vectors such that
$$|\langle x^*, T^n x \rangle| \geq 1 - \epsilon, \quad n = 0, 1, ..., k_{N-1}.$$
From $|\chi_{\Omega_N}| \leq (1-\epsilon)^{-1} |f_{x,x^*}|$, we see that $\chi_{\Omega_N} \in E$ and $\varphi_E(N) = \|\chi_{\Omega_N}\|_E \leq (1-\epsilon)^{-1} K$. Since this holds for all $N$, we have $\sup_N \varphi_E(N) \leq (1-\epsilon)^{-1} K$, a contradiction. ////

The theorem obviously fails for the space $E = l^\infty$. This shows that the condition $\lim_{n\to\infty} \varphi_E(n) = \infty$ cannot be omitted.

For a Banach function space $E(\mathbb{R}_+)$ over $\mathbb{R}_+$ we define
$$\varphi_{E(\mathbb{R})}(t) = \|\chi_{[0,t)}\|_{E(\mathbb{R}_+)}, \qquad t \geq 0,$$
provided $\chi_{(0,t)} \in E$; otherwise we put $\varphi_{E(\mathbb{R})}(t) = \infty$.

Let **T** be a $C_0$–semigroup on $X$. By applying Theorem 3.1.4 to $T(1)$ we obtain the following result.

**Theorem 3.1.5.** Let **T** be a $C_0$–semigroup on a Banach space $X$ and let $E = E(\mathbb{R}_+)$ be a Banach function space over $\mathbb{R}_+$ with $\lim_{t\to\infty} \varphi_E(t) = \infty$. If, for all $x \in X$, the maps $t \mapsto \|T(t)x\|$ belong to $E$, then **T** is uniformly exponentially stable.

*Proof:* Let $E(\mathbb{N})$ be the Banach function space over $\mathbb{N}$ consisting of all sequences $(\alpha_n)$ such that $\sum_n \alpha_n \chi_{[n,n+1)} \in E$, with the norm

$$\|(\alpha_n)\|_{E(\mathbb{N})} := \left\| \sum_n \alpha_n \chi_{[n,n+1)} \right\|_E.$$

Note that $\lim_{n\to\infty} \varphi_{E(\mathbb{N})}(n) = \infty$.

Let $M := \sup_{0 \le t \le 1} \|T(t)\|$. For all $x \in X$ and $t \ge 0$, $t = t_0 + t_1$ with $t_0 \in \mathbb{N}$ and $0 \le t_1 < 1$, we have $\|T(t_0+1)x\| \le M\|T(t)x\|$. Therefore

$$\|T(\cdot)x\| \ge M^{-1} \sum_n \|T(n+1)\| \chi_{[n,n+1)}.$$

It follows that for all $x \in X$ the map $n \mapsto \|T(n+1)x\|$ defines an element of $E(\mathbb{N})$. Then certainly $n \mapsto \langle x^*, T(n+1)x \rangle \in E(\mathbb{N})$ for all $x \in X$ and $x^* \in X^*$, and the theorem follows from Theorem 3.1.4 applied to $T = T(1)$ and $k_n = n+1$.  ////

We are going to apply this theorem to certain weighted $L^p$-spaces. For $1 \le p \le \infty$ and a strictly positive function $0 < \alpha \in L^1_{loc}(\mathbb{R}_+)$, we define $L^p_\alpha(\mathbb{R}_+)$ as the set of all functions $g$ such that $\alpha g$ (the pointwise a.e. product) is in $L^p(\mathbb{R}_+)$. With the norm

$$\|g\|_{L^p_\alpha(\mathbb{R}_+)} := \|\alpha g\|_{L^p(\mathbb{R}_+)},$$

it is easy to check that $L^p_\alpha(\mathbb{R}_+)$ is a Banach function space over $\mathbb{R}_+$. Note that the strict positivity implies that only the zero function has norm zero. We have

$$\varphi_{L^p_\alpha(\mathbb{R}_+)}(t) = \left( \int_0^t (\alpha(s))^p \, ds \right)^{\frac{1}{p}}.$$

Using this space, we have the following corollary to Theorem 3.1.5.

**Corollary 3.1.6.** Let $0 \le \beta \in L^1_{loc}(\mathbb{R}_+)$ be any function such that

$$\int_0^\infty \beta(t) \, dt = \infty.$$

Let $1 \le p < \infty$. If **T** is a $C_0$–semigroup on a Banach space $X$ such that

$$\int_0^\infty \beta(t) \|T(t)x\|^p \, dt < \infty, \qquad \forall x \in X,$$

then **T** is uniformly exponentially stable.

*Proof:* If $T(t) = 0$ for some $t \geq 0$, then **T** is eventually zero and there is nothing to prove. Therefore, we may assume that $T(t) \neq 0$ for all $t \geq 0$. Define the function $\alpha \in L^1_{loc}(\mathbb{R}_+)$ by $\alpha(t) := (\beta(t))^{\frac{1}{p}}$. We are going to modify $\alpha$ to a strictly positive function as follows. Define

$$\tilde{\alpha}(t) := \begin{cases} \alpha(t), & \text{if } \alpha(t) \neq 0; \\ e^{-t}\|T(t)\|^{-1}, & \text{if } \alpha(t) = 0. \end{cases}$$

Since $\tilde{\alpha} \geq \alpha = \beta^{\frac{1}{p}} \geq 0$ we have

$$\lim_{t \to \infty} \varphi_{L^p_{\tilde{\alpha}}(\mathbb{R}_+)}(t) \geq \lim_{t \to \infty} \left( \int_0^t \beta(s)\,ds \right)^{\frac{1}{p}} = \infty.$$

Also, by the triangle inequality for $L^p(\mathbb{R}_+)$, for all $x \in X$ we have

$$\big\| \|T(\cdot)x\| \big\|_{L^p_{\tilde{\alpha}}(\mathbb{R}_+)} \leq \left( \int_0^\infty \big((\alpha(t) + e^{-t}\|T(t)\|^{-1})\|T(t)x\|\big)^p\, dt \right)^{\frac{1}{p}}$$
$$\leq \left( \int_0^\infty \beta(t)\|T(t)x\|^p\, dt \right)^{\frac{1}{p}} + \left( \int_0^\infty e^{-pt}\|x\|^p\, dt \right)^{\frac{1}{p}} < \infty.$$

Therefore, for all $x \in X$ the map $t \mapsto T(t)x$ defines an element of $L^p_{\tilde{\alpha}}(\mathbb{R}_+)$, and the desired conclusion follows from Theorem 3.1.5. ////

The special case $\beta := 1$ is usually referred to as the *Datko-Pazy theorem*. Explicitly, it states that **T** is uniformly exponentially stable if and only if there exist $1 \leq p < \infty$ such that

$$\int_0^\infty \|T(t)x\|^p\, dt < \infty, \quad \forall x \in X. \tag{3.1.1}$$

If (3.1.1) holds, the closed graph theorem applied to the map $X \to L^p(\mathbb{R}_+, X)$, $x \mapsto T(\cdot)x$, implies the existence of a constant $C > 0$ such that

$$\int_0^\infty \|T(t)x\|^p\, dt \leq C\|x\|^p, \quad \forall x \in X. \tag{3.1.2}$$

Our next objective is to prove the following quantitative version of the Datko-Pazy theorem: If (3.1.2) holds, then $\omega_0(\mathbf{T}) \leq -1/(pC)$. The proof is based on the following non-trivial result due to V. Müller: Let $T$ be a bounded operator on a Banach space $X$ with spectral radius $r(T) \geq 1$. Then for each $0 < \epsilon < 1$ and each sequence $1 \geq \alpha_0 \geq \alpha_1 \geq \ldots \downarrow 0$ there exists a norm one vector $x \in X$ such that

$$\|T^n x\| \geq (1-\epsilon)\alpha_n, \quad \forall n = 0, 1, 2, \ldots$$

We have the following analogue for semigroups.

**Lemma 3.1.7.** Let $\mathbf{T}$ be a $C_0$-semigroup on a Banach space $X$ with $\omega_0(\mathbf{T}) \geq 0$. Then for each $0 < \epsilon < 1$ and each non-increasing function $\alpha : [0, \infty) \to [0, 1]$ with $\lim_{t \to \infty} \alpha(t) = 0$ there exists a norm one vector $x \in X$ such that
$$\|T(t)x\| \geq (1 - \epsilon)\alpha(t), \quad \forall t \geq 0.$$

*Proof:* Let the non-increasing function $\alpha : [0, \infty) \to [0, 1]$, $0 \leq \alpha \leq 1$, $\alpha(t) \downarrow 0$, be fixed.

*Step 1.* Define $\beta : [0, \infty) \to [0, 1]$ by
$$\beta(t) = \begin{cases} \alpha(0), & 0 \leq t < 1; \\ \alpha(t-1), & t \geq 1. \end{cases}$$
Then $\beta$ is non-increasing and $\lim_{t \to \infty} \beta(t) = 0$. Put $T := T(1)$. Because of the identity $r(T(t)) = e^{\omega_0(\mathbf{T})t}$, we have $r(T) \geq 1$. By Müller's theorem, we can choose a vector $x_0 \in X$ of norm one such that $\|T^k x_0\| \geq \frac{1}{2}\beta(k)$ for all $k$. For $t \geq 0$ we let $[t]$ be the integer part of $t$. With $M := \sup_{0 \leq t \leq 1} \|T(t)\|$, for all $t \geq 0$ we have
$$\|T(t)x_0\| \geq \frac{1}{M}\|T([t]+1)x_0\| \geq \frac{1}{2M}\beta([t]+1)$$
$$= \frac{1}{2M}\alpha([t]) \geq \frac{1}{2M}\alpha(t). \tag{3.1.3}$$

*Step 2.* Fix $0 < \epsilon < 1$. We will now show that the constant $(2M)^{-1}$ in (3.1.3) can actually be replaced by $1 - \epsilon$. Fix $\delta > 0$ such that $(1-\delta)/(1+\delta) \geq 1 - \epsilon$. We start by choosing integers $0 = M_0 < M_1 < ...$ such that $0 \leq \alpha(t) \leq (1+\delta)^{-n}$ whenever $t \geq M_n$. Next, choose integers $0 = N_0 < N_1 < ...$ in such a way that $N_n \geq M_n$ for each $n$ and $N_m + N_n \leq N_{m+n}$ for all $m, n$. Define the non-increasing function $\gamma : [0, \infty) \to [0, 1]$ by $\gamma(t) := (1+\delta)^{-n}$ whenever $N_n \leq t < N_{n+1}$. Note that $0 \leq \alpha \leq \gamma \leq 1$ and $\lim_{t \to \infty} \gamma(t) = 0$.

We claim that $\gamma(t+s) \geq (1+\delta)^{-1}\gamma(t)\gamma(s)$ for all $t, s \geq 0$. Indeed, choose integers $k_t$ and $k_s$ such that $N_{k_t} \leq t < N_{k_t+1}$ and $N_{k_s} \leq s < N_{k_s+1}$. Then $\gamma(t) = (1+\delta)^{-k_t}$ and $\gamma(s) = (1+\delta)^{-k_s}$, whereas from $t + s < N_{k_t+1} + N_{k_s+1} \leq N_{k_t+k_s+2}$ we have $\gamma(t+s) \geq (1+\delta)^{-k_t-k_s-1}$. This proves the claim.

Now choose a norm one vector $x_0 \in X$ such that $\|T(t)x_0\| \geq (2M)^{-1}\gamma(t)$ for all $t \geq 0$ using Step 1 applied to the function $\gamma$. Let
$$\eta := \inf_{t \geq 0} \frac{\|T(t)x_0\|}{\gamma(t)}.$$
Note that $\eta \geq (2M)^{-1}$; moreover, for all $t \geq 0$ we have $\|T(t)x_0\| \geq \eta\gamma(t)$. Choose $t_0 \geq 0$ such that
$$\frac{\eta\gamma(t_0)}{\|T(t_0)x_0\|} \geq 1 - \delta$$
and put $x := \|T(t_0)x_0\|^{-1}T(t_0)x_0$. Then for all $t \geq 0$ we have
$$\|T(t)x\| = \frac{\|T(t+t_0)x_0\|}{\|T(t_0)x_0\|} \geq \frac{\eta\gamma(t+t_0)}{\|T(t_0)x_0\|} \geq \frac{1-\delta}{1+\delta}\gamma(t) \geq (1-\epsilon)\gamma(t) \geq (1-\epsilon)\alpha(t).$$
////

**Theorem 3.1.8.** *Let* **T** *be a* $C_0$-*semigroup on a Banach space* $X$. *If there exist* $1 \leq p < \infty$ *and* $C > 0$ *such that*

$$\int_0^\infty \|T(t)x\|^p\, dt \leq C\|x\|^p, \quad \forall x \in X,$$

*then* $\omega_0(\mathbf{T}) \leq -1/(pC)$.

*Proof:* By Lemma 3.1.7 applied to the $C_0$–semigroup $\mathbf{T}_\omega := \{e^{-\omega t}T(t)\}_{t\geq 0}$, where $\omega := \omega_0(\mathbf{T})$, for all $\delta > 0$ and $0 < \epsilon < 1$ there exists a norm one vector $x = x_{\delta,\epsilon} \in X$ such that

$$\|e^{-\omega t}T(t)x\| \geq (1-\epsilon)e^{-\delta t}, \quad \forall t \geq 0.$$

Hence, for this $x$ we have

$$\int_0^\infty \|T(t)x\|^p\, dt \geq (1-\epsilon)^p \int_0^\infty e^{(\omega-\delta)pt}\, dt = -\frac{(1-\epsilon)^p}{p(\omega-\delta)}.$$

Combining this with the assumption of the theorem, we see that

$$-\frac{(1-\epsilon)^p}{p(\omega-\delta)} \leq C.$$

Since $\delta > 0$ and $\epsilon > 0$ were arbitrary, it follows that $-(\omega p)^{-1} \leq C$, so $\omega \leq -(pC)^{-1}$ as was to be proved. ////

The estimate is the best possible in the sense that $\omega_0(\mathbf{T})$ is the infimum of the numbers $-(pC)^{-1}$, where $C > 0$ runs over all constants for which (3.1.2) holds with regard to some equivalent norm. To see this, let $\omega_0(\mathbf{T}) < \omega < 0$ be arbitrary and choose a constant $M > 0$ such that $\|T(t)\| \leq Me^{\omega t}$ for all $t \geq 0$. Define the equivalent norm $\|\|\cdot\|\|$ by

$$\|\|x\|\| := \sup_{t\geq 0} e^{-\omega t}\|T(t)x\|.$$

Then $\|\|T(t)\|\| \leq e^{\omega t}$ for all $t \geq 0$, and for all $1 \leq p < \infty$ and $x \in X$ we have

$$\int_0^\infty \|\|T(t)x\|\|^p\, dt \leq \int_0^\infty e^{\omega pt}\|\|x\|\|^p \leq -\frac{1}{p\omega}\|\|x\|\|^p.$$

Thus, (3.1.2) holds for the norm $\|\|\cdot\|\|$ with $C = -(p\omega)^{-1}$.

The consideration of equivalent norms is necessary because $\omega_0(\mathbf{T})$ is an isomorphic quantity whereas the assumption (3.1.2) is isometric.

The (qualitative) Datko-Pazy theorem leads to a short alternative proof of Corollary 2.2.5.

*Second proof of Corollary 2.2.5:* Fix $\omega_0 > \omega_0(\mathbf{T})$ arbitrary. By the Hilbert space-valued Plancherel theorem applied to the maps $t \mapsto e^{-\omega t}T(t)x \cdot \chi_{\mathbb{R}_+}(t)$, $\omega \geq \omega_0$, it follows that

$$\sup_{\omega \geq \omega_0} \int_{-\infty}^{\infty} \|R(\omega + it, A)x\|^2 \, dt \leq K^2 \|x\|^2, \quad \forall x \in H, \qquad (3.1.4)$$

where $K > 0$ is a constant independent of $x$. By the resolvent identity, for all $\omega > 0$ we have

$$\|R(\omega + it, A)\| \leq \|(I + (\omega_0 - \omega)R(\omega + it, A))\| \, \|R(\omega_0 + it, A)\|$$

and hence by (3.1.4) and the uniform boundedness of the resolvent in the right half-plane there is a constant $k > 0$ such that

$$\sup_{0 < \omega < \omega_0} \int_{-\infty}^{\infty} \|R(\omega + it, A)x\|^2 \, dt \leq k^2 K^2 \|x\|^2, \quad \forall x \in H. \qquad (3.1.5)$$

Combining (3.1.4) and (3.1.5), the Hilbert space-valued Paley-Wiener theorem now implies that for each $x \in H$ there exists a function $g_x \in L^2(\mathbb{R}_+, H)$ such that $R(z, A)x$ is the Laplace transform of $g_x$. Since the Laplace transforms of $g_x$ and $T(\cdot)x$ agree on $\{\operatorname{Re} \lambda > \omega_0(\mathbf{T})\}$, by the uniqueness of the Laplace transform it follows that $g_x(t) = T(t)x$ for almost all $t \geq 0$. We have proved that

$$\int_0^\infty \|T(t)x\|^2 \, dt < \infty, \quad \forall x \in H.$$

Therefore, $\mathbf{T}$ is uniformly exponentially stable. ////

We conclude this section with an orbitwise analogue of Theorem 3.1.5 that will be useful later.

**Lemma 3.1.9.** *Let $\mathbf{T}$ be a $C_0$-semigroup on a Banach space $X$. Let $E = E(\mathbb{R}_+)$ be a rearrangement invariant Banach function space over $\mathbb{R}_+$ with $\lim_{t \to \infty} \varphi_E(t) = \infty$. If, for some $x_0 \in X$, the map $t \mapsto \|T(t)x_0\|$ belongs to $E$, then*

$$\lim_{t \to \infty} \|T(t)x_0\| = 0.$$

*Proof:* Suppose the contrary. Then there exists an $\epsilon > 0$ and a sequence $t_n \to \infty$ such that $\|T(t_n)x_0\| \geq \epsilon$ for all $n \in \mathbb{N}$. By passing to a subsequence, we may assume that $t_0 \geq 1$ and $t_{n+1} - t_n \geq 1$ for all $n$. Let $M := \sup_{0 \leq t \leq 1} \|T(t)\|$. For $t \in [t_n - 1, t_n]$ we have $\|T(t_n)x_0\| \leq M \|T(t)x_0\|$, and hence $\|T(t)x_0\| \geq M^{-1}\epsilon$. It follows that

$$\|T(\cdot)x_0\| \geq M^{-1}\epsilon \cdot \chi_{\bigcup_{n=0}^{N-1}[t_n-1,t_n]}, \quad N = 1, 2, ...,$$

and therefore, using the rearrangement invariance,

$$\big\| \|T(\cdot)x_0\| \big\|_E \geq M^{-1}\epsilon \|\chi_{\bigcup_{n=0}^{N-1}[t_n-1,t_n]}\|_E$$
$$= M^{-1}\epsilon \|\chi_{[0,N]}\|_E$$
$$= M^{-1}\epsilon \varphi_E(N), \quad N = 1, 2, ...$$

This contradicts the assumption that $\lim_{t \to \infty} \varphi_E(t) = \infty$. ////

## 3.2. The theorem of Rolewicz

In this section, we apply Theorem 3.1.5 to Orlicz spaces. We refer the reader to Appendix A.4 for the definition of this class of spaces and for the notation used here.

The key to applying Theorem 3.1.5 is the following construction.

**Lemma 3.2.1.** *Let $\phi : \mathbb{R}_+ \to \mathbb{R}_+$ be a non-decreasing function. Then there exists an Orlicz space $E = E(\mathbb{R}_+)$ with $\lim_{t \to \infty} \varphi_E(t) = \infty$, which has the property that $f \in E$ for all $f \in L^\infty(\mathbb{R}_+)$ that satisfy*

$$\int_0^\infty \phi(|f(s)|)\, ds < \infty. \tag{3.2.1}$$

*Proof:* Upon replacing $\phi(t)$ by $\lim_{s \uparrow t} \phi(s)$, we may assume that $\phi$ is left-continuous. Also, upon replacing $\phi$ by some multiple of itself, we may assume that $\phi(1) = 1$. Define

$$\tilde{\phi}(t) := \begin{cases} \phi(t), & 0 < t \leq 1, \\ 1, & t > 1. \end{cases}$$

Let $\tilde{\Phi}$ be its indefinite integral,

$$\tilde{\Phi}(t) = \int_0^t \tilde{\phi}(s)\, ds, \quad t > 0.$$

Fix a bounded measurable function $f$ such that (3.2.1) is satisfied. Without loss of generality we may assume that $\|f\|_\infty \leq 1$. Since $\tilde{\Phi}(t) \leq \tilde{\phi}(t) = \phi(t)$ for all $0 \leq t \leq 1$, we have

$$\int_{\mathbb{R}_+} \tilde{\Phi}(|f(s)|)\, ds \leq \int_{\mathbb{R}_+} \phi(|f(s)|)\, ds < \infty.$$

It follows that

$$M^{\tilde{\Phi}}(f) = \int_{\mathbb{R}_+} \tilde{\Phi}(|f(s)|)\, ds < \infty,$$

and hence $f \in L^{\tilde{\Phi}}$, the Orlicz space associated to $\tilde{\Phi}$. Finally, $\tilde{\phi}(t) > 0$ for all $t > 0$ and hence $\lim_{t \to \infty} \varphi_{L^{\tilde{\Phi}}}(t) = \infty$. This proves the lemma, with $E = L^{\tilde{\Phi}}$.  ////

**Theorem 3.2.2.** *Let $\phi : \mathbb{R}_+ \to \mathbb{R}_+$ be a non-decreasing function with $\phi(t) > 0$ for all $t > 0$. If $\mathbf{T}$ is a $C_0$-semigroup on a Banach space $X$ such that*

$$\int_0^\infty \phi(\|T(t)x\|)\, dt < \infty, \quad \forall x \in X,\; \|x\| \leq 1,$$

*then $\mathbf{T}$ is uniformly exponentially stable.*

*Proof:* Let $E$ be the Orlicz space of Lemma 3.2.1.

We claim that **T** is uniformly bounded. In fact, we have $\lim_{t\to\infty} \|T(t)x\| = 0$. Indeed, if there were $x \in X$, $\epsilon > 0$, and a sequence $t_n \to \infty$ such that $\|T(t_n)x\| \geq \epsilon$ for all $n$, then for all $n$ and all $t \in [t_n - 1, t_n]$ we have $\|T(t)x\| \geq M^{-1}\epsilon$, where $M = \sup_{0\leq s \leq 1} \|T(s)\|$. Assuming without loss of generality that $t_0 \geq 1$ and $t_{n+1} - t_n \geq 1$ for all $n$, it follows that

$$\int_0^\infty \phi(\|T(t)x\|)\, dt \geq \sum_{n=0}^\infty \phi(\epsilon M^{-1}) = \infty,$$

a contradiction.

Since **T** is uniformly bounded, for each $x \in X$ of norm $\leq 1$ the function $f_x(t) := \|T(t)x\|$ satisfies the assumptions of Lemma 3.2.1, and therefore it belongs to $E$. By linearity, the same then holds for arbitrary $x \in X$. Since $\lim_{t\to\infty} \varphi_E(t) = \infty$ by Proposition A.4.2, we can now apply Theorem 3.1.5. ////

For continuous $\phi$, this result is due to S. Rolewicz.

By imposing stronger conditions on $\phi$, it is possible to improve this result and obtain an analogue of Corollary 3.1.6. We say that $\phi$ satisfies a $\Delta_2$-*condition* if there is a constant $K > 0$ such that for all $t > 0$ we have $\phi(t) \leq K\phi(\frac{1}{2}t)$.

**Theorem 3.2.3.** *Let **T** be a $C_0$-semigroup on a Banach space $X$. Let $\phi : \mathbb{R}_+ \to \mathbb{R}_+$ be a non-decreasing function satisfying a $\Delta_2$-condition. Let $\alpha$ be a non-negative measurable function on $\mathbb{R}_+$ such that $\phi \circ \alpha \in L^1_{loc}(\mathbb{R}_+)$ and*

$$\int_0^\infty \phi(\alpha(t))\, dt = \infty.$$

*If*

$$\int_0^\infty \phi(\alpha(t)\|T(t)x\|)\, dt < \infty, \quad \forall x \in X,\ \|x\| \leq 1,$$

*then **T** is uniformly exponentially stable.*

*Proof:* Let $\phi(t) \leq K\phi(\frac{1}{2}t)$ for all $t > 0$. Put $t_0 := 0$ and let $t_1 > 0$ be so large that $\int_0^{t_1} \phi(\alpha(t))\, dt \geq 1$. Inductively, suppose $t_0 < ... < t_{n-1}$ have been chosen such that

$$\int_{t_{k-1}}^{t_k} \phi(2^{-k+1}\alpha(t))\, dt \geq 1, \quad k = 1, ..., n-1.$$

Since $\phi \circ \alpha \in L^1_{loc}(\mathbb{R}_+)$, we have $\int_{t_{n-1}}^\infty \phi(\alpha(t))\, dt = \infty$. Hence also

$$\int_{t_{n-1}}^\infty \phi(2^{-n+1}\alpha(t))\, dt = \infty$$

by the $\Delta_2$-condition. Therefore, for $t_n > t_{n-1}$ large enough,
$$\int_{t_{n-1}}^{t_n} \phi(2^{-n+1}\alpha(t))\,dt \geq 1.$$
This completes the induction step.

Suppose, for a contradiction, that $\omega_0(\mathbf{T}) \geq 0$. Let $\gamma \in C_0[0,\infty)$ be a norm one function such that $\gamma(t) \geq 2^{-n+1}$ for $t \in [t_{n-1}, t_n); n = 1, 2, \ldots$

By Lemma 3.1.7, there is a norm one vector $x \in X$ such that $\|T(t)x\| \geq \frac{1}{2}\gamma(t)$ for all $t \geq 0$. But then, using the $\Delta_2$-condition,
$$\int_0^\infty \phi\big(\alpha(t)\|T(t)x\|\big)\,dt \geq K^{-1} \int_0^\infty \phi\big(\alpha(t)\gamma(t)\big)\,dt$$
$$\geq K^{-1} \sum_{n=1}^\infty \int_{t_{n-1}}^{t_n} \phi(2^{-n+1}\alpha(t))\,dt = \infty.$$
This contradiction concludes the proof. ////

Note that Corollary 3.1.6 corresponds to the special case $\phi(t) = t^p$ and $\beta = \alpha^{\frac{1}{p}}$.

## 3.3. Characterization by convolutions

Consider the inhomogeneous abstract Cauchy problem
$$\frac{du}{dt}(t) = Au(t) + f(t), \quad t \geq 0,$$
$$u(0) = x,$$
where $A$ is the generator of a $C_0$-semigroup $\mathbf{T}$ on a Banach space $X$ and $f \in L^p(\mathbb{R}_+, X)$ or $C_0(\mathbb{R}_+, X)$. As is well-known, for all $x \in X$ this problem has a unique mild solution given by
$$u(t) = T(t)x + (\mathbf{T} * f)(t),$$
where the convolution $\mathbf{T} * f$ is defined by
$$(\mathbf{T} * f)(t) := \int_0^t T(s)f(t-s)\,ds, \quad t \geq 0.$$
This observation motivates us to study the action by convolution of a $C_0$-semigroup on vector-valued function spaces over $\mathbb{R}_+$.

The main result of this section, which relies on Latushkin - Montgomery-Smith theory, is a characterization of uniform exponential stability in terms of convolutions. In the proof we use *Vitali's theorem*: If $\Omega_1 \subset \Omega_0 \subset \mathbb{C}$ are simply connected open sets, $(f_n)$ is a uniformly bounded sequence of bounded $X$-valued holomorphic functions on $\Omega_0$, and $f$ is an $X$-valued holomorphic function on $\Omega_1$ such that $f_n \to f$ on $\Omega_1$ pointwise, then $f$ admits a holomorphic extension to $\Omega_0$ and we have $f_n \to f$ locally uniformly on $\Omega_0$.

**Theorem 3.3.1.** Let **T** be a $C_0$–semigroup on a Banach space $X$ and let $1 \le p < \infty$. Then the following assertions are equivalent:

(i) **T** is uniformly exponentially stable;
(ii) $\mathbf{T} * f \in L^p(\mathbb{R}_+, X)$ for all $f \in L^p(\mathbb{R}_+, X)$;
(iii) $\mathbf{T} * f \in C_0(\mathbb{R}_+, X)$ for all $f \in C_0(\mathbb{R}_+, X)$.

*Proof:* For $1 \le p < \infty$, we define the $C_0$–semigroup $\mathbf{S}_p$ on $L^p(\mathbb{R}_+, X)$ by

$$(S_p(t)f)(s) = \begin{cases} T(t)f(s-t), & s \ge t; \\ 0, & \text{else.} \end{cases}$$

Similarly, we define the $C_0$–semigroup $\mathbf{S}_\infty$ on $C_{00}(\mathbb{R}_+, X)$, the space of $X$-valued continuous functions vanishing at infinity satisfying $f(0) = 0$, by

$$(S_\infty(t)f)(s) = \begin{cases} T(t)f(s-t), & s \ge t; \\ 0, & \text{else.} \end{cases}$$

We denote the generators of these semigroups by $B_p$ and $B_\infty$, respectively.

First we prove (ii)$\Rightarrow$(i).

By Lemma 2.5.5, we have to prove that $s(B_p) < 0$. For this, it is enough to prove that the resolvent of $B_p$ exists and is uniformly bounded in the right half-plane. Indeed, once this is established, Lemma 2.3.4 shows that $s(B_p) < 0$.

We start by observing that there is a constant $M > 0$ such that

$$\|\mathbf{T} * f\|_{L^p(\mathbb{R}_+, X)} \le M \|f\|_{L^p(\mathbb{R}_+, X)}, \qquad \forall f \in L^p(\mathbb{R}_+, X).$$

To see this, we claim that the map $f \mapsto \mathbf{T} * f$ is closed as a map of $L^p(\mathbb{R}_+, X)$ into itself. Indeed, assume $f_n \to f$ and $\mathbf{T} * f_n \to g$ in $L^p(\mathbb{R}_+, X)$. Then it is immediate that $(\mathbf{T} * f_n)(s) \to (\mathbf{T} * f)(s)$ for all $s > 0$. On the other hand, since a norm convergent sequence in $L^p(\mathbb{R}_+, X)$ contains a subsequence that converges pointwise a.e., it follows that $(\mathbf{T} * f_{n_k})(s) \to g(s)$ for some sequence $(n_k)$ and almost all $s$. Therefore, $\mathbf{T} * f = g$, as was to be shown. The existence of the constant $M$ now follows from the closed graph theorem.

For $T_0 > 0$, we define $\pi_{T_0}: L^p(\mathbb{R}_+, X) \to L^p([0, T_0], X)$ by restriction: $\pi_{T_0} f = f|_{[0, T_0]}$. For $T > 0$, $T_0 > 0$ and $f \in L^p(\mathbb{R}_+, X)$, we define the entire $L^p(\mathbb{R}_+, X)$-valued function $F_{T,f}$ and the entire $L^p([0, T_0], X)$-valued function $F_{T,T_0,f}$ by

$$F_{T,f}(z) = \int_0^T e^{-zt} S_p(t) f \, dt,$$
$$F_{T,T_0,f}(z) = \pi_{T_0}(F_{T,f}(z)).$$

For each $z$, the map $f \mapsto F_{T,T_0,f}(z)$ is bounded as a map from $L^p(\mathbb{R}_+, X)$ into $L^p([0, T_0], X)$. A trivial estimate shows that each of the functions $F_{T,f}$ and $F_{T,T_0,f}$ is bounded in each vertical strip $\{0 < \operatorname{Re} z < c\}$, $c > 0$.

For $\lambda \in \mathbb{R}$ and $f \in L^p(\mathbb{R}_+, X)$, let $f_\lambda(s) := e^{i\lambda s} f(s)$, $s \geq 0$. The restriction of $\mathbf{S}_p$ to the invariant subspace $C_{00}(\mathbb{R}_+, X) \cap L^p(\mathbb{R}_+, X)$ extends to the $C_0$−semigroup $\mathbf{S}_\infty$ on $C_{00}(\mathbb{R}_+, X)$. Since point evaluations on the latter space are continuous, for $f \in C_{00}(\mathbb{R}_+, X) \cap L^p(\mathbb{R}_+, X)$, $T \geq T_0$, and $0 \leq s \leq T_0$ we have

$$\left(\int_0^T e^{-i\lambda t} S_p(t) f \, dt\right)(s) = \int_0^T e^{-i\lambda t} S_p(t) f(s) \, dt$$

$$= \int_0^s e^{-i\lambda t} T(t) f(s-t) \, dt$$

$$= e^{-i\lambda s} \int_0^s T(t) f_\lambda(s-t) \, dt.$$

Therefore, for $T \geq T_0$,

$$\|F_{T,T_0,f}(i\lambda)\|_{L^p([0,T_0],X)} = \left\|\pi_{T_0}\left(e^{-i\lambda(\cdot)}\int_0^{(\cdot)} T(t) f_\lambda(\cdot - t) \, dt\right)\right\|_{L^p([0,T_0],X)}$$

$$\leq \|\pi_{T_0}\| \cdot \|\mathbf{T} * f_\lambda\|_{L^p(\mathbb{R}_+, X)} \leq \|\mathbf{T} * f_\lambda\|_{L^p(\mathbb{R}_+, X)}$$

$$\leq M \|f_\lambda\|_{L^p(\mathbb{R}_+, X)} = M \|f\|_{L^p(\mathbb{R}_+, X)}.$$

Since $C_{00}(\mathbb{R}_+, X) \cap L^p(\mathbb{R}_+, X)$ is dense in $L^p(\mathbb{R}_+, X)$, it follows that

$$\|F_{T,T_0,f}(i\lambda)\|_{L^p([0,T_0],X)} \leq M \|f\|_{L^p(\mathbb{R}_+, X)} \qquad (3.3.1)$$

for all $\lambda \in \mathbb{R}$, $T \geq T_0$ and $f \in L^p(\mathbb{R}_+, X)$. Also, if we choose constants $N > 0$ and $\omega \geq 0$ such that $\|S_p(t)\| \leq N e^{\omega t}$ for all $t \geq 0$, then for $\mathrm{Re}\, z = \omega + 1$ we have

$$\|F_{T,T_0,f}(z)\|_{L^p([0,T_0],X)} \leq \|\pi_{T_0}\| \int_0^T e^{-(\omega+1)t} N e^{\omega t} \|f\|_{L^p(\mathbb{R}_+, X)} \, dt$$

$$\leq N(1 - e^{-T}) \|f\|_{L^p(\mathbb{R}_+, X)} \leq N \|f\|_{L^p(\mathbb{R}_+, X)}.$$

It follows that for each $f \in L^p(\mathbb{R}_+, X)$ and $T_0 > 0$ fixed, the functions $z \mapsto F_{T,T_0,f}(z)$ are bounded on the line $\mathrm{Re}\, z = \omega + 1$, uniformly with respect to $T > 0$, with bound $N \|f\|_{L^p(\mathbb{R}_+, X)}$. Therefore, by (3.3.1) and the Phragmen-Lindelöf theorem, for each $f$ and $T_0$ fixed we have

$$\|F_{T,T_0,f}(z)\|_{L^p([0,T_0],X)} \leq \max\{M, N\} \|f\|_{L^p(\mathbb{R}_+, X)} \qquad (3.3.2)$$

for all $0 < \mathrm{Re}\, z < \omega + 1$ and $T \geq T_0$. Also, for $\mathrm{Re}\, z > \omega$ we have

$$\lim_{T \to \infty} F_{T,T_0,f}(z) = \pi_{T_0} R(z, B_p) f.$$

Combining these facts, by Vitali's theorem for each $f \in L^p(\mathbb{R}_+, X)$ and $T_0 > 0$ the function $z \mapsto \pi_{T_0} R(z, B_p)f$ has an holomorphic extension $F_{\infty, T_0, f}$ to $\{0 < \text{Re}\, z < \omega + 1\}$, and that for $0 < \text{Re}\, z < \omega + 1$,

$$F_{\infty, T_0, f}(z) = \lim_{T \to \infty} F_{T, T_0, f}(z)$$

uniformly on compacta. Moreover, by (3.3.2),

$$\|F_{\infty, T_0, f}(z)\|_{L^p([0, T_0], X)} \leq \max\{M, N\} \|f\|_{L^p(\mathbb{R}_+, X)} \tag{3.3.3}$$

for all $0 < \text{Re}\, z < \omega + 1$ and $T_0 > 0$. By regarding $L^p([0, T_0], X)$ as a closed subspace of $L^p(\mathbb{R}_+, X)$, for all $\omega < \text{Re}\, z < \omega + 1$ we have

$$\lim_{T_0 \to \infty} F_{\infty, T_0, f}(z) = \lim_{T_0 \to \infty} \pi_{T_0} R(z, B_p)f = R(z, B_p)f, \tag{3.3.4}$$

the convergence being with respect to the norm of $L^p(\mathbb{R}_+, X)$. Again by Vitali's theorem, now using (3.3.3), it follows that $R(z, B_p)f$ admits a holomorphic extension $F_{\infty, \infty, f}$ to $\{0 < \text{Re}\, z < \omega + 1\}$, and that for all $0 < \text{Re}\, z < \omega + 1$,

$$\lim_{T_0 \to \infty} \lim_{T \to \infty} F_{T, T_0, f}(z) = \lim_{T_0 \to \infty} F_{\infty, T_0, f}(z) = F_{\infty, \infty, f}(z) \tag{3.3.5}$$

uniformly on compacta. By Proposition 1.1.6, $\{0 < \text{Re}\, z < \omega + 1\} \subset \varrho(B_p)$ and $F_{\infty, \infty, f}(z) = R(z, B_p)f$.

Therefore, by (3.3.3), (3.3.5), and the uniform boundedness theorem, it follows that $R(z, B_p)$ is uniformly bounded in $\{0 < \text{Re}\, z < \omega + 1\}$. By the Hille-Yosida theorem, $R(z, B_p)$ is also uniformly bounded in $\{\text{Re}\, z \geq \omega + 1\}$. Thus, $R(z, B_p)$ exists and is uniformly bounded in $\{\text{Re}\, z > 0\}$. This completes the proof of (ii)⇒(i).

Next, we prove (i) ⇒(ii). Assume $\omega_0(\mathbf{T}) < 0$, and choose $M > 0$ and $\mu > 0$ such that $\|T(t)\| \leq M e^{-\mu t}$ for all $t \geq 0$. Let $1 \leq p < \infty$ be arbitrary and fixed. By applying Jensen's inequality to the probability measure $\mu(1 - e^{-\mu s})^{-1} e^{-\mu t}\, dt$ on $[0, s]$, we have, noting that $\mu^{-(p-1)}(1 - e^{-\mu s})^{p-1} \leq \mu^{-(p-1)}$ for all $s > 0$,

$$\int_0^\infty \left\| \int_0^s T(t) f(s-t)\, dt \right\|^p ds \leq \int_0^\infty \left( \int_0^s M e^{-\mu t} \|f(s-t)\|\, dt \right)^p ds$$

$$\leq M^p \int_0^\infty \mu^{-p}(1 - e^{-\mu s})^p \int_0^s \|f(s-t)\|^p \mu(1 - e^{-\mu s})^{-1} e^{-\mu t}\, dt\, ds$$

$$\leq M^p \mu^{-(p-1)} \int_0^\infty \int_0^s e^{-\mu t} \|f(s-t)\|^p\, dt\, ds$$

$$= M^p \mu^{-(p-1)} \int_0^\infty e^{-\mu t} \int_t^\infty \|f(s-t)\|^p\, ds\, dt$$

$$\leq M^p \mu^{-p} \|f\|^p.$$

This proves (i)⇒(ii).

It remains to prove the equivalence of (i) and (iii). The implication (i)⇒(iii) is proved as follows. Choose $M > 0$ and $\mu > 0$ such that $\|T(t)\| \leq Me^{-\mu t}$ for all $t \geq 0$. Fix $\epsilon > 0$ and $f \in C_0(\mathbb{R}_+, X)$ arbitrary. Choose $N$ so large that $se^{-\mu s} \leq \epsilon$ and $\|f(s)\| \leq \epsilon \|f\|_{C_0(\mathbb{R}_+, X)}$ for all $s \geq N$. Then, for $s \geq 2N$,

$$\left\| \int_0^s T(t) f(s-t) dt \right\| \leq \int_{\frac{s}{2}}^s Me^{-\mu \frac{s}{2}} \|f\|_{C_0(\mathbb{R}_+, X)} dt + \int_0^{\frac{s}{2}} Me^{-\mu t} \epsilon \|f\|_{C_0(\mathbb{R}_+, X)} dt$$
$$\leq M(1 + \mu^{-1}) \epsilon \|f\|_{C_0(\mathbb{R}_+, X)}.$$

Since $\mathbf{T} * f$ also is continuous, we obtain the desired conclusion.

Finally, we have to prove (iii)⇒(i). We do this by modifying the proof of (ii)⇒(i). First we note that there exists a constant $M > 0$ such that

$$\|\mathbf{T} * f\|_{C_0(\mathbb{R}_+, X)} \leq M \|f\|_{C_0(\mathbb{R}_+, X)} \qquad \forall f \in C_0(\mathbb{R}_+, X).$$

Indeed, this follows from applying the uniform boundedness theorem to the operators $T_s : f \mapsto \int_0^s T(t) f(s-t) \, dt$.

Let $f \in C_{00}(\mathbb{R}_+, X)$ be arbitrary. Since $\mathbf{T} * f \in C_0(\mathbb{R}_+, X)$ by assumption and $(\mathbf{T} * f)(0) = 0$, it follows that $\mathbf{T}$ acts boundedly on $C_{00}(\mathbb{R}_+, X)$ by convolution, with norm at most $M$.

Let $C_{00}([0, T_0], X)$ be the closed subspace of $C_{00}(\mathbb{R}_+, X)$ consisting of all functions vanishing on $[T_0, \infty)$. For each $T_0 \geq 1$, define the piecewise linear function $g$ on $\mathbb{R}_+$ by

$$g_{T_0}(t) = \begin{cases} 1, & 0 \leq t \leq T_0 - 1; \\ T_0 - t, & T_0 - 1 \leq t \leq T_0; \\ 0, & \text{else} \end{cases}$$

and the operator $\Pi_{T_0} : C_{00}(\mathbb{R}_+, X) \to C_{00}([0, T_0], X)$ by

$$(\Pi_{T_0} f)(t) = g_{T_0}(t) f(t), \qquad 0 \leq t \leq T_0.$$

Note that for all $f \in C_{00}(\mathbb{R}_+, X)$, $\|\Pi_{T_0} f\|_{C_{00}([0, T_0], X)} \leq \|f\|_{C_{00}(\mathbb{R}_+, X)}$. With $\pi_{T_0}$ replaced by $\Pi_{T_0}$, the proof of now follows along the lines of (ii)⇒(i). ////

The reason of introducing the operators $\Pi_{T_0}$ is as follows. If we simply truncate a function in $C_{00}(\mathbb{R}_+, X)$ with $\pi_{T_0}$, the resulting function need not define an element of $C_{00}([0, T_0], X)$, so that we cannot perform the limiting operation (3.3.4). With the operator $\Pi_{T_0}$, this poses no problems.

In the next section, we will improve part of Theorem 3.3.1 by showing that $\omega_0(\mathbf{T}) < 0$ if (and only if) $\mathbf{T} * f$ is merely *bounded* for all $f \in C_0(\mathbb{R}_+, X)$.

Theorem 3.3.1 remains valid if the role of $L^p$ is taken over by any rearrangement invariant Banach function space $E = E(\mathbb{R}_+)$ over $\mathbb{R}_+$ with order continuous norm. By the order continuity of the norm, translation is strongly continuous on $E(\mathbb{R}_+)$. The proof of (i) ⇒ (ii) then uses the fact that Lemma 2.5.4 admits a straightforward generalization to this setting, and (ii) ⇒ (i) is proved as follows. For $\mu > 0$ and $g \in L^1_{loc}(\mathbb{R}_+)$, define

$$(T_\mu(g))(s) := \int_0^s e^{-\mu t} g(s-t)\, dt, \qquad s \geq 0.$$

The proof of Theorem 3.3.1, (i)⇒(ii), shows that this defines a bounded operator $T_\mu : L^1(\mathbb{R}_+) \to L^1(\mathbb{R}_+)$ of norm $\leq \mu^{-1}$. Also, it is trivial that $T_\mu$ is bounded as an operator $L^\infty(\mathbb{R}_+) \to L^\infty(\mathbb{R}_+)$, of norm $\leq \mu^{-1}$. By Proposition A.4.1, every rearrangement invariant Banach function space over $\mathbb{R}_+$ with order continuous norm is an interpolation space between $L^1(\mathbb{R}_+)$ and $L^\infty(\mathbb{R}_+)$. Therefore, $T_\mu$ is bounded as an operator $E(\mathbb{R}_+) \to E(\mathbb{R}_+)$.

Now let $f \in E(\mathbb{R}_+, X)$ be arbitrary. Since $\omega_0(\mathbf{T}) < 0$, there are constants $M > 0$ and $\mu > 0$ such that $\|T(t)\| \leq M e^{-\mu t}$ for all $t \geq 0$. Since $\|f(\cdot)\| \in E(\mathbb{R}_+)$, we have

$$\|\mathbf{T} * f\|_{E(\mathbb{R}_+, X)} = \Big\|\|(\mathbf{T}*f)(\cdot)\|\Big\|_{E(\mathbb{R}_+)} \leq \Big\|\int_0^{(\cdot)} M e^{-\mu t} \|f(\cdot - t)\|\, dt\Big\|_{E(\mathbb{R}_+)}$$

$$= M\Big\|T_\mu(\|f(\cdot)\|)\Big\|_{E(\mathbb{R}_+)} \leq M\|T_\mu\|_{E(\mathbb{R}_+)} \Big\|\|f(\cdot)\|\Big\|_{E(\mathbb{R}_+)}$$

$$= M\|T_\mu\|_{E(\mathbb{R}_+)} \|f\|_{E(\mathbb{R}_+, X)}.$$

This concludes the proof. ////

In the remainder of this section we study the weak analogue of Theorem 3.3.1. If $E(\mathbb{R}_+)$ is a given Banach function space over $\mathbb{R}_+$, we want to characterize those semigroups $\mathbf{T}$ on $X$ for which $\langle x^*, \mathbf{T} * f\rangle$ defines an element of $E(\mathbb{R}_+)$ for all $x^* \in X^*$ and $f \in E(\mathbb{R}_+, X)$. Here, and in the following, for a $g \in L^1_{loc}(\mathbb{R}_+, X)$ and a functional $x^* \in X^*$, the function $\langle x^*, g\rangle \in L^1_{loc}(\mathbb{R}_+)$ is defined in the natural way: $\langle x^*, g\rangle(s) = \langle x^*, g(s)\rangle$; $s \geq 0$.

For $E = L^1$ we solve this problem as follows. A semigroup $\mathbf{T}$ is said to be *scalarly integrable* if

$$\int_0^\infty |\langle x^*, T(t)x\rangle|\, dt < \infty, \qquad \forall x \in X, x^* \in X^*.$$

**Theorem 3.3.2.** *Let $\mathbf{T}$ be a $C_0$–semigroup on a Banach space $X$. Then the following assertions are equivalent:*

(i) $\mathbf{T}$ *is scalarly integrable;*
(ii) $\langle x^*, \mathbf{T} * f\rangle \in L^1(\mathbb{R}_+)$ *for all $f \in L^1(\mathbb{R}_+, X)$ and $x^* \in X^*$.*

*Proof:* Assume (ii). The bilinear map $T : X^* \times L^1(\mathbb{R}_+, X) \to L^1(\mathbb{R}_+)$ defined by $T(x^*, f) = \langle x^*, \mathbf{T} * f\rangle$ is separately continuous by the closed graph theorem. Therefore, by Lemma 3.1.1, there exists a constant $M > 0$ such that

$$\|\langle x^*, \mathbf{T} * f\rangle\|_{L^1(\mathbb{R}_+)} \leq M \|f\|_{L^1(\mathbb{R}_+, X)} \|x^*\|, \qquad \forall f \in L^1(\mathbb{R}_+, X), x^* \in X^*.$$

Fix $x \in X$, $x^* \in X^*$ and $s_0 > 1$ arbitrary. Choose $0 < \tau_0 < 1$ such that

$$\frac{1}{\tau_0}\left|\int_0^{\tau_0} \langle x^*, T(s-t)x\rangle\, dt\right| \geq \frac{1}{2}|\langle x^*, T(s)x\rangle|, \qquad \forall 1 \leq s \leq s_0.$$

Then,

$$\begin{aligned}
\int_1^{s_0} |\langle x^*, T(s)x\rangle|\, ds &\leq 2\int_1^{s_0} \frac{1}{\tau_0}\left|\int_0^{\tau_0} \langle x^*, T(s-t)x\rangle\, dt\right| ds \\
&= 2\int_1^{s_0} \frac{1}{\tau_0}\left|\int_0^s \langle x^*, T(t)x\rangle\, \chi_{[0,\tau_0]}(s-t)\, dt\right| ds \\
&\leq \frac{2}{\tau_0}\|\langle x^*, \mathbf{T} * (\chi_{[0,\tau_0]} \otimes x)\rangle\|_{L^1(\mathbb{R}_+)} \\
&\leq \frac{2M}{\tau_0}\|\chi_{[0,\tau_0]} \otimes x\|_{L^1(\mathbb{R}_+,X)}\|x^*\| = 2M\|x\|\,\|x^*\|.
\end{aligned}$$

Since $s_0 > 1$ is arbitrary, it follows that

$$\int_1^\infty |\langle x^*, T(s)x\rangle|\, ds \leq 2M\|x\|\,\|x^*\|, \qquad \forall x \in X, x^* \in X^*.$$

Therefore, $\int_0^\infty |\langle x^*, T(s)x\rangle|\, ds < \infty$ for all $x \in X$ and $x^* \in X^*$, which proves (i).

Conversely, assume (i). By Lemma 3.1.2, there exists a constant $C > 0$ such that

$$\int_0^\infty |\langle x^*, T(t)x\rangle|\, dt \leq C\|x\|\,\|x^*\|, \qquad \forall x \in X, x^* \in X^*.$$

Let $N := \sup_{0 \leq s \leq 1} \|T(s)\|$. Fix $x \in X$, $x^* \in X^*$ and real numbers $0 \leq t_0 < t_1$ with $t_1 - t_0 \leq 1$. Then,

$$\begin{aligned}
\int_0^\infty &\left|\left\langle x^*, \int_0^s T(\tau)(\chi_{(t_0,t_1)} \otimes x)(s-\tau)\, d\tau\right\rangle\right| ds \\
&= \int_0^\infty \left|\left\langle x^*, \int_{\max\{s-t_1,0\}}^{\max\{s-t_0,0\}} T(\tau)x\, d\tau\right\rangle\right| ds \\
&= \int_{t_0}^{t_1} \left|\left\langle x^*, \int_0^{s-t_0} T(\tau)x\, d\tau\right\rangle\right| ds + \int_{t_1}^\infty \left|\left\langle x^*, \int_{s-t_1}^{s-t_0} T(\tau)x\, d\tau\right\rangle\right| ds \\
&\leq \int_{t_0}^{t_1} (s-t_0)N\|x\|\,\|x^*\|\, ds + \int_0^\infty \left|\left\langle x^*, T(s)\int_0^{t_1-t_0} T(\tau)x\, d\tau\right\rangle\right| ds \\
&\leq (t_1-t_0)N\|x\|\,\|x^*\| + C\left\|\int_0^{t_1-t_0} T(\tau)x\, d\tau\right\|\|x^*\| \\
&\leq (t_1-t_0)N(1+C)\|x\|\,\|x^*\|.
\end{aligned}$$

Therefore, with $M = N(1+C)$, we have

$$\|\langle x^*, \mathbf{T} * (\chi_{(t_0,t_1)} \otimes x)\rangle\|_{L^1(\mathbb{R}_+)} \leq M(t_1 - t_0)\|x\|\,\|x^*\|.$$

Next, let $f$ be a stepfunction of the form $f = \sum_{k=0}^{n-1} \chi_{(t_k,t_{k+1})} \otimes x_k$ with the intervals $(t_k, t_{k+1})$ pairwise disjoint. By splitting large intervals into finitely many smaller ones, we may assume that $0 < t_{k+1} - t_k \leq 1$ for all $k = 0, ..., n-1$. By the above estimate we have

$$\begin{aligned}\|\langle x^*, \mathbf{T} * f\rangle\|_{L^1(\mathbb{R}_+)} &\leq \sum_{k=0}^{n-1} \|\langle x^*, \mathbf{T} * (\chi_{(t_k,t_{k+1})} \otimes x_k)\rangle\|_{L^1(\mathbb{R}_+)} \\ &\leq M \sum_{k=0}^{n-1} (t_{k+1} - t_k)\|x_k\|\,\|x^*\| = M\|f\|_{L^1(\mathbb{R}_+,X)}\|x^*\|.\end{aligned} \quad (3.3.6)$$

Since the stepfunctions supported by finite unions of disjoint intervals of length $\leq 1$ are dense in $L^1(\mathbb{R}_+, X)$, (3.3.6) holds for arbitrary $f \in L^1(\mathbb{R}_+, X)$. This proves the implication (i)$\Rightarrow$(ii). ////

Scalarly integrable semigroups will be studied in more detail in Section 4.6. Since there exist scalarly integrable semigroups whose uniform growth bound is positive (combine Theorem 4.6.6 below and Example 1.4.4), the theorem shows that condition (ii) does not characterize uniform exponential stability. In Hilbert space however, a $C_0$-semigroup $\mathbf{T}$ is scalarly integrable if and only if it is uniformly exponentially stable; this will be proved in Section 4.6. In combination with Theorem 3.3.2, this leads to the following result.

**Corollary 3.3.3.** *Let $\mathbf{T}$ be a $C_0$-semigroup on a Hilbert space $H$. Then the following assertions are equivalent:*

(i) $\mathbf{T}$ *is uniformly exponentially stable;*
(ii) $\langle y, \mathbf{T} * f\rangle \in L^1(\mathbb{R}_+)$ *for all $f \in L^1(\mathbb{R}_+, H)$ and $y \in H$.*

## 3.4. Characterization by almost periodic functions

In this section we apply Theorem 3.3.1 in order to prove that a $C_0$-semigroup $\mathbf{T}$ on a Banach space $X$ is uniformly exponentially stable if and only if convolution with $\mathbf{T}$ maps certain closed subspaces of $BUC(\mathbb{R}_+, X)$ into $L^\infty(\mathbb{R}_+, X)$. This, in turn, leads to a characterization of uniform exponential stability in terms of almost periodic functions.

Let $BUC(\mathbb{R}_+, X)$ denote the space of all $X$-valued, bounded, uniformly continuous functions on $\mathbb{R}_+$. A linear subspace $E$ of $BUC(\mathbb{R}_+, X)$ will be called *locally* dense in $BUC(\mathbb{R}_+, X)$ if for every $\epsilon > 0$, every bounded closed interval $I \subset \mathbb{R}_+$, and every $f \in C(I)$ there exists a function $f_{\epsilon,I} \in E$ such that

$$\sup_{t \in I} \|f(t) - f_{\epsilon,I}(t)\| \leq \epsilon.$$

If, in addition, there is a constant $K > 0$, independent of $I$ and $\epsilon$, such that for every $f \in C(I)$ the function $f_{\epsilon,I}$ can be chosen in such a way that $\|f_{\epsilon,I}\|_{BUC(\mathbb{R}_+,X)} \leq K\|f\|_{C(I)}$, we say that $E$ is *boundedly locally dense* in $BUC(\mathbb{R}_+, X)$.

**Theorem 3.4.1.** *Let $\mathbf{T}$ be a $C_0$-semigroup a Banach space $X$ and let $E$ be a closed, boundedly locally dense subspace of $BUC(\mathbb{R}_+, X)$. Then the following assertions are equivalent:*

*(i) $\mathbf{T}$ is uniformly exponentially stable;*

*(ii) $\sup_{s>0} \left\| \int_0^s T(t)g(t)\, dt \right\| < \infty, \quad \forall g \in E.$*

*Proof:* The implication (i)⇒(ii) is trivial. We will prove (ii)⇒(i). By the uniform boundedness theorem, there is a constant $C > 0$ such that

$$\sup_{s>0} \left\| \int_0^s T(t)g(t)\, dt \right\| \leq C\|g\|_{BUC(\mathbb{R}_+,X)}, \quad \forall g \in E. \tag{3.4.1}$$

For a given $f \in C_0(\mathbb{R}_+, X)$ and $s > 0$, let $M_s = \sup_{0 \leq t \leq s} \|T(t)\|$ and let $f_s \in E$ be any function such that

$$\sup_{0 \leq t \leq s} \|f(s-t) - f_s(t)\| \leq \frac{1}{sM_s}\|f\|_{C_0(\mathbb{R}_+,X)},$$

and

$$\|f_s\|_{BUC(\mathbb{R}_+,X)} \leq K\|f\|_{C_0(\mathbb{R}_+,X)}.$$

Such an $f_s$ exists by the definition of a boundedly locally dense subspace; $K$ is the constant from the definition. Then, by (3.4.1),

$$\|(\mathbf{T} * f)(s)\| = \left\| \int_0^s T(t)\left(f(s-t) - f_s(t) + f_s(t)\right) dt \right\|$$

$$\leq \|f\|_{C_0(\mathbb{R}_+,X)} + \left\| \int_0^s T(t)f_s(t)\, dt \right\|$$

$$\leq \|f\|_{C_0(\mathbb{R}_+,X)} + C\|f_s\|_{BUC(\mathbb{R}_+,X)} \leq (1 + CK)\|f\|_{C_0(\mathbb{R}_+,X)}. \tag{3.4.2}$$

It follows that $\mathbf{T} * f$ is a bounded continuous function. If we can prove that

$$\lim_{s \to \infty} \int_0^s T(t)f(s-t)\, dt = 0,$$

it follows that convolution with $\mathbf{T}$ maps $C_0(\mathbb{R}_+, X)$ into $C_0(\mathbb{R}_+, X)$ and (3.4.2) shows that this map is bounded. Then we can apply Theorem 3.3.1 to conclude that $\omega_0(\mathbf{T}) < 0$.

Fix $\epsilon > 0$ arbitrary and choose $N$ so large that $\|f(s)\| \leq (1 + CK)^{-1}\epsilon$ for all $s \geq N$. Write $f = f_0 + f_1$, where $f_0 \in C_0(\mathbb{R}_+, X)$ is chosen in such a way that $f_0(s) = f(s)$ for all $s \geq N$ and $\|f_0\|_{C_0(\mathbb{R}_+, X)} \leq (1 + CK)^{-1}\epsilon$. Note that the support of $f_1$ is contained in the interval $[0, N]$. By (3.4.2), for all $s \geq 0$ we have

$$\left\| \int_0^s T(t) f_0(s - t) \, dt \right\| \leq (1 + CK)\|f_0\|_{C_0(\mathbb{R}_+, X)} \leq \epsilon.$$

It follows that it is sufficient to prove that

$$\lim_{s \to \infty} T(s - N) \left( \int_0^N T(t) f_1(N - t) \, dt \right) = \lim_{s \to \infty} \int_0^s T(t) f_1(s - t) \, dt = 0. \quad (3.4.3)$$

Since $\lim_{\lambda \to \infty} \lambda^2 R(\lambda, A)^2 f_1(\cdot) \to f_1(\cdot)$ uniformly on $[0, N]$, and hence on all of $\mathbb{R}_+$, (3.4.2) shows that it is even sufficient to prove that

$$\lim_{s \to \infty} T(s - N) \left( \int_0^N T(t) \lambda^2 R(\lambda, A)^2 f_1(N - t) \, dt \right) = 0 \quad (3.4.4)$$

for all $\lambda$ sufficiently large. Note that, for each such $\lambda$,

$$\int_0^N T(t) \lambda^2 R(\lambda, A)^2 f_1(N - t) \, dt \in D(A^2). \quad (3.4.5)$$

In order to prove (3.4.4), we claim that the resolvent of $A$ exists and is uniformly bounded in the right half-plane.

To prove this, fix $\mu > 0$, $g \in E$, and $s > 0$ arbitrary. Choose a function $g_{\mu,s} \in E$ such that

$$\sup_{0 \leq t \leq s} \|g_{\mu,s}(t) - e^{-\mu t} g(t)\| \leq \frac{1}{sM_s} \|g\|_{BUC(\mathbb{R}_+, X)}$$

and $\|g_{\mu,s}\|_{BUC(\mathbb{R}_+, X)} \leq K\|g\|_{BUC(\mathbb{R}_+, X)}$. Then, by (3.4.1),

$$\left\| \int_0^s e^{-\mu t} T(t) g(t) \, dt \right\| \leq \|g\|_{BUC(\mathbb{R}_+, X)} + C\|g_{\mu,s}\|_{BUC(\mathbb{R}_+, X)} \quad (3.4.6)$$
$$\leq (1 + CK)\|g\|_{BUC(\mathbb{R}_+, X)}.$$

Next, let $\nu \in \mathbb{R}$ and $x \in X$ be arbitrary and fixed. Let $g_{\nu,x,s} \in E$ be a function such that

$$\sup_{0 \leq t \leq s} \|g_{\nu,x,s}(t) - e^{-i\nu t} \otimes x\| \leq \frac{1}{sM_s} \|x\|$$

and $\|g_{\nu,x,s}\|_{BUC(\mathbb{R}_+,X)} \leq K\|x\|$. Then, by (3.4.6),

$$\left\|\int_0^s e^{-(\mu+i\nu)t}T(t)x\,dt\right\| \leq \|x\| + \left\|\int_0^s e^{-\mu t}T(t)g_{\nu,x,s}(t)\,dt\right\|$$
$$\leq \|x\| + (1+CK)\|g_{\nu,x,s}\|_{BUC(\mathbb{R}_+,X)} \qquad (3.4.7)$$
$$\leq (1 + (1+CK)K)\|x\|.$$

Since $s > 0$ and $\mu > 0, \nu > 0$ are arbitrary, (3.4.7) shows that the entire $X$-valued functions $z \mapsto \int_0^s e^{-zt}T(t)x\,dt$ are bounded in the right half-plane, uniformly in $s > 0$. Moreover, for $\operatorname{Re} z$ sufficiently large, they converge to $R(z,A)x$ as $s \to \infty$. An application of Vitali's theorem and an Proposition 1.1.6 show that the right half-plane is contained in $\varrho(A)$ and that the resolvent of $A$ is uniformly bounded there. This concludes the proof of the claim.

We next borrow a result from Chapter 4, where it is proved that the uniform boundedness of the resolvent in the right half-plane implies exponential stability. All we need here is the weaker (and much easier to prove) fact that there are constants $k > 0$ and $\delta > 0$ such that $\|T(t)y\| \leq ke^{-\delta t}\|y\|_{D(A^2)}$ for all $y \in D(A^2)$ and $t \geq 0$.

Using this, it is immediate that (3.4.4) is a consequence of (3.4.5). This concludes the proof. ////

It is clear that $E = C_0(\mathbb{R}_+, X)$ is a closed, boundedly locally dense subspace of $BUC(\mathbb{R}_+, X)$. Therefore, Theorem 3.4.1 implies:

**Corollary 3.4.2.** *Let $\mathbf{T}$ be a $C_0$–semigroup on a Banach space $X$. Then $\mathbf{T}$ is uniformly exponentially stable if and only if*

$$\sup_{s>0}\left\|\int_0^s T(t)g(t)\,dt\right\| < \infty, \qquad \forall g \in C_0(\mathbb{R}_+, X).$$

As another application, we define $AP(\mathbb{R}_+, X)$ as the closure in $BUC(\mathbb{R}_+, X)$ of the linear span of the functions $\{e^{i\lambda(\cdot)} \otimes x : \lambda \in \mathbb{R}, x \in X\}$. It is easy to see that $AP(\mathbb{R}_+, X)$ is boundedly locally dense in $BUC(\mathbb{R}_+, X)$. Indeed, if $I \subset \mathbb{R}_+$ is a bounded closed interval and $f \in C(I)$ is given, we choose $N$ so large that $I \subset [0, N]$ and fix an arbitrary continuous function $f_N \in C([0, N+1])$ that coincides with $f$ on $I$ and satisfies $f(0) = f(N+1)$. Then we approximate $f_N$ uniformly in $[0, N+1]$ by linear combinations of functions $e^{i\delta t} \otimes x$, $\delta \in \{k/(2\pi(N+1)) : k \in \mathbb{Z}\}$, $x \in X$. Since these functions are $(N+1)$-periodic, their sup-norms on $\mathbb{R}_+$ are the same as their sup-norms in $[0, N+1]$. Therefore, $AP(\mathbb{R}_+, X)$ is boundedly locally dense in $BUC(\mathbb{R}_+, X)$. Since $AP(\mathbb{R}_+, X)$ is also closed in $BUC(\mathbb{R}_+, X)$, we obtain:

**Corollary 3.4.3.** *Let $\mathbf{T}$ be a $C_0$–semigroup on a Banach space $X$. Then $\mathbf{T}$ is uniformly exponentially stable if and only if*

$$\sup_{s>0}\left\|\int_0^s T(t)g(t)\,dt\right\| < \infty, \qquad \forall g \in AP(\mathbb{R}_+, X). \qquad (3.4.8)$$

The interest of this result is as follows. If (3.4.8) is fulfilled, the uniform boundedness theorem implies the existence of a constant $M > 0$ such that

$$\sup_{s>0} \left\| \int_0^s T(t)g(t)\,dt \right\| \leq M\|g\|_{BUC(\mathbb{R}_+,X)}, \qquad \forall g \in AP(\mathbb{R}_+, X).$$

Since for each $\lambda \in \mathbb{R}$ the function $t \mapsto e^{-i\lambda t} \otimes x$ is almost periodic, it follows that

$$\sup_{\lambda \in \mathbb{R}} \sup_{s>0} \left\| \int_0^s e^{-i\lambda t} T(t) x\,dt \right\| \leq M\|x\|, \qquad \forall x \in X. \tag{3.4.9}$$

Thus, the Laplace transform of **T** at the imaginary axis is interpreted as pointwise evaluation of **T** at the almost periodic functions $e^{-i\lambda t} \otimes x$. In Section 4.5 we shall prove that condition (3.4.9) implies *exponential stability* rather than uniform exponential stability.

We conclude this section with an improvement of Theorem 3.3.1 for the case $C_0(\mathbb{R}_+, X)$.

**Theorem 3.4.4.** *Let **T** be a $C_0$-semigroup a Banach space $X$ and let $E$ be a closed, boundedly locally dense subspace of $BUC(\mathbb{R}_+, X)$. Then the following assertions are equivalent:*

*(i) **T** is uniformly exponentially stable ;*
*(ii) $\mathbf{T} * f \in L^\infty(\mathbb{R}_+, X)$ for all $f \in E$.*

*Proof:* The implication (i)⇒(ii) is trivial. We will prove (ii)⇒(i). By the uniform boundedness theorem applied to the operators $T_s : f \mapsto \int_0^s T(t)f(s-t)\,dt$, there is a constant $C > 0$ such that

$$\sup_{s>0} \left\| \int_0^s T(t)f(s-t)\,dt \right\| \leq C\|f\|_{BUC(\mathbb{R}_+,X)}, \qquad \forall f \in E. \tag{3.4.10}$$

For a given $f \in E$ and $s > 0$, let $M_s = \sup_{0 \leq t \leq s} \|T(t)\|$ and let $f_s \in E$ be any function such that

$$\sup_{0 \leq t \leq s} \|f(s-t) - f_s(t)\| \leq \frac{1}{sM_s} \|f\|_{BUC(\mathbb{R}_+,X)},$$

and

$$\|f_s\|_{BUC(\mathbb{R}_+,X)} \leq K\|f\|_{BUC(\mathbb{R}_+,X)},$$

where $K$ is the constant from the definition of a boundedly locally dense subspace. Then, by (3.4.10),

$$\left\| \int_0^s T(t)f(t)\,dt \right\| \leq \|f\|_{BUC(\mathbb{R}_+,X)} + \left\| \int_0^s T(t)f_s(s-t)\,dt \right\|$$
$$\leq \|f\|_{BUC(\mathbb{R}_+,X)} + C\|f_s\|_{BUC(\mathbb{R}_+,X)}$$
$$\leq (1 + CK)\|f\|_{BUC(\mathbb{R}_+,X)}.$$

Therefore, $\omega_0(\mathbf{T}) < 0$ by Theorem 3.4.1. ////

## 3.5. Positive semigroups on $L^p$-spaces

In Section 1.4 we saw an example of a positive $C_0$-semigroup on the space $L^p \cap L^q$, $1 \leq p < q < \infty$, whose spectral bound is strictly smaller than the uniform growth bound. In this section we prove the theorem of L. Weis that this cannot happen for positive $C_0$-semigroups on $L^p(\mu)$. The proof is based on two key ingredients: the use of Latushkin - Montgomery-Smith theory, which allows us to deal with the spectral bound of $s(B)$ rather than with $\omega_0(\mathbf{T})$ itself, where $B$ is the generator of the induced semigroup on $L^p(\Gamma, L^p(\mu))$ as in Section 2.5; and an extrapolation technique which allows us to reduce the theorem to the 'easy' cases $p = 1$ and $p = 2$.

In view of the complexity of the proof, let us sketch the main line of argument. Let $\mathbf{T}$ be a positive $C_0$-semigroup on $L^p(\mu)$. By duality, it suffices to consider the case $1 \leq p \leq 2$ and by rescaling, it suffices to prove that $s(A) < 0$ implies $\omega_0(\mathbf{T}) < 0$.

So let $1 \leq p \leq 2$ and $s(A) < 0$. First, we show that it is possible to extrapolate the operator $R(0, A)$ to bounded operators $R_0^{(r)}$ on each of the spaces $L^r(\mu_g)$, $1 \leq r \leq \infty$, where $\mu_g = g\,d\mu$ for a certain strictly positive function $0 < g \in L^1(\mu)$. The positivity of $\mathbf{T}$ is used to extend the other resolvent operators $R(\lambda, A)$, $\operatorname{Re}\lambda \geq 0$, to the spaces $L^r(\mu_g)$ as well. One obtains a family $\{R_\lambda^{(r)}\}_{\operatorname{Re}\lambda \geq 0}$ of operators on each $L^r(\mu_g)$ satisfying the resolvent identity. For $r = p$, $R_\lambda^{(p)}$ is just the resolvent of $\mathbf{T}$, which by a similarity transformation is thought of as a semigroup on $L^p(\mu_g)$. For $r \neq p$, the operators $R_\lambda^{(r)}$ need not be the resolvent of some operator $A^{(r)}$, however. To overcome this problem, we now formally imitate the Yosida approximation construction. Recall that if $A$ is the generator of a $C_0$-semigroup, its *Yosida approximation* is the operator $A_\lambda := \lambda^2 R(\lambda, A) - \lambda$, where $\lambda > 0$ is sufficiently large. The operators $A_\lambda$ are bounded and have the property that $\lim_{\lambda \to \infty} A_\lambda x = Ax$ for all $x \in D(A)$. Accordingly, we define the bounded operators $A_\lambda^{(r)} := \lambda^2 R_\lambda^{(r)} - \lambda$ on $L^r(\mu_g)$. Then $\{e^{tA_\lambda^{(r)}}\}_{t \geq 0}$ is a positive uniformly continuous semigroup on $L^r(\mu_g)$ with generator $A_\lambda^{(r)}$. We then prove that the resolvent of $A_\lambda^{(r)}$ is uniformly bounded in the right half-plane, uniformly in $r \geq 1$ and $\lambda \geq 1$. Now Latushkin - Montgomery-Smith theory comes in: for $r = 1$ and $r = 2$ it is easy to prove that also the resolvent of $B_\lambda^{(r)}$ is uniformly bounded in the right half-plane, where $B_\lambda^{(r)}$ is the generator of the semigroup on $L^r(\Gamma, L^r(\mu_g))$ induced by $A_\lambda^{(r)}$ as in Section 2.5. This being done, we go back to the case $r = p$ by interpolation between $L^1(\Gamma, L^1(\mu_g))$ and $L^2(\Gamma, L^2(\mu_g))$ and conclude that the resolvent of $B_\lambda^{(p)}$ is uniformly bounded in the right half-plane, uniformly for $\lambda \geq 1$. Now the operators $B_\lambda^{(p)}$ are indeed the Yosida approximation of a generator, viz. of $B$ on $L^p(\Gamma, L^p(\mu_g))$, where $A$ is the generator of $\mathbf{T}$ as a semigroup on $L^p(\mu_g)$. By letting $\lambda \to \infty$ and using Vitali's theorem, we obtain that the resolvent of $B$

is uniformly bounded in the right half-plane. Then, by the results of Section 2.5, $\omega_0(\mathbf{T}) = s(B) < 0$ and the theorem is proved.

We now start with the details of the proof. Let $(\Omega, \mu)$ be a positive $\sigma$-finite measure space. We shall use the following notation. For a function $0 \leq g \in L^1(\mu)$, we denote by $\mu_g$ the finite measure defined by

$$\mu_g(E) := \int_E g \, d\mu.$$

By $\rho_{g,p} : L^p(\mu_g) \to L^p(\mu)$ we denote the multiplication operator $\rho_{g,p}(f) := fg^{\frac{1}{p}}$. Note that $\rho_{g,p}$ is invertible if and only if $g$ is strictly positive, i.e. if $g$ vanishes on a set of $\mu$-measure zero. In that case the inverse $\rho_{g,p}^{-1} : L^p(\mu) \to L^p(\mu_g)$ is given by $\rho_{g,p}^{-1} f = fg^{-\frac{1}{p}}$. Also, a trivial calculation shows that the restriction to $L^q(\mu)$ ($\frac{1}{p} + \frac{1}{q} = 1$) of the adjoint of $\rho_{g,p}$ is given by $\rho_{g,q}^{-1}$.

**Theorem 3.5.1.** *Let $1 \leq p < \infty$ and let $(\Omega, \mu)$ be a positive $\sigma$-finite measure space. Let $T$ be a positive operator on $L^p(\mu)$. Then there exists a strictly positive function $g \in L^1(\mu)$, $\|g\|_{L^1(\mu)} = 1$, such that for each $1 \leq r \leq \infty$ the following assertion is true: The operator $\rho_{g,p}^{-1} T \rho_{g,p}$ maps $L^p(\mu_g) \cap L^r(\mu_g)$ into itself and the restriction of $\rho_{g,p}^{-1} T \rho_{g,p}$ to $L^p(\mu_g) \cap L^r(\mu_g)$ extends to a bounded operator $T^{(r)}$ on $L^r(\mu_g)$ with norm*

$$\|T^{(r)}\|_{L^r(\mu_g)} \leq 2\|T\|_{L^p(\mu)}.$$

*Proof:* By rescaling, we may assume that $\|T\|_{L^p(\mu)} = 1$.

First assume $1 < p < \infty$. Define the non-linear mapping $S : L^1_+(\mu) \to L^1_+(\mu)$ by

$$Sh := \frac{1}{2}(T(h^{\frac{1}{p}}))^p + \frac{1}{2}(T^*(h^{\frac{1}{q}}))^q \qquad (\frac{1}{p} + \frac{1}{q} = 1).$$

It follows from the concavity of the functions $t \mapsto t^{\frac{1}{r}}$, $r = p, q$, that $S$ is continuous. Also, $\|Sh\|_{L^1(\mu)} \leq \|h\|_{L^1(\mu)}$, and $0 \leq g \leq h$ implies $0 \leq Sg \leq Sh$. Fix a strictly positive $f_0 \in L^1(\mu)$, $\|f_0\|_{L^1(\mu)} = 1$, and choose $a > 1$ so small that $(2a)^{\frac{1}{p}} \leq 2$ and $(2a)^{\frac{1}{q}} \leq 2$. Define

$$f_{n+1} := f_0 + \frac{1}{a} S(f_n) \qquad n = 0, 1, 2, \ldots$$

Then $0 \leq f_0 \leq f_1 \leq f_2 \leq \ldots$, and for each $n$ we have $\|f_n\|_{L^1(\mu)} \leq \sum_{k=0}^n a^{-k} \leq a/(a-1)$. An easy application of the monotone convergence theorem shows that $(f_n)$ converges in $L^1(\mu)$ to $f := \sup_n f_n \in L^1(\mu)$. It follows that

$$f = f_0 + \frac{1}{a} S(f)$$

and hence $S(f) \leq af$. For the strictly positive function $g := \|f\|_{L^1(\mu)}^{-1} f$ we obtain

$$T(g^{\frac{1}{p}}) \leq (2S(g))^{\frac{1}{p}} \leq 2g^{\frac{1}{p}};$$
$$T^*(g^{\frac{1}{q}}) \leq (2S(g))^{\frac{1}{q}} \leq 2g^{\frac{1}{q}}.$$

Using this, for all $f \in L^p(\mu_g)$ satisfying $|f| \leq 1$ it follows that

$$|\rho_{g,p}^{-1} T \rho_{g,p}(f)| \leq \rho_{g,p}^{-1} T \rho_{g,p}(1) = g^{-\frac{1}{p}} T(g^{\frac{1}{p}}) \leq 2;$$
$$|\rho_{g,q}^{-1} T^* \rho_{g,q}(f)| \leq \rho_{g,q}^{-1} T^* \rho_{g,q}(1) = g^{-\frac{1}{q}} T^*(g^{\frac{1}{q}}) \leq 2.$$

The first inequality shows that $\rho_{g,p}^{-1} T \rho_{g,p}$ maps $L^\infty(\mu_g)$ into itself, with norm $\leq 2$. The second inequality shows that also $\rho_{g,q}^{-1} T^* \rho_{g,q}$ maps $L^\infty(\mu_g)$ into itself, also with norm $\leq 2$.

Next, let $f \in L^p(\mu_g)$. Then $\rho_{g,p} f \in L^p(\mu)$, $T\rho_{g,p} f \in L^p(\mu)$, and hence $\rho_{g,p}^{-1} T \rho_{g,p} f \in L^p(\mu_g) \subset L^1(\mu_g)$, using that $\mu_g$ is a finite measure. For all $h \in L^\infty(\mu_g)$ of norm $\leq 1$ we have

$$|\langle \rho_{g,p}^{-1} T \rho_{g,p} f, h \rangle| = \left| \int_\Omega g^{-\frac{1}{p}} T(fg^{\frac{1}{p}}) h \, d\mu_g \right|$$
$$= \left| \int_\Omega T(fg^{\frac{1}{p}}) h g^{\frac{1}{q}} \, d\mu \right|$$
$$= \left| \int_\Omega fg^{\frac{1}{p}} T^*(hg^{\frac{1}{q}}) \, d\mu \right|$$
$$= \left| \int_\Omega fg^{-\frac{1}{q}} T^*(hg^{\frac{1}{q}}) \, d\mu_g \right|$$
$$= |\langle f, \rho_{g,q}^{-1} T^* \rho_{g,q} h \rangle|.$$

Therefore,

$$\|\rho_{g,p}^{-1} T \rho_{g,p} f\|_{L^1(\mu_g)} \leq \|\rho_{g,q}^{-1} T^* \rho_{g,q}\|_{L^\infty(\mu_g)} \|f\|_{L^1(\mu_g)} \leq 2\|f\|_{L^1(\mu_g)}.$$

Since $L^p(\mu_g)$ is dense in $L^1(\mu_g)$, it follows that $\rho_{g,p}^{-1} T \rho_{g,p}$ extends to a bounded operator from $L^1(\mu_g)$ into itself, of norm $\leq 2$.

Thus, we have shown that $\rho_{g,p}^{-1} T \rho_{g,p}$ extends to bounded operators on $L^\infty(\mu_g)$ and $L^1(\mu_g)$. The theorem now follows by interpolation.

If $p = 1$, we argue as above, now taking $S(h) := Th$ and $a = \frac{1}{2}$. ////

The following lemma concerns the Yosida approximation $A_\lambda := \lambda^2 R(\lambda, A) - \lambda$ of a generator $A$. As we already noted, these operators have the property that

$$\lim_{\lambda \to \infty} A_\lambda x = Ax, \quad \forall x \in D(A);$$

this is an easy consequence of (1.1.6).

**Lemma 3.5.2.** Let $\mathbf{T}$ be a $C_0$–semigroup on a Banach space $X$, with generator $A$. For all $\lambda > 2\max\{0, \omega_0(\mathbf{T})\}$ and $\operatorname{Re} z > 2\max\{0, \omega_0(\mathbf{T})\}$ we have $z \in \varrho(A_\lambda)$. Moreover,

$$\lim_{\lambda \to \infty} R(z, A_\lambda)x = R(z, A)x, \qquad \forall x \in X,\ \operatorname{Re} z > 2\max\{0, \omega_0(\mathbf{T})\}.$$

*Proof:* Fix $\omega > \omega_0(\mathbf{T})$ arbitrary. By the Hille-Yosida theorem, there is a constant $M$ such that

$$\|R(\lambda, A)^n\| \leq \frac{M}{(\lambda - \omega)^n}, \qquad \lambda > \omega,\ n = 1, 2, \ldots$$

Therefore, by the definition of $A_\lambda$, for $\lambda > \omega$ we have

$$\|e^{tA_\lambda}\| \leq e^{-\lambda t} \sum_{k=0}^{\infty} \frac{1}{k!} t^k \lambda^{2k} \|R(\lambda, A)^k\|$$

$$\leq M e^{-\lambda t} \sum_{k=0}^{\infty} \frac{1}{k!} \frac{t^k \lambda^{2k}}{(\lambda - \omega)^k}$$

$$= M e^{\lambda \omega (\lambda - \omega)^{-1} t}.$$

For $\lambda \geq 2\max\{0, \omega\}$ it follows that

$$\|e^{tA_\lambda}\| \leq M e^{2\max\{0,\omega\} t}, \qquad t \geq 0.$$

This implies that $\{\operatorname{Re} z > 2\max\{0, \omega\}\} \subset \varrho(A_\lambda)$ for all $\lambda > 2\max\{0, \omega\}$ and

$$\|R(z, A_\lambda)\| \leq \frac{M}{\operatorname{Re} z - 2\max\{0, \omega\}}, \qquad \operatorname{Re} z > 2\max\{0, \omega\}.$$

Let $x \in X$ and $\operatorname{Re} z_0 > 2\max\{0, \omega\}$ be arbitrary, and put $y = R(z_0, A)x$. For $\lambda$ large enough, let $x_\lambda = (z_0 - A_\lambda)y$. Since $\lim_{\lambda \to \infty} A_\lambda y = Ay$ we have $\lim_{\lambda \to \infty} x_\lambda = x$, and the above estimate on the resolvent of $A_\lambda$ implies that $\lim_{\lambda \to \infty} R(z_0, A_\lambda)(x - x_\lambda) = 0$. Therefore,

$$\lim_{\lambda \to \infty} R(z_0, A_\lambda)x = \lim_{\lambda \to \infty} (y + R(z_0, A_\lambda)(x - x_\lambda)) = y = R(z_0, A)x.$$

////

After these preparations we can state and prove Weis's theorem on the stability of positive semigroups on $L^p$. The proof presented here is a modification, due to W. Arendt, of Weis's original argument.

We recall from Section 2.2 that the adjoint $\mathbf{T}^*$ of a $C_0$–semigroup $\mathbf{T}$ on a reflexive Banach space is strongly continuous. Since $\sigma(A) = \sigma(A^*)$ and $\|T(t)\| = \|T^*(t)\|$ for all $t \geq 0$, we have $s(A) = s(A^*)$ and $\omega_0(\mathbf{T}) = \omega_0(\mathbf{T}^*)$.

**Theorem 3.5.3.** Let $(\Omega, \mu)$ be a $\sigma$-finite measure space and let $1 \leq p < \infty$. Let $\mathbf{T}$ be a positive $C_0$–semigroup on $L^p(\mu)$, with generator $A$. Then $s(A) = \omega_0(\mathbf{T})$.

*Proof:* Since for $2 \leq p < \infty$ the spaces $L^p(\mu)$ are reflexive, by applying the above duality procedure it follows that it suffices to prove the theorem for $1 \leq p \leq 2$.

By rescaling, it is enough to show that $s(A) < 0$ implies $\omega_0(\mathbf{T}) < 0$. So we shall assume that $s(A) < 0$.

By Theorem 1.4.1, $-A^{-1} = R(0, A)$ is a positive operator. By Theorem 3.5.1 applied to $T = -A^{-1}$, there exists a strictly positive function $g \in L^1(\mu)$ of norm one such that the following assertion is true. For each $1 \leq r \leq \infty$, the operator $\rho_{g,p}^{-1} A^{-1} \rho_{g,p}$ maps $L^p(\mu_g) \cap L^r(\mu_g)$ into itself, and the restriction of $-\rho_{g,p}^{-1} A^{-1} \rho_{g,p}$ to $L^p(\mu_g) \cap L^r(\mu_g)$ extends to a bounded operator $R_0^{(r)}$ on $L^r(\mu_g)$ such that

$$\|R_0^{(r)}\|_{L^r(\mu_g)} \leq 2\|A^{-1}\|_{L^p(\mu)}.$$

We claim that for all $\lambda \geq 0$, the operator $\rho_{g,p}^{-1} R(\lambda, A) \rho_{g,p}$ maps $L^p(\mu_g) \cap L^r(\mu_g)$ into itself, and that the restriction of $\rho_{g,p}^{-1} R(\lambda, A) \rho_{g,p}$ to $L^p(\mu_g) \cap L^r(\mu_g)$ extends to a bounded positive operator $R_\lambda^{(r)}$ on $L^r(\mu_g)$ of norm $\leq 2\|A^{-1}\|_{L^p(\mu)}$.

To see this, fix $1 \leq r \leq \infty$ and $f \in L^p(\mu_g) \cap L^r(\mu_g)$ arbitrary. For all $\lambda \geq 0$, by Theorem 1.4.1 we have

$$|R(\lambda, A)\rho_{g,p} f| = |R(\lambda, A)(fg^{\frac{1}{p}})| = \left|\lim_{\tau \to \infty} \int_0^\tau e^{-\lambda t} T(t)(fg^{\frac{1}{p}})\, dt\right|$$

$$\leq \lim_{\tau \to \infty} \int_0^\tau T(t)(|f|g^{\frac{1}{p}})\, dt = -A^{-1}(|f|g^{\frac{1}{p}}) = -A^{-1}\rho_{g,p}|f|.$$

Therefore,

$$|\rho_{g,p}^{-1} R(\lambda, A)\rho_{g,p} f| \leq \rho_{g,p}^{-1}(-A^{-1})\rho_{g,p}|f| = R_0^{(r)}|f|.$$

Since $R_0^{(r)}$ maps $L^p(\mu_g) \cap L^r(\mu_g)$ into itself, it follows that $|\rho_{g,p}^{-1} R(\lambda, A)\rho_{g,p} f| \in L^p(\mu_g) \cap L^r(\mu_g)$ and

$$\|\rho_{g,p}^{-1} R(\lambda, A)\rho_{g,p} f\|_{L^r(\mu_g)} \leq \|R_0^{(r)}|f|\|_{L^r(\mu_g)} \leq 2\|A^{-1}\|_{L^p(\mu)} \|f\|_{L^r(\mu_g)}.$$

This proves the claim. Note that for $r = p$ we have

$$R_\lambda^{(p)} = \rho_{g,p}^{-1} R(\lambda, A)\rho_{g,p}, \qquad \forall \lambda \geq 0.$$

For all $1 \leq r < \infty$ and any two $\lambda_0, \lambda_1 \geq 0$ we have

$$R_{\lambda_0}^{(r)} - R_{\lambda_1}^{(r)} = (\lambda_1 - \lambda_0) R_{\lambda_0}^{(r)} R_{\lambda_1}^{(r)} \tag{3.5.1}$$

Indeed, by the resolvent identity this holds for the restrictions to $L^p(\mu_g) \cap L^r(\mu_g)$, and by a density argument it follows for all of $L^r(\mu_g)$.

We now define, for $1 \le r < \infty$ and $\lambda \ge 0$,

$$A_\lambda^{(r)} := \lambda^2 R_\lambda^{(r)} - \lambda.$$

These are bounded operators on $L^r(\mu_g)$. From the identity

$$T_\lambda^{(r)}(t) := e^{tA_\lambda^{(r)}} = e^{-\lambda t} e^{t\lambda^2 R_\lambda^{(r)}}$$

and the positivity of $R_\lambda^{(r)}$ it follows that the uniformly continuous semigroup $\mathbf{T}_\lambda^{(r)}$ generated by $A_\lambda^{(r)}$ is positive. Also, for $\lambda > 0$, by (3.5.1) we see that $\lambda^{-1} + R_0^{(r)}$ is a two-sided, positive inverse of $-A_\lambda^{(r)}$. Therefore, in view of Theorem 1.4.1 (ii) and the boundedness of $A_\lambda^{(r)}$,

$$\omega_0(\mathbf{T}_\lambda^{(r)}) = s(A_\lambda^{(r)}) < 0.$$

It follows that $\{\operatorname{Re} z > 0\} \subset \varrho(A_\lambda^{(r)})$, and for $\operatorname{Re} z > 0$ and $f \in L^r(\mu_g)$ we have

$$|R(z, A_\lambda^{(r)})f| = \left| \int_0^\infty e^{-zt} e^{tA_\lambda^{(r)}} f \, dt \right| \le \int_0^\infty e^{tA_\lambda^{(r)}} |f| \, dt = R(0, A_\lambda^{(r)})|f|,$$

the integrals being in the improper sense. Therefore, for all $\lambda > 0$ and $\operatorname{Re} z > 0$,

$$\begin{aligned} \|R(z, A_\lambda^{(r)})\|_{L^r(\mu_g)} &\le \|R(0, A_\lambda^{(r)})\|_{L^r(\mu_g)} \\ &\le \lambda^{-1} + \|R_0^{(r)}\|_{L^r(\mu_g)} \le \lambda^{-1} + 2\|A^{-1}\|_{L^p(\mu)}. \end{aligned} \qquad (3.5.2)$$

We now apply Corollary 2.5.3 to the operators $A_\lambda^{(r)}$. It follows that the generator $B_\lambda^{(r)}$ of the positive semigroup $\mathbf{S}_\lambda^{(r)}$ on $L^r(\Gamma, L^r(\mu_g))$ defined by

$$S_\lambda^{(r)}(t) f(e^{i\theta}) = e^{tA_\lambda^{(r)}} f(e^{i(\theta-t)})$$

satisfies $s(B_\lambda^{(r)}) = \omega_0(\mathbf{T}_\lambda^{(r)}) < 0$. This implies that $\{\operatorname{Re} z > 0\} \subset \varrho(B_\lambda^{(r)})$, and for all $\operatorname{Re} z > 0$ and $f \in L^r(\Gamma, L^r(\mu_g))$ we have

$$R(z, B_\lambda^{(r)}) f = \int_0^\infty e^{-zt} S_\lambda^{(r)}(t) f \, dt,$$

the integral being in the improper sense. The next claim is that

$$\|R(z, B_\lambda^{(r)})\|_{L^r(\Gamma, L^r(\mu_g))} \le 2(\lambda^{-1} + 2\|A^{-1}\|_{L^p(\mu)})$$

for all $\operatorname{Re} z > 0$, $\lambda > 0$, and $1 \le r \le 2$. First we show this for $r = 1$ and $r = 2$.

($r = 1$): Using (3.5.2), for all $0 \le f \in L^1(\Gamma, L^1(\mu_g))$ we have

$$\|R(z, B_\lambda^{(1)})f\|_{L^1(\Gamma, L^1(\mu_g))}$$
$$= \int_\Gamma \left\| \int_0^\infty e^{-zt} e^{tA_\lambda^{(1)}} f(e^{i(\theta-t)}) \, dt \right\|_{L^1(\mu_g)} d\theta$$
$$\le \int_\Gamma \int_0^\infty \left\| e^{tA_\lambda^{(1)}} f(e^{i(\theta-t)}) \right\|_{L^1(\mu_g)} dt \, d\theta$$
$$= \left\| \int_0^\infty e^{tA_\lambda^{(1)}} \left( \int_\Gamma f(e^{i(\theta-t)}) \, d\theta \right) dt \right\|_{L^1(\mu_g)}$$
$$\le \|R(0, A_\lambda^{(1)})\|_{L^1(\mu_g)} \left\| \int_\Gamma f(e^{i(\theta-t)}) \, d\theta \right\|_{L^1(\mu_g)}$$
$$\le \|R(0, A_\lambda^{(1)})\|_{L^1(\mu_g)} \|f\|_{L^1(\Gamma, L^1(\mu_g))}$$
$$\le (\lambda^{-1} + 2\|A^{-1}\|_{L^p(\mu)}) \|f\|_{L^1(\Gamma, L^1(\mu_g))}.$$

By splitting an arbitrary $f \in L^1(\Gamma, L^1(\mu_g))$ into real and imaginary part and each of these into positive and negative part and using the additivity of the $L^1$-norm, the case $r = 1$ follows from this.

($r = 2$): We denote by $e_k$ the function $e_k(e^{i\theta}) := e^{ik\theta}$. Using the orthogonality of the $e_k$ in $L^2(\Gamma)$, for all finite sequences $f_{-n}, ..., f_n \in L^2(\mu_g)$ we have, using (3.5.2),

$$\left\| R(z, B_\lambda^{(2)}) \sum_{k=-n}^n e_k \otimes f_k \right\|_{L^2(\Gamma, L^2(\mu_g))}^2$$
$$= \left\| \sum_{k=-n}^n e_k \otimes R(z+ik, A_\lambda^{(2)}) f_k \right\|_{L^2(\Gamma, L^2(\mu_g))}^2$$
$$= 2\pi \sum_{k=-n}^n \left\| R(z+ik, A_\lambda^{(2)}) f_k \right\|_{L^2(\mu_g)}^2$$
$$\le 2\pi \left\| R(z+ik, A_\lambda^{(2)}) \right\|_{L^2(\mu_g)}^2 \sum_{k=-n}^n \|f_k\|_{L^2(\mu_g)}^2$$
$$\le 2\pi (\lambda^{-1} + 2\|A^{-1}\|_{L^p(\mu)})^2 \sum_{k=-n}^n \|f_k\|_{L^2(\mu_g)}^2$$
$$= (\lambda^{-1} + 2\|A^{-1}\|_{L^p(\mu)})^2 \left\| \sum_{k=-n}^n e_k \otimes f_k \right\|_{L^2(\Gamma, L^2(\mu_g))}^2$$

Since the finite linear combinations $\sum_{k=-n}^n e_k \otimes f_k$ are dense in $L^2(\Gamma, L^2(\mu_g))$, the desired conclusion for $r = 2$ follows from this.

By Appendix A.2, Proposition A.2.2, the claim follows from these two cases by complex interpolation. Taking $r = p$ in the claim, we obtain
$$\|R(z, B_\lambda^{(p)})\|_{L^p(\Gamma, L^p(\mu_g))} \leq 2 + 4\|A^{-1}\|_{L^p(\mu)}, \qquad \forall \operatorname{Re} z > 0, \quad \lambda \geq 1. \qquad (3.5.3)$$
The next step is to prove that
$$\lim_{\lambda \to \infty} R(z, B_\lambda^{(p)}) f = R(z, B^{(p)}) f \qquad (3.5.4)$$
for all $\operatorname{Re} z > 2 \max\{0, \omega_0(\mathbf{T})\}$ and $f \in L^p(\Gamma, L^p(\mu_g))$, where $A^{(p)}$ denotes the generator of the induced semigroup $\rho_{g,p}^{-1} \mathbf{T} \rho_{g,p}$ on $L^p(\mu_g)$ and $B^{(p)}$ denotes the generator of the semigroup $\mathbf{S}^{(p)}$ on $L^p(\Gamma, L^p(\mu_g))$ induced by $\rho_{g,p}^{-1} \mathbf{T} \rho_{g,p}$:
$$\mathbf{S}^{(p)} f(e^{i\theta}) := \rho_{g,p}^{-1} T(t) \rho_{g,p} f(e^{i(\theta-t)}).$$
To this end, we note that
$$A_\lambda^{(p)} = \lambda^2 R_\lambda^{(p)} - \lambda = \lambda^2 \rho_{g,p}^{-1} R(\lambda, A) \rho_{g,p} - \lambda = \rho_{g,p}^{-1} A_\lambda \rho_{g,p},$$
where $A_\lambda$ is the Yosida approximation of $A$. Hence $\varrho(A_\lambda^{(p)}) = \varrho(A_\lambda)$ and for all $z \in \varrho(A_\lambda)$ we have
$$R(z, A_\lambda^{(p)}) = \rho_{g,p}^{-1} R(z, A_\lambda) \rho_{g,p}.$$
Regarding the $e_k$ as functions in $L^p(\Gamma)$, for all finite sequences $f_{-n}, \ldots, f_n \in L^p(\mu_g)$ we have, using Lemma 3.5.2,
$$\lim_{\lambda \to \infty} R(z, B_\lambda^{(p)}) \sum_{k=-n}^n e_k \otimes f_k = \lim_{\lambda \to \infty} \sum_{k=-n}^n e_k \otimes R(z + ik, A_\lambda^{(p)}) f_k$$
$$= \lim_{\lambda \to \infty} \sum_{k=-n}^n e_k \otimes \rho_{g,p}^{-1} R(z + ik, A_\lambda) \rho_{g,p} f_k$$
$$= \sum_{k=-n}^n e_k \otimes \rho_{g,p}^{-1} R(z + ik, A) \rho_{g,p} f_k$$
$$= \sum_{k=-n}^n e_k \otimes R(z + ik, A^{(p)}) f_k$$
$$= R(z, B^{(p)}) \sum_{k=-n}^n e_k \otimes f_k.$$
Since the functions $\sum_{k=-n}^n e_k \otimes f_k$ are dense in $L^p(\Gamma, L^p(\mu_g))$, (3.5.4) follows from the uniform boundedness of the operators $\{R(z, B_\lambda^{(p)}), \lambda \geq 1, \operatorname{Re} z > 0\}$.

In view of (3.5.3) and (3.5.4), we are in a position to apply Vitali's theorem (cf. Section 3.3) to the functions $z \mapsto R(z, B_\lambda^{(p)}) f$, $f \in L^p(\Gamma, L^p(\mu_g))$, $\lambda \geq 1$. It follows that the maps $z \mapsto R(z, B^{(p)}) f$ admit bounded holomorphic extensions to $\{\operatorname{Re} z > 0\}$, with bounds $(2 + 4\|A^{-1}\|_{L^p(\mu)}) \|f\|_{L^p(\Gamma, L^p(\mu_g))}$. Proposition 1.1.6 now shows that $\{\operatorname{Re} z > 0\} \subset \varrho(B^{(p)})$ (and that the resolvent of $B^{(p)}$ exists and is uniformly bounded there with bound $2 + 4\|A^{-1}\|_{L^p(\mu)}$). Therefore, $\omega_0(\mathbf{T}) = \omega_0(\rho_{g,p}^{-1} \mathbf{T} \rho_{g,p}) = s(B^{(p)}) < 0$. ////

It is not a coincidence that we have dealt explicitly with the cases $r = 1$ and $r = 2$ in the proof above. In fact, the theorem admits a much simpler proof for $p = 1$ and $p = 2$. If $p = 2$, then $L^2(\mu)$ is a Hilbert space. Therefore, $\omega_0(\mathbf{T})$ coincides with the abscissa of uniform boundedness of the resolvent of its generator, and this abscissa is equal to $s(A)$ by Theorem 1.4.1. If $p = 1$ we argue as follows: For all $\tau > 0$ and $f \in L^1(\mu)$ we have

$$\int_0^\tau \|T(t)f\|\,dt \leq \int_0^\tau \|T(t)|f|\,\|\,dt = \int_0^\tau \langle 1, T(t)|f|\rangle\,dt.$$

Here, 1 denotes the constant one function in $L^\infty(\mu)$. If $s(A) < 0$, then by Theorem 1.2.3 $\omega_1(\mathbf{T}) < 0$ and the right hand side converges to $\langle 1, R(0,A)|f|\rangle$ as $\tau \to \infty$. Therefore,

$$\int_0^\infty \|T(t)f\|\,dt < \infty, \quad \forall f \in L^1(\mu),$$

and we can apply the Datko-Pazy theorem.

We conclude this section with the case of positive $C_0$–semigroups on $C_0(\Omega)$, where $\Omega$ is a locally compact Hausdorff space.

If $\mathbf{T}$ is a $C_0$–semigroup, then $\omega_0(\mathbf{T}) = \omega_0(\mathbf{T}^\odot)$, where $\mathbf{T}^\odot$ is the strongly continuous part of the adjoint of $\mathbf{T}$ as defined in Section 2.1. Indeed, it is obvious that $\|T^\odot(t)\| \leq \|T^*(t)\| = \|T(t)\|$ for all $t \geq 0$; on the other hand, with $M := \limsup_{\lambda \to \infty} \|\lambda R(\lambda, A)\|$ the identity

$$|\langle x^*, T(t)x\rangle| = \lim_{\lambda \to \infty} |\langle T^\odot(t)\lambda R(\lambda, A^*)x^*, x\rangle|$$

shows that $|\langle x^*, T(t)x\rangle| \leq M\|T^\odot(t)\|\,\|x^*\|\,\|x\|$ for all $x^* \in X^*$ and $x \in X$, so $\|T^\odot(t)\| \geq M^{-1}\|T(t)\|$ for all $t \geq 0$. Hence

$$M^{-1}\|T(t)\| \leq \|T^\odot(t)\| \leq \|T(t)\|, \quad t \geq 0,$$

and the identity $\omega_0(\mathbf{T}) = \omega_0(\mathbf{T}^\odot)$ follows.

**Theorem 3.5.4.** *Let $\mathbf{T}$ be a positive $C_0$–semigroup with generator $A$ on $C_0(\Omega)$, $\Omega$ locally compact Hausdorff. Then $s(A) = \omega_0(\mathbf{T})$.*

*Proof.* If $\omega_0(\mathbf{T}) = -\infty$ there is nothing to prove; hence without loss of generality we may assume that $\omega_0(\mathbf{T}) = 0$. It suffices to prove that $0 \in \sigma(A)$.

Suppose the contrary. For each $\tau > 0$, consider the bounded operator $S_\tau$ on $(C_0(\Omega))^*$ defined by

$$\langle S_\tau \nu, f\rangle := \int_0^\tau \langle \nu, T(t)f\rangle\,dt, \quad f \in C_0(\Omega).$$

For all $f \in C_0(\Omega)$ and $\nu \in (C_0(\Omega))^*$ we have, using Theorem 1.4.1 and the assumption that $0 \notin \sigma(A)$,

$$\lim_{\tau \to \infty} \langle S_\tau \nu, f\rangle = \lim_{\tau \to \infty} \int_0^\tau \langle \nu, T(t)f\rangle\,dt = \langle \nu, R(0,A)f\rangle = \langle R(0,A)^*\nu, f\rangle.$$

It follows that $S_\tau \nu$ converges weak* to $R(0, A)^* \nu$ for all $\nu$. In particular,
$$\sup_{\tau > 0} \|S_\tau\| =: N < \infty.$$

Now we use the fact that for any two $0 \leq \nu_0, \nu_1 \in (C_0(\Omega))^*$ we have
$$\|\nu_0 + \nu_1\| = \|\nu_0\| + \|\nu_1\|.$$

By splitting the integral $\int_0^\tau T^\odot(t) \lambda R(\lambda, A^*) \mu \, dt$ into Riemann sums, for all $0 \leq \mu \in (C_0(\Omega))^*$ and $\lambda > 0$ we have
$$\int_0^\tau \|T^\odot(t) \lambda R(\lambda, A^*) \mu\| \, dt = \left\| \int_0^\tau T^\odot(t) \lambda R(\lambda, A^*) \mu \, dt \right\| = \|\lambda R(\lambda, A^*) S_\tau \mu\|.$$

Take an arbitrary real $\nu \in (C_0(\Omega))^\odot$ and let $\nu_+$ and $\nu_-$ denote its positive and negative part. Then, for all $\tau > 0$,

$$\begin{aligned}
\int_0^\tau \|T^\odot(t) \nu\| \, dt &= \lim_{\lambda \to \infty} \int_0^\tau \|T^\odot(t) \lambda R(\lambda, A^\odot) \nu\| \, dt \\
&\leq \limsup_{\lambda \to \infty} \int_0^\tau \|T^\odot(t) \lambda R(\lambda, A^*) \nu_+\| \, dt \\
&\quad + \limsup_{\lambda \to \infty} \int_0^\tau \|T^\odot(t) \lambda R(\lambda, A^*) \nu_-\| \, dt \\
&= \limsup_{\lambda \to \infty} \|\lambda R(\lambda, A^*) S_\tau \nu_+\| \\
&\quad + \limsup_{\lambda \to \infty} \|\lambda R(\lambda, A^*) S_\tau \nu_-\| \\
&\leq M(\|S_\tau \nu_+\| + \|S_\tau \nu_-\|) \\
&\leq MN(\|\nu_+\| + \|\nu_-\|) = MN\|\nu\|,
\end{aligned}$$

where $M := \limsup_{\lambda \to \infty} \|\lambda R(\lambda, A)\|$. By splitting an arbitrary $\nu \in (C_0(\Omega))^\odot$ into real and imaginary parts, and noting that these are in $(C_0(\Omega))^\odot$ again, we obtain
$$\int_0^\infty \|T^\odot(t) \nu\| \, dt \leq 2MN\|\nu\|, \quad \forall \nu \in (C_0(\Omega))^\odot.$$

Therefore we can apply the Datko-Pazy theorem to $\mathbf{T}^\odot$ and conclude that $\omega_0(\mathbf{T}) = \omega_0(\mathbf{T}^\odot) < 0$. This contradiction concludes the proof.   ////

For spaces $C(K)$ with $K$ compact Hausdorff, a much stronger result is valid. For its proof we need the *Krein-Rutman theorem* which we state next. By $M(K)$ we denote the space of all regular bounded Borel measures on $K$; so $M(K) = (C(K))^*$.

**Proposition 3.5.5.** *Let $T$ be a positive bounded operator on the space $C(K)$, $K$ compact Hausdorff. Then there exists a non-zero $0 \leq \mu \in M(K)$ such that $T^*\mu = r(T)\mu$.*

*Proof:* For all $f \in C(K)$ and all $\mu \in \mathbb{C}$ with $|\mu| > r := r(T)$ we have

$$|R(\mu, T)f| = \left|\sum_{n=0}^{\infty} \frac{T^n f}{\mu^{n+1}}\right| \leq \sum_{n=0}^{\infty} \frac{T^n |f|}{|\mu|^{n+1}} = R(|\mu|, T)|f|$$

and hence $\|R(\mu, T)\| \leq \|R(|\mu|, T)\|$. By this and Proposition 1.1.5, we see that $\lim_{\lambda \downarrow r} \|R(\lambda, T)^*\| = \lim_{\lambda \downarrow r} \|R(\lambda, T)\| = \infty$. By the uniform boundedness theorem, there exists a sequence $\lambda_n \downarrow r$ and a $\nu \in M(K)$ such that $\lim_{n \to \infty} \|R(\lambda_n, T)^*\nu\| = \infty$; by replacing $\nu$ by its modulus we may take $\nu$ to be positive. Define

$$\mu_n := \|R(\lambda_n, T)^*\nu\|^{-1} R(\lambda_n, T)^*\nu.$$

Then,

$$\lim_{n \to \infty} (r - T^*)\mu_n = \lim_{n \to \infty} \left((r - \lambda_n) + (\lambda_n - T^*)\right)\mu_n$$
$$= \lim_{n \to \infty} (r - \lambda_n)\mu_n + \|R(\lambda_n, T)^*\nu\|^{-1}\nu = 0.$$

By the Banach-Alaoglu theorem, the dual unit ball of any Banach space is weak*-compact. Therefore, the sequence $(\mu_n)$ has a weak*-cluster point $\mu$. Clearly, $\mu \geq 0$. Since the operator $r - T^*$ is weak*-continuous, it follows that $(r - T^*)\mu$ is a weak*-cluster point of the sequence $((r - T^*)\mu_n)$, and therefore $(r - T^*)\mu = 0$. Also, since $K$ is compact, the constant one function $1$ defines an element of $C(K)$, and there is a subsequence $(n_k)$ such that

$$\|\mu\| = \langle \mu, 1 \rangle = \lim_{k \to \infty} \langle \mu_{n_k}, 1 \rangle = \lim_{k \to \infty} \|\mu_{n_k}\| = 1.$$

This proves that $\mu$ is non-zero. ////

**Theorem 3.5.6.** *Let $\mathbf{T}$ be a positive $C_0$-semigroup, with generator $A$, on $X = C(K)$ with $K$ a compact Hausdorff space. Then the following assertions are equivalent:*

(i) $s(A) < 0$;
(ii) $\mathbf{T}$ *is uniformly exponentially stable;*
(iii) $\lim_{t \to \infty} \langle x^*, T(t)y \rangle = 0$ *for all $y \in D(A)$ and $x^* \in X^*$.*

*Proof:* In view of Theorem 3.5.4, all we have to prove is that (iii) implies (i).

From the uniform boundedness theorem, for all $x \in D(A)$ there is a constant $M_x$ such that that $\|T(t)x\| \leq M_x$ for all $t \geq 0$. Hence, $s(A) = \omega_1(\mathbf{T}) \leq 0$. In order to prove that $s(A) < 0$, by Theorem 1.4.1 it suffices to prove that $0 \notin \sigma(A)$.

Suppose, for a contradiction, that $0 \in \sigma(A)$. By the spectral inclusion theorem, $1 \in \sigma(T(t))$ for all $t \geq 0$. We also have $\omega_0(\mathbf{T}) = s(A) \leq 0$ by Theorem 3.5.4 and hence $r(T(t)) \leq 1$. It follows that $r(T(t)) = 1$ and therefore by the Krein-Rutman theorem there exists a non-zero $\mu \in M(K)$ such that $T^*(t)\mu = \mu$. Let $f \in D(A)$ be any function such that $\langle \mu, f \rangle \neq 0$; such $f$ exists since $D(A)$ is dense. Then, $\langle \mu, T(t)f \rangle = \langle T^*(t)\mu, f \rangle = \langle \mu, f \rangle$ and the latter fails to converge to zero. ////

The equivalences (i)⇔(iii) and (ii)⇔(iii) break down in the locally compact case: the translation semigroup on $C_0(\mathbb{R}_+)$ defined by $(T(t)f)(s) := f(s+t)$ is uniformly stable and therefore (iii) holds, but $\mathbf{T}$ is not uniformly exponentially stable and we have $s(A) = 0$.

## 3.6. The essential spectrum

In this section we study the essential growth bound $\omega_0^{ess}(\mathbf{T})$ of a $C_0$–semigroup $\mathbf{T}$ with generator $A$ and prove that

$$\omega_0(\mathbf{T}) = \max\{s(A), \omega_0^{ess}(\mathbf{T})\}.$$

Thus, the failure of the identity $s(A) = \omega_0(\mathbf{T})$ can be interpreted in terms of the essential spectrum of $\mathbf{T}$. We also give necessary and sufficient conditions for a $C_0$–semigroup to converge uniformly to a rank one projection. This theorem assumes a particularly nice form for irreducible positive $C_0$–semigroups, which we discuss briefly.

We start with some definitions. The two-sided ideal in $\mathcal{L}(X)$ of all compact operators is denoted by $\mathcal{K}(X)$. The *Calkin algebra* $\mathcal{C}(X)$ is the quotient $\mathcal{L}(X)/\mathcal{K}(X)$. With the quotient norm, $\mathcal{C}(X)$ is a Banach algebra with unit.

The *essential spectrum*, notation $\sigma^{ess}(T)$, of a bounded operator $T$ is defined as the spectrum of $T + \mathcal{K}(X)$ in the Banach algebra $\mathcal{C}(X)$. Explicitly, it consists of the set of all $\lambda \in \mathbb{C}$ for which no bounded operator $S$ can be found such that $(\lambda - T)S - I$ and $S(\lambda - T) - I$ are compact. The *essential spectral radius* $r^{ess}(T)$ of $T$ is defined as $\sup\{|\lambda| : \lambda \in \sigma^{ess}(T)\}$. We shall need the following fact from Fredholm theory; the proof may be found in the book [CPY]. If $r^{ess}(T) < r(T)$, then $\sigma(T) \cap \{|\lambda| > r^{ess}(T)\}$ consists of at most finitely many isolated eigenvalues (each of which is a pole of the resolvent of $T$).

We define the *essential growth bound* $\omega_0^{ess}(\mathbf{T})$ of a $C_0$–semigroup $\mathbf{T}$ as the growth bound of the quotient semigroup $\mathbf{T} + \mathcal{K}(X)$ on $\mathcal{C}(X)$, i.e. as the infimum of all $\omega \in \mathbb{R}$ for which there exists an $M > 0$ such that

$$\|T(t) + \mathcal{K}(X)\|_{\mathcal{C}(X)} \leq M e^{\omega t}, \quad \forall t \geq 0.$$

As in Proposition 1.2.2, for all $t_0 > 0$ we have

$$\omega_0^{ess}(\mathbf{T}) = \frac{\log r^{ess}(T(t_0))}{t_0} = \lim_{t \to \infty} \frac{\log \|T(t) + \mathcal{K}(X)\|_{\mathcal{C}(X)}}{t}.$$

The following theorem describes the relation between $\omega_0^{ess}(\mathbf{T})$ and $\omega_0(\mathbf{T})$.

**Theorem 3.6.1.** *Let* **T** *be a $C_0$–semigroup on a Banach space $X$. Then either $\omega_0^{ess}(\mathbf{T}) = \omega_0(\mathbf{T})$ or the line $\operatorname{Re}\lambda = \omega_0(\mathbf{T})$ contains an eigenvalue of $A$. In particular,*
$$\omega_0(\mathbf{T}) = \max\{s(A), \omega_0^{ess}(\mathbf{T})\}.$$

*Proof:* Assume that $\omega_0^{ess}(\mathbf{T}) < \omega_0(\mathbf{T})$. Then also $r^{ess}(T(1)) < r(T(1))$. Let $\lambda \in \sigma(T(1))$ with $|\lambda| = r(T(1))$ be arbitrary. Then $\lambda$ is an eigenvalue of $T(1)$ and by the spectral mapping theorem for the point spectrum there exists a $\mu \in \sigma_p(A)$ such that $e^\mu = \lambda$. Clearly, $\operatorname{Re}\mu = \omega_0(\mathbf{T})$, so $s(A) = \omega_0(\mathbf{T})$. ////

We say that $A$ has a *strictly dominant eigenvalue* if $\sigma(A) \cap \{\operatorname{Re}\lambda = s(A)\}$ consists of a single point $\{\omega\}$ which is an eigenvalue of $A$. An isolated point $\lambda \in \sigma(A)$ is called *algebraically simple* if the corresponding spectral projection is a rank one operator.

The following theorem describes the asymptotic behaviour of a semigroup with a strictly dominant, algebraically simple eigenvalue.

**Theorem 3.6.2.** *Let* **T** *be a $C_0$–semigroup on a Banach space $X$ and let $\omega \in \mathbb{C}$. Then the following assertions are equivalent:*

(i) *There exists a rank one projection $P$ such that*
$$\lim_{t \to \infty} \|e^{-\omega t}T(t) - P\| = 0.$$

(ii) $\omega_0^{ess}(\mathbf{T}) < \omega_0(\mathbf{T})$ *and $\omega$ is a strictly dominant, algebraically simple eigenvalue of $A$.*

*In this case, the rank one projection is the spectral projection corresponding to $\{\omega\}$. Moreover, there are constants $M > 0$ and $\delta > 0$ such that*
$$\|e^{-\omega t}T(t) - P\| \leq Me^{-\delta t}, \qquad t \geq 0,$$
*and $\sigma(A) \cap \{\operatorname{Re}\lambda > \omega - \delta\} = \{\omega\}$.*

*Proof:* By rescaling **T** along the vertical axis we may assume that $\omega \in \mathbb{R}$.

Assume (i). Then,
$$T(t)P = \lim_{s \to \infty} e^{-\omega s}T(t+s) = e^{\omega t}P$$

and similarly $PT(t) = e^{\omega t}P$. Hence, $T(t)P = PT(t) = e^{\omega t}P$, $Px \in D(A)$ for all $x \in X$ and $APx = \omega Px$, showing that $\omega$ is an eigenvalue of $A$.

Since $P$ is a projection commuting with **T**, we have a direct sum decomposition $X = X_0 \oplus X_1$, where $X_0 = PX$ and $X_1 = (I - P)X$ are **T**-invariant. Let $\mathbf{T}_i$ be the corresponding restrictions and let $A_i$ be their generators, $i = 0, 1$. From
$$\lim_{t \to \infty} \|e^{-\omega t}T_1(t)\| = \lim_{t \to \infty} \|e^{-\omega t}T(t)(I - P)\| = \lim_{t \to \infty} \|e^{-\omega t}T(t) - P\| = 0$$

it follows that

$$\lim_{t\to\infty} e^{(\omega_0(\mathbf{T}_1)-\omega)t} = \lim_{t\to\infty} e^{-\omega t} r(T_1(t)) \leq \limsup_{t\to\infty} e^{-\omega t} \|T_1(t)\| = 0$$

and therefore $\omega_0(\mathbf{T}_1) < \omega$. Hence there exists a $\delta > 0$ and a constant $M = M_\delta > 0$ such that

$$\|T_1(t)\| \leq M e^{(\omega-\delta)t}, \quad t \geq 0.$$

This implies that

$$\|e^{-\omega t} T(t) - P\| = \|e^{-\omega t} T_1(t)\| \leq M e^{-\delta t}.$$

Since $P$ is rank one, hence compact, $T(t)P$ is compact for all $t \geq 0$ and

$$\begin{aligned}
\|T(t) + \mathcal{K}(X)\|_{\mathcal{C}(X)} &= \|T(t)(I-P) + \mathcal{K}(X)\|_{\mathcal{C}(X)} \\
&= \|T_1(t) + \mathcal{K}(X)\|_{\mathcal{C}(X)} \\
&\leq \|T_1(t)\| \leq M e^{(\omega-\delta)t}, \quad t \geq 0.
\end{aligned}$$

This implies that $\omega_0^{ess}(\mathbf{T}) \leq \omega - \delta$. On the other hand, $\omega_0(\mathbf{T}) = \omega$, and it follows that $\omega_0^{ess}(\mathbf{T}) < \omega_0(\mathbf{T})$. We claim that $\sigma(A) \cap \{\operatorname{Re}\lambda > \omega - \delta\} = \{\omega\}$.

Indeed, let $\lambda \in \sigma(A)$ with $\operatorname{Re}\lambda > \omega - \delta$. Then $e^\lambda \in \sigma(T(1))$, $|e^\lambda| > e^{\omega-\delta} \geq r^{ess}(T(1))$ and therefore $e^\lambda$ is an eigenvalue of $T(1)$. If $y$ is an eigenvector, then

$$0 = \lim_{n\to\infty} \|e^{-\omega n} T(n) y - Py\| = \lim_{n\to\infty} \|e^{(-\omega+\lambda)n} y - Py\|$$

which is only possible if $\lambda = \omega$ (and $y = Py$). This proves the claim and at the same time shows that $\omega$ is strictly dominant.

It remains to prove that $\omega$ is algebraically simple. For this it suffices to prove that $P = P_\omega$, the spectral projection corresponding to $\omega$. To this end, we fix $x \in X$ arbitrary. Let $\Gamma = \Gamma_{\frac{\delta}{2},\omega}$ be the counterclockwise oriented circle of radius $\frac{\delta}{2}$ and centre $\omega$. By what we have already proved, $\Gamma \subset \varrho(A)$ and the only element of $\sigma(A)$ enclosed by $\Gamma$ is $\omega$. Since $APx = \omega Px$, for all $\lambda \in \Gamma$ we have $R(\lambda, A)Px = (\lambda - \omega)^{-1} Px$. On the other hand, for all $\lambda \in \Gamma$ we have

$$R(\lambda, A)(I-P)x = R(\lambda, A_1)(I-P)x = \int_0^\infty e^{-\lambda t} T_1(t) x \, dt,$$

the integral being absolutely convergent since $\omega_0(\mathbf{T}_1) \leq \omega - \delta < \operatorname{Re}\lambda$. Therefore, by Fubini's theorem and Cauchy's theorem,

$$\begin{aligned}
P_\omega x &= \frac{1}{2\pi i} \int_\Gamma R(\lambda, A) x \, d\lambda \\
&= \frac{1}{2\pi i} \int_\Gamma \frac{Px}{\lambda - \omega} d\lambda + \int_0^\infty \left( \frac{1}{2\pi i} \int_\Gamma e^{-\lambda t} d\lambda \right) T_1(t) x \, dt \\
&= Px + 0 = Px.
\end{aligned}$$

This concludes the proof of (i)⇒(ii).

Conversely, assume (ii). We claim that there exists $\delta > 0$ such that $\omega_0^{ess}(\mathbf{T}) \leq \omega - \delta$ and $\operatorname{Re} \lambda \leq \omega - \delta$ for all $\lambda \in \sigma(A) \setminus \{\omega\}$. Indeed, otherwise there would be a sequence $(\lambda_n) \subset \sigma(A)$ such that $\operatorname{Re} \lambda_n \uparrow \omega$. By the spectral inclusion theorem, $e^{\lambda_n} \in \sigma(T(1))$ for all $n$. Since $|e^{\lambda_n}| \uparrow e^\omega$, it follows that the circle with radius $e^\omega$ contains an accumulation point $\mu$ of the sequence $(e^{\lambda_n})$. Since the spectrum of $T(1)$ is closed, we have $\mu \in \sigma(T(1))$. But $\sigma(T(1)) \cap \{|\lambda| > r^{ess}(T(1))\}$ consists of isolated points, and therefore we must have $r^{ess}(T(1)) \geq e^\omega$ and hence $\omega_0^{ess}(\mathbf{T}) \geq \omega$. Since on the other hand $\omega$ is an eigenvalue of $A$, $e^\omega$ is an eigenvalue of $T(1)$ and hence $\omega_0(\mathbf{T}) \geq \omega$. It follows that $\omega_0(\mathbf{T}) \leq \omega_0^{ess}(\mathbf{T})$, contradicting the assumptions of (ii). This proves the claim.

Let $P = P_\omega$ the spectral projection corresponding to $\{\omega\}$; by assumption this projection is rank one. We have a direct sum decomposition $X = X_0 \oplus X_1$ of $\mathbf{T}$-invariant closed subspaces, $X_0 = PX$, $X_1 = (I-P)X$. Let $\mathbf{T}_i$ be the restriction of $\mathbf{T}$ to $X_i$, and let $A_i$ be its generator, $i = 0, 1$. Since $\omega_0^{ess}(\mathbf{T}_1) \leq \omega_0^{ess}(\mathbf{T}) \leq \omega - \delta$ and $s(A_1) \leq \omega - \delta$ by the claim, Theorem 3.6.1 implies that

$$\omega_0(\mathbf{T}_1) = \max\{s(A_1), \omega_0^{ess}(\mathbf{T}_1)\} \leq \omega - \delta.$$

It follows that for all $0 < \epsilon < \delta$ there is a constant $M = M_\epsilon > 0$ such that $\|T_1(t)\| \leq M e^{(\omega-\epsilon)t}$, $t \geq 0$, and hence

$$\|e^{-\omega t} T(t) - P\| = \|e^{-\omega t} T_1(t)\| \leq M e^{-\epsilon t}.$$

Here we used the fact that $T(t)P = e^{\omega t} P$, $t \geq 0$; this follows from the fact that the one-dimensional range of $P$ is spanned by an eigenvector of $A$ with eigenvalue $\omega$. ////

A positive $C_0$-semigroup on a Banach lattice $X$ is called *irreducible* if $X$ has no proper $\mathbf{T}$-invariant ideal other than $\{0\}$. Theorem 3.6.2 has the following consequence.

**Theorem 3.6.3.** *Let $\mathbf{T}$ be an irreducible semigroup on a Banach lattice $X$. If $\omega_0^{ess}(\mathbf{T}) < \omega_0(\mathbf{T})$, in particular if $\mathbf{T}$ is eventually compact and $\omega_0(\mathbf{T}) > -\infty$, there exist $0 \leq x_0 \in X$, $0 \leq x_0^* \in X^*$ and an $\delta > 0$ such that*

$$\|e^{-s(A)t} T(t) - x_0^* \otimes x_0\| \leq e^{-\delta t}, \quad t \geq 0.$$

That Theorem 3.6.3 follows from 3.6.2 is far from obvious. We shall not give the complete proof, which is beyond the scope of these notes, but only sketch the main line of the argument. The detailed proof can be found in the book [Na, Chapter C-III].

From the assumption $\omega_0^{ess}(\mathbf{T}) < \omega_0(\mathbf{T})$ it follows that $s(A) = \omega_0(\mathbf{T})$ and $r^{ess}(T(t)) < r(T(t))$ for all $t > 0$. Hence by Fredholm theory, for all $t > 0$, $r(T(t))$ is a pole of the resolvent of the operator $T(t)$. This easily implies that $s(A)$ is a pole of the resolvent of $A$. Next, one shows that $s(A)$ is a strictly dominant eigenvalue. For this one uses the fact that $\sigma(A) \cap \{\operatorname{Re}\lambda = s(A)\}$ is additively cyclic if $s(A)$ is a pole of the resolvent. Thus, if $s(A)$ is not strictly dominant, there is an $\alpha > 0$ such that $s(A) + i\alpha\mathbb{Z} \subset \sigma(A)$. The spectral inclusion theorem then implies that for all but countable many $t$ the circle with radius $r(T(t))$ is contained in the spectrum of $T(t)$. In particular, none of the points on this circle is isolated in $\sigma(T(t))$. By Fredholm theory, this implies that $r^{ess}(T(t)) = r(T(t))$ and hence $\omega_0^{ess}(\mathbf{T}) = \omega_0(\mathbf{T})$, contradicting the assumptions. Therefore $s(A)$ is strictly dominant. Now the irreducibility of $\mathbf{T}$ comes in to show that $s(A)$ is an algebraically simple pole whenever it is a pole, and that in this case the corresponding spectral projection is of the form $x_0^* \otimes x_0$ with $\langle x^*, x_0 \rangle = 1$, where $0 \leq x_0$ and $0 \leq x_0^*$ are eigenvectors with eigenvalue $s(A)$ of $A$ and $A^*$, respectively.

**Notes.** Lemma 3.1.2 is taken from [NSW] but various forms of it have been observed earlier by a number of authors.

R. Datko [Dk] proved Corollary 3.1.6 for $\alpha = 1$ and $p = 2$, and A. Pazy [Pz] obtained the case $\alpha = 1$ and $1 \leq p < \infty$.

The general approach to the Datko-Pazy theorem presented here not only leads to stronger results, but also to a unified treatment of Rolewicz's theorem. A direct proof of the Datko-Pazy theorem is given in [Pz, Theorem 4.4.1].

Theorems 3.1.4 and 3.1.5 are taken from [Ne2]. For $E = l^p$, Theorem 3.1.4 is due to G. Weiss [Ws2]. Earlier, special cases have been obtained by J. Zabczyk [Zb1] and K.M. Przyluski [Pl].

Müller's theorem is proved in [Mü] and depends of Fredholm theory. The proof is valid only for complex Banach spaces. An elementary proof for power bounded operators, valid for real and complex Banach spaces, is given in [Ne3]. Related results can be found in the book [Be].

The first step of Lemma 3.1.7 was proved in [Ne2]; the technique to derive the sharper result in Step 2 is used in [Ne3]. Theorem 3.1.8 seems to be new and answers a question raised by Q. Zheng [Zh2]. It shows that the result in [Tg] is not sharp. For Hilbert spaces, Theorem 3.1.8 follows from Gearhart's theorem and the following result of G. Weiss (see Section 4.6): Let $\mathbf{T}$ be a $C_0$-semigroup on a Banach space $X$. If there exist $1 \leq p < \infty$ and $C > 0$ such that

$$\int_0^\infty |\langle x^*, T(t)x \rangle|^p \, dt \leq C \|x\|^p \|x^*\|^p, \quad \forall x \in X, \, x^* \in X^*,$$

then $s_0(A) \leq -1/(pC)$.

Theorems 3.2.2 and 3.2.3 are taken from [Ne2]. Note that the proof of Theorem 3.2.3 depends on Step 1 of Lemma 3.1.7 only. The special case of Theorem 3.2.2 for continuous $\phi$ is due to S. Rolewicz [Ro]. He actually proves a more slightly more general result valid for so-called evolution families $\{U(t,s)\}_{t\geq s\geq 0}$. A shorter proof of Rolewicz's theorem is given by Q. Zheng [Zh1] who also removed the continuity assumption. Another proof of (the semigroup case of) Rolewicz's result was offered by W. Littman [Li]. All these proofs are direct and ad hoc. Theorem 3.2.3 with $\alpha = 1$ leads to yet another proof.

The material of Sections 3.3 and 3.4 is taken from [Ne4]. A proof of Vitali's theorem is given in [HP, Thm 3.14.1].

Theorem 3.5.1 was proved by L. Weis [We1], thereby extending the case $p = 1$ which had been obtained earlier by W.B. Johnson and L. Jones [JJ]. Theorem 3.5.3 is also due to L. Weis [We2]. His proof is based on interpolation between $L^1(\Gamma, L^1(\mu_g))$ and $L^\infty(\Gamma, L^\infty(\mu_g))$ and duality arguments. Some difficulties in the proof can be avoided if one replaces the role of $L^\infty(\Gamma, L^\infty(\mu_g))$ by $L^2(\Gamma, L^2(\mu_g))$. This was discovered by W. Arendt (unpublished manuscript); we follow his modification of Weis's proof here.

A second and much shorter proof of Weis's theorem has been announced by S. Montgomery-Smith [Mo]. He verifies the conditions of Theorem 2.5.4 directly by first approximating a function in $L^p(\Gamma, L^p(\mu))$ by stepfunctions and then proving an inequality for $L^p$-valued stepfunctions. At the moment this book went to print, there was one subtle measure-theoretic difficulty in the proof which had not been resolved yet.

The equality $s(A) = \omega_0(\mathbf{T})$ for positive semigroups on $L^p$ has been a long-standing open problem; before its final solution partial results have been obtained by J. Voigt [Vo] and M. Hieber [Hb].

The identity $s(A) = \omega_0(\mathbf{T})$ for positive $C_0$-semigroups on $L^1(\mu)$ and $L^2(\mu)$ is due to R. Derndinger [De] and to G. Greiner and R. Nagel [GNa], respectively.

For the spaces $C(K)$ and $C_0(\Omega)$, the identity $s(A) = \omega_0(\mathbf{T})$ was proved by R. Derndinger [De] and C.J.K. Batty and E.B. Davies [BD], respectively. U. Groh and F. Neubrander [GNe] proved the result for $C^*$-algebras with unit. W. Arendt and G. Greiner [AG] proved that for positive $C_0$-groups on $C_0(\Omega)$ the weak spectral mapping theorem holds. This is no longer true for positive $C_0$-groups on arbitrary Banach lattices; an example of an irreducible positive $C_0$-group with $s(A) < \omega_0(\mathbf{T})$ was given by M. Wolff [Wo].

Theorems 3.6.1 and 3.6.2 are due to J. Prüss [Pr1] and G. Webb [Wb], respectively. Theorem 3.6.3 is due to G. Greiner and has applications, e.g., to the theory of age-structured populations; see for instance [Hk].

# Chapter 4

## Boundedness of the resolvent

In this chapter we study to what extent the exponential type of certain orbits of a $C_0$-semigroup is determined by boundedness properties of the resolvent in a given right half-plane. The prototype of such results has already been proved in Section 2.2, where we showed that $\omega_0(\mathbf{T}) = s_0(A)$ for $C_0$-semigroups on Hilbert spaces. We shall prove a generalization of this result for semigroups in arbitrary Banach spaces.

We start in Section 4.1 by introducing the fractional growth bounds $\omega_\alpha(\mathbf{T})$, $\alpha \geq 0$. These are, roughly speaking, defined by the exponential growth of orbits originating from the domains of the fractional powers of the generator. We prove the convexity theorem of Weis and Wrobel, which states that the function $\alpha \mapsto \omega_\alpha(\mathbf{T})$ is convex.

This result is applied in Section 4.2 to prove the inequality

$$\omega_{\frac{1}{p}-\frac{1}{q}}(\mathbf{T}) \leq s_0(A) \tag{4.0.1}$$

for $C_0$-semigroups $\mathbf{T}$ on a Banach space with Fourier type $p \in [1, 2]$. Since Hilbert spaces have Fourier type 2, the identity $\omega_0(\mathbf{T}) = s_0(A)$ is recovered. On the other hand, every Banach space has Fourier type 1 and we obtain the inequality

$$\omega_1(\mathbf{T}) \leq s_0(A)$$

for arbitrary $C_0$-semigroups. We also present an example which shows that (4.0.1) is the best possible.

In Section 4.3 we prove the following individual version of (4.0.1): If $A$ generates a $C_0$-semigroup on a Banach space $X$ with Fourier type $p \in (1, 2]$ and $x_0 \in X$ is such that the map $\lambda \mapsto R(\lambda, A)x_0$ admits a bounded holomorphic extension to $\{\operatorname{Re} \lambda > 0\}$, then for all $\alpha > \frac{1}{p}$ and $\operatorname{Re} \lambda_0 > \omega_0(\mathbf{T})$ we have

$$\lim_{t \to \infty} \|T(t)(\lambda_0 - A)^{-\alpha} x_0\| = 0. \tag{4.0.2}$$

By modifying the proof, in Section 4.4 we show that this extends to the case $p = 1$ if $X$ has the analytic Radon-Nikodym property.

In arbitrary Banach spaces, (4.0.2) is no longer true. Nevertheless, even in the limit case $\alpha = 1$ the norm of $T(t)R(\lambda_0, A)x_0$ cannot grow very fast: we prove in Section 4.5 that there exists a constant $M > 0$ such that

$$\|T(t)R(\lambda_0, A)x_0\| \leq M(1+t), \quad t \geq 0.$$

In Section 4.6 we apply some of the results of the previous sections to semigroups enjoying certain scalar integrability properties. Under fairly general assumptions it is shown that such semigroups have uniformly bounded resolvent. This leads to the theorem of Huang, Liu, and Weiss that a $C_0$−semigroup $\mathbf{T}$ on a Hilbert space $H$ is uniformly exponentially stable if and only if all orbits are scalarly $p$-integrable for some $1 \leq p < \infty$, i.e.,

$$\int_0^\infty |\langle x^*, T(t)x \rangle|^p \, dt < \infty, \quad \forall x, x^* \in H.$$

By means of an example we show that uniform boundedness of the resolvent need not imply scalar $p$-integrability for any $p > 1$. Finally, we present some results giving sufficient conditions for scalar $p$-integrability of the semigroup in terms of the behaviour along vertical lines of both the resolvent and its adjoint.

## 4.1. The convexity theorem of Weis and Wrobel

In this section we introduce the fractional growth bounds $\omega_\alpha(\mathbf{T})$ and study their basic properties. For $0 \leq \alpha \leq 1$, these quantities interpolate between $\omega_0(\mathbf{T})$ and $\omega_1(\mathbf{T})$. The relevant definitions and facts concerning fractional powers and interpolation theory are collected in Appendices A.1 and A.2, respectively.

Let $A$ be the generator of a $C_0$−semigroup $\mathbf{T}$ on a Banach space $X$ and let $\operatorname{Re} \omega > \omega_0(\mathbf{T})$. Then for $A_\omega := A - \omega$ we have $\|R(\lambda, A_\omega)\| \leq M/(1+\lambda)$ for some $M > 0$ and all $\lambda > 0$. Consequently, the fractional powers $(-A_\omega)^\alpha$ are defined. By Proposition A.1.2 in Appendix A.1, for real $\omega$ the domains $D((-A_\omega)^\alpha)$ are independent of $\omega > \omega_0(\mathbf{T})$. We will denote this common domain by $X_\alpha$.

For $\alpha \geq 0$ we define the *fractional growth bound* $\omega_\alpha(\mathbf{T})$ as the infimum of all $\omega \in \mathbb{R}$ such that for all $x \in X_\alpha$ there exists a constant $M = M_{\omega,x} > 0$ such that $\|T(t)x\| \leq Me^{\omega t}$ for all $t \geq 0$. Note that for $\alpha = 0$ and $\alpha = 1$ this reduces to the definitions of $\omega_0(\mathbf{T})$ and $\omega_1(\mathbf{T})$ given before.

From property (iii) in Appendix A.1 it follows that the map $\alpha \mapsto \omega_\alpha(\mathbf{T})$ is non-increasing. In fact, much more is true: this map is *convex*.

**Theorem 4.1.1.** *The function $\alpha \mapsto \omega_\alpha(\mathbf{T})$ is convex on $[0, \infty)$.*

*Proof:* Let $0 \leq \alpha_0 \leq \alpha_1$ and $0 < \theta < 1$ be arbitrary, and put $\alpha_\theta := (1-\theta)\alpha_0 + \theta\alpha_1$. We have to show that

$$\omega_{\alpha_\theta}(\mathbf{T}) \leq (1-\theta)\omega_{\alpha_0}(\mathbf{T}) + \theta\omega_{\alpha_1}(\mathbf{T}).$$

By Proposition A.2.1 of Appendix A.2, for all $0 < \epsilon < \theta$ we have a continuous inclusion

$$D((-A)^{\alpha_\theta}) \subset [D((-A)^{\alpha_0}), D((-A)^{\alpha_1})]_{\theta-\epsilon}. \tag{4.1.1}$$

Fix $\alpha_i > \omega_{\alpha_i}(\mathbf{T})$, $i = 0, 1$, $\alpha_0 \geq \alpha_1$. By definition of the growth bounds $\omega_\alpha(\mathbf{T})$, for each $x \in D((-A)^{\alpha_i})$ we have

$$T(\cdot)x \in L^1(\mathbb{R}_+, e^{-\alpha_i t}dt, X), \qquad i = 0, 1,$$

and by the closed graph theorem the maps

$$T : D((-A)^{\alpha_i}) \to L^1(\mathbb{R}_+, e^{-\alpha_i t}dt, X),$$
$$Tx := T(\cdot)x,$$

are bounded, $i = 0, 1$. Using Proposition A.2.3, by interpolation it follows that $T$ is bounded as a map

$$T : [D((-A)^{\alpha_0}), D((-A)^{\alpha_1})]_{\theta-\epsilon}$$
$$\to [L^1(\mathbb{R}_+, e^{-\alpha_0 t}dt, X), L^1(\mathbb{R}_+, e^{-\alpha_1 t}dt, X)]_{\theta-\epsilon} = L^1(\mathbb{R}_+, e^{-\alpha_{\theta-\epsilon} t}dt, X),$$

where $\alpha_{\theta-\epsilon} = (1 - (\theta - \epsilon))\alpha_0 + (\theta - \epsilon)\alpha_1$. Therefore, by (4.1.1), $T$ is bounded as a map

$$T : D((-A)^{\alpha_\theta}) \to L^1(\mathbb{R}_+, e^{-\alpha_{\theta-\epsilon} t}dt, X).$$

This means that for all $x \in D((-A)^{\alpha_\theta})$ the map $t \mapsto T(t)x$ defines an element of $L^1(\mathbb{R}_+, e^{-\alpha_{\theta-\epsilon} t}dt, X)$. By Lemma 3.1.9 applied to the $C_0$-semigroup $\mathbf{T}_{\alpha_{\theta-\epsilon}} = \{e^{-\alpha_{\theta-\epsilon} t}T(t)\}_{t \geq 0}$, this implies that the map $t \mapsto e^{-\alpha_{\theta-\epsilon} t}T(t)x$ is bounded for all $x \in D((-A)^{\alpha_\theta})$. Hence,

$$\omega_{\alpha_\theta}(\mathbf{T}) \leq \alpha_{\theta-\epsilon} = (1 - (\theta - \epsilon))\alpha_0 + (\theta - \epsilon)\alpha_1.$$

Since this holds for all $0 < \epsilon < \theta$ and $\alpha_i > \omega_{\alpha_i}(\mathbf{T})$, $i = 0, 1$, it follows that

$$\omega_{\alpha_\theta}(\mathbf{T}) \leq (1 - \theta)\omega_{\alpha_0}(\mathbf{T}) + \theta \omega_{\alpha_1}(\mathbf{T}).$$

////

**Corollary 4.1.2.** *The function $\alpha \mapsto \omega_\alpha(\mathbf{T})$ is continuous on $(0, \infty)$.*

## 4.2. Exponential stability and boundedness of the resolvent

In this section, as an application of the Weis-Wrobel convexity theorem we will discuss the asymptotic behaviour of $C_0$-semigroups whose resolvent is uniformly bounded in a given right half-plane.

In order to treat the Banach space case and the Hilbert space case simultaneously, we include the so-called Fourier type of the underlying Banach space into our considerations. Let $1 \leq p \leq 2$. A Banach space $X$ has *Fourier type p* if the Fourier transform $\mathcal{F}$ extends to a bounded operator

$$\mathcal{F} : L^p(\mathbb{R}, X) \to L^q(\mathbb{R}, X), \quad \frac{1}{p} + \frac{1}{q} = 1.$$

In other words, it is assumed that the Hausdorff-Young theorem holds for the space $L^p(\mathbb{R}, X)$. Trivially, every Banach space has Fourier type 1. Only a Hilbert space has Fourier type 2 [Kw]. The Banach space $L^p(\Omega, \mu)$ has Fourier type $\min\{p, q\}$, $\frac{1}{p} + \frac{1}{q} = 1$ [Pe]. Every uniformly convex space has Fourier type $p$ for some $p$ with $1 < p \leq 2$ [Bo]. If a Banach space has Fourier type $p$ for some $1 \leq p \leq 2$, then it has Fourier type $r$ for all $1 \leq r \leq p$.

Let $A$ be a linear operator whose resolvent is uniformly bounded in the right half-plane. Then by Lemma 2.3.4, the resolvent is even bounded in $\{\operatorname{Re} \lambda > -\epsilon\}$ for $\epsilon > 0$ small enough. For all $\omega \in (-\epsilon, 0)$, $\alpha > 0$, and $x \in D(A)$, the integral

$$(S_\alpha x)(t) := \frac{1}{2\pi i} \int_{\operatorname{Re} \lambda = \omega} e^{\lambda t} (-\lambda)^{-\alpha} R(\lambda, A) x \, d\lambda \qquad (4.2.1)$$

is absolutely convergent and independent of the choice of $\omega$. The absolute convergence follows by writing $x = R(\mu, A)y$ and applying the resolvent identity. We also define

$$(S_{\alpha,\omega} x)(t) := e^{-\omega t} (S_\alpha x)(t).$$

**Lemma 4.2.1.** *Let $X$ be a Banach space with Fourier type $p$ for some $1 \leq p \leq 2$. Let $A$ be the generator of a $C_0$-semigroup $\mathbf{T}$ on $X$ whose resolvent is uniformly bounded in the right half-plane. Then, for all $\alpha > \frac{1}{p} - \frac{1}{q}$, $\epsilon > 0$ small enough, and $\omega \in (-\epsilon, 0)$, $S_{\alpha,\omega}$ extends to a bounded linear operator $S_{\alpha,\omega} : X \to L^q(\mathbb{R}_+, X)$.*

*Proof:* Let $\epsilon > 0$ be such that the resolvent is uniformly bounded in $\{\operatorname{Re} \lambda > -\epsilon\}$. Fix $\omega \in (-\epsilon, 0)$ and $x \in D(A)$, and define $g : \mathbb{R} \to X$ by

$$g(s) = (-\omega + is)^{-\alpha} R(\omega - is, A) x.$$

Then $g \in L^1(\mathbb{R}, X)$, and from the definition of $S_{\alpha,\omega} x$ and (4.2.1) we have

$$(S_{\alpha,\omega} x)(t) = \frac{1}{2\pi} \int_{-\infty}^{\infty} e^{-ist} g(s) \, ds = \frac{1}{2\pi} \mathcal{F} g(t). \qquad (4.2.2)$$

We are going to show that $g \in L^p(\mathbb{R}, X)$. Choose $M > 0$ such that $\|R(\lambda, A)\| \leq M$ for all $\operatorname{Re} \lambda \geq \omega$. Let $\omega_0 > \max\{0, \omega_0(\mathbf{T})\}$ be arbitrary and fixed. By the resolvent identity, for all $s \in \mathbb{R}$ we have

$$\|R(\omega - is, A)x\| = \|(I + (\omega_0 - \omega)R(\omega - is, A))R(\omega_0 - is, A)x\|$$
$$\leq (1 + M|\omega_0 + \epsilon|) \|R(\omega_0 - is, A)x\|.$$

Since $\frac{1}{r} := \frac{1}{p} - \frac{1}{q} < \alpha$, we can apply Hölder's inequality and obtain

$$\left(\int_{-\infty}^{\infty} \|g(s)\|^p \, ds\right)^{\frac{1}{p}} \leq \left(\int_{-\infty}^{\infty} |\omega - is|^{-\alpha r} (1 + M|\omega_0 + \epsilon|)^r \, ds\right)^{\frac{1}{r}}$$

$$\cdot \left(\int_{-\infty}^{\infty} \|R(\omega_0 - is, A)x\|^q ds\right)^{\frac{1}{q}}.$$

The first of the integrals on the right hand side is absolutely convergent; let $N$ denote its value. As to the second integral, since $\omega_0 > \omega_0(\mathbf{T})$ there exists a constant $C > 0$ independent of $x$ such that

$$\left(\int_{-\infty}^{\infty} \|R(\omega_0 - is, A)x\|^q \, ds\right)^{\frac{1}{q}} = \left(\int_{-\infty}^{\infty} \|R(\omega_0 + is, A)x\|^q \, ds\right)^{\frac{1}{q}}$$

$$\leq c_p \left(\int_{0}^{\infty} \|e^{-\omega_0 t} T(t)x\|^p \, dt\right)^{\frac{1}{p}} \leq c_p C \|x\|.$$

Here, $c_p$ is the norm of the Fourier transform as a map $L^p(\mathbb{R}, X) \to L^q(\mathbb{R}, X)$. Summarizing, we find that $g \in L^p(\mathbb{R}, X)$ and

$$\|g\|_p \leq N c_p C \|x\|.$$

Using once more the Fourier type of $X$, from (4.2.2) we obtain

$$\|(S_{\alpha,\omega} x)(\cdot)\|_{L^q(\mathbb{R}_+, X)} \leq \frac{c_p}{2\pi} \|g\|_p \leq \frac{N c_p^2 C}{2\pi} \|x\|.$$

Since $D(A)$ is dense in $X$, the lemma follows from this. ////

We can express $(S_{\alpha,\omega} x)(t)$ in terms of the semigroup:

**Lemma 4.2.2.** Let $A$ be the generator of a $C_0$–semigroup $\mathbf{T}$ on a Banach space $X$ and assume that the resolvent of $A$ is uniformly bounded in the right half-plane. Then for all $\alpha > 0$, $\epsilon > 0$ small enough, $\omega \in (-\epsilon, 0)$, and $x \in D(A^2)$, we have

$$(S_{\alpha,\omega} x)(t) = e^{-\omega t} T(t)(-A)^{-\alpha} x, \qquad \forall t \geq 0.$$

*Proof:* Let $\epsilon > 0$ be as in Lemma 4.2.1. By Cauchy's theorem,

$$\frac{1}{2\pi i} \int_{\text{Re } \lambda = \omega} e^{\lambda t} (-\lambda)^{-\alpha - 1} \, d\lambda = 0. \tag{4.2.3}$$

Using this and writing $-\lambda = (-\lambda + A) - A$, we see that

$$(S_\alpha x)(t) = \frac{1}{2\pi i} \int_{\text{Re } \lambda = \omega} e^{\lambda t} (-\lambda)^{-\alpha - 1} R(\lambda, A)(-Ax) \, d\lambda.$$

Since $-Ax \in D(A)$, we may differentiate under the integral sign and obtain

$$\frac{d}{dt}(S_\alpha x)(t) = \frac{1}{2\pi i} \int_{\operatorname{Re}\lambda=\omega} \lambda e^{\lambda t}(-\lambda)^{-\alpha-1} R(\lambda, A)(-Ax)\, d\lambda$$
$$= A\Big(\frac{1}{2\pi i} \int_{\operatorname{Re}\lambda=\omega} e^{\lambda t}(-\lambda)^{-\alpha-1} R(\lambda, A)(-Ax)\, d\lambda\Big) = A((S_\alpha x)(t)).$$

To obtain the second identity we used (4.2.3) once more.

By the resolvent identity, $\|R(\lambda, A)x\| = O(1 + |\lambda|)^{-1}$ as $|\lambda| \to \infty$ in the half-plane $\operatorname{Re}\lambda \geq \omega$. Therefore, by Cauchy's theorem we can deform the path of integration in (4.2.1) to the path in (A.1.2) of Appendix A.1. It follows that $(S_\alpha x)(0) = (-A)^{-\alpha}x$.

We have shown that $(S_\alpha x)(\cdot)$ is a solution to the abstract Cauchy problem

$$\frac{du}{dt}(t) = Au(t), \quad t \geq 0,$$
$$u(0) = (-A)^{-\alpha}x.$$

By uniqueness of solutions, we must have $(S_\alpha x)(t) = T(t)(-A)^{-\alpha}x$. In view of the definition of $S_{\alpha,\omega}x$, this proves the lemma.    ////

By an easy density argument it can be shown that the lemma actually holds for all $x \in D(A)$; we will not need this fact.

**Lemma 4.2.3.** *Let $X$ be a Banach space with Fourier type $p$ for some $1 \leq p \leq 2$. Let $A$ be the generator of a $C_0$–semigroup $\mathbf{T}$ on $X$ and assume that the resolvent of $A$ is uniformly bounded in the right half-plane. Then for all $\alpha > \frac{1}{p} - \frac{1}{q}$ and $x \in D((-A)^\alpha)$, the map $t \mapsto T(t)x$ belongs to $L^q(\mathbb{R}_+, X)$.*

Proof: Let $\epsilon > 0$ be as before and fix $\omega \in (-\epsilon, 0)$.

Let $x \in D((-A)^\alpha)$ be arbitrary and put $y := (-A)^\alpha x$. Choose a sequence $y_k \to y$ with $y_k \in D(A^2)$ for each $k$. By Lemma 4.2.1, $S_{\alpha,\omega}y_k \to S_{\alpha,\omega}y$ in $L^q(\mathbb{R}_+, X)$. Upon passing to a subsequence we may assume that $(S_{\alpha,\omega}y_k)(t) \to (S_{\alpha,\omega}y)(t)$ for almost all $t \geq 0$. For any such $t$, by Lemma 4.2.2 we have

$$e^{-\omega t}T(t)x = e^{-\omega t}T(t)(-A)^{-\alpha}y$$
$$= \lim_{k\to\infty} e^{-\omega t}T(t)(-A)^{-\alpha}y_k$$
$$= \lim_{k\to\infty} (S_{\alpha,\omega}y_k)(t) = (S_{\alpha,\omega}y)(t).$$

Therefore, $e^{-\omega(\cdot)}T(\cdot)x$ is equal a.e. to the function $(S_{\alpha,\omega}y)(\cdot) \in L^q(\mathbb{R}_+, X)$. Since $\omega < 0$, the lemma follows from this.    ////

We now come to our main stability result for semigroups on Banach spaces with Fourier type.

**Theorem 4.2.4.** *Let $1 \leq p \leq 2$ and let $X$ be a Banach space with Fourier type $p$. Let $A$ be the generator of a $C_0$-semigroup $\mathbf{T}$ on $X$. Then,*

$$\omega_{\frac{1}{p}-\frac{1}{q}}(\mathbf{T}) \leq s_0(A), \quad \frac{1}{p}+\frac{1}{q}=1.$$

*Proof:* By rescaling it is enough to prove the following: if the resolvent of $A$ is uniformly bounded in the right half-plane, then $\omega_{\frac{1}{p}-\frac{1}{q}}(\mathbf{T}) \leq 0$.

If $\frac{1}{p} - \frac{1}{q} = 0$, then $p=2$ and we are in the case of a generator on a Hilbert space. By Corollary 2.2.5, in that case we have $\omega_0(\mathbf{T}) = s_0(A)$. Therefore, in the rest of the proof we assume that $\frac{1}{p} - \frac{1}{q} > 0$.

Let $\alpha > \frac{1}{p} - \frac{1}{q}$ be arbitrary and fix $x \in D((-A)^\alpha)$. By Lemma 4.2.3, $T(\cdot)x \in L^q(\mathbb{R}_+, X)$. By Lemma 3.1.9, this implies that $T(\cdot)x$ is bounded. As this holds for all $x \in D((-A)^\alpha)$, it follows that $\omega_\alpha(\mathbf{T}) \leq 0$. Since $\alpha \mapsto \omega_\alpha(\mathbf{T})$ is continuous on $(0, \infty)$ by Corollary 4.1.2, the theorem is proved.    ////

Note that the Weis-Wrobel theorem was only used to improve the inequality

$$\omega_\alpha(\mathbf{T}) \leq s_0(A), \quad \forall \alpha > \frac{1}{p} - \frac{1}{q},$$

to

$$\omega_{\frac{1}{p}-\frac{1}{q}} \leq s_0(A).$$

We record some interesting special cases of Theorem 4.2.4, using what is known about Fourier type of the spaces in question.

**Corollary 4.2.5.** *Let $A$ be the generator of a $C_0$-semigroup $\mathbf{T}$ on a uniformly convex Banach space. Then there exists $0 < \epsilon \leq 1$ such that $\omega_{1-\epsilon}(\mathbf{T}) \leq s_0(A)$.*

**Corollary 4.2.6.** *Let $A$ be the generator of a $C_0$-semigroup $\mathbf{T}$ on a space $L^p(\mu)$, $1 \leq p < \infty$. Then $\omega_{|\frac{1}{p}-\frac{1}{q}|}(\mathbf{T}) \leq s_0(A)$.*

**Corollary 4.2.7.** *Let $A$ be the generator of a $C_0$-semigroup $\mathbf{T}$ on an arbitrary Banach space. Then $\omega_1(\mathbf{T}) \leq s_0(A)$.*

**Corollary 4.2.8.** *Let $\mathbf{T}$ be a $C_0$-semigroup on an arbitrary Banach space. If the resolvent of the generator $A$ exists and is uniformly bounded in the right half-plane, then there exists $0 < \epsilon \leq 1$ such that $\omega_{1-\epsilon}(\mathbf{T}) < 0$. In particular, $\mathbf{T}$ is exponentially stable.*

*Proof:* By Lemma 2.3.4, $s_0(A) < 0$, so $\omega_1(\mathbf{T}) < 0$ by Corollary 4.2.7, and the result follows from Corollary 4.1.2.    ////

Theorem 4.2.4 and its corollaries can be considered as analogues for $C_0$-semigroups of the classical Lyapunov stability theorem for matrices.

The following example shows that Theorem 4.2.4 is the best possible.

**Example 4.2.9.** Let $1 \leq p \leq 2$, $\frac{1}{p} + \frac{1}{q} = 1$, and let $\mathbf{T}$ be the $C_0$–semigroup on $X := L^p(1, \infty) \cap L^q(1, \infty)$ defined by

$$(T(t)f)(s) = f(se^t), \quad s > 1.$$

Since both $L^p(1, \infty)$ and $L^q(1, \infty)$ have Fourier type $p$, so does their intersection $X$. As we have seen in Section 1.4,

$$s(A) = \omega_1(\mathbf{T}) = -\frac{1}{p}; \qquad \omega_0(\mathbf{T}) = -\frac{1}{q}.$$

We are going to prove that

$$\omega_\alpha(\mathbf{T}) = \begin{cases} -\frac{1}{q} - \alpha, & 0 \leq \alpha \leq \frac{1}{p} - \frac{1}{q}; \\ -\frac{1}{p}, & \frac{1}{p} - \frac{1}{q} \leq \alpha \leq 1. \end{cases} \tag{4.2.4}$$

Let $f : (1, \infty) \to \mathbb{C}$ be a smooth function with $\operatorname{supp} f \subset (1, 2)$. For $n \in \mathbb{N}$ we define the function $f_n \in X$ by

$$f_n(s) := \begin{cases} f(s - n), & s - n > 1; \\ 0, & \text{else}. \end{cases}$$

Then $f_n \in D(A)$ for all $n$ and $\operatorname{supp} f_n \subset (n + 1, n + 2)$. Since $(Af_n)(s) = sf'_n(s)$, for all $n$ we have

$$\|Af_n\| \leq (n + 2)\|f'_n\| = (n + 2)\|f'\| \leq (n + 2)\|Af\|.$$

Fix $0 < \alpha < 1$. By (A.2.5) in Appendix A.2 there is a constant $C > 0$ such that for all $n \geq 1$,

$$\|f_n\|_{D((-A)^\alpha)} \leq C\|f_n\|^{1-\alpha}\|Af_n\|^\alpha$$
$$\leq C(n+2)^\alpha\|f\|^{1-\alpha}\|Af\|^\alpha$$
$$\leq 3Cn^\alpha\|f\|^{1-\alpha}\|Af\|^\alpha.$$

Since $\operatorname{supp} f_n \subset (n + 1, n + 2)$, by (1.4.3) for all $t \leq \log n$ we have

$$\|T(t)f_n\| = \max\{\|T_p(t)f_n\|_p, \|T_q(t)f_n\|_q\} = \max\{e^{-\frac{t}{p}}\|f_n\|_p, e^{-\frac{t}{q}}\|f_n\|_q\},$$

where $\mathbf{T}_p$ and $\mathbf{T}_q$ have the obvious meaning. By the uniform boundedness theorem, for each $\omega > \omega_\alpha(\mathbf{T})$ there exists an $M > 0$ such that

$$\|T(t)f_n\| \leq Me^{\omega t}\|f_n\|_{D((-A)^\alpha)}, \quad \forall n \in \mathbb{N}.$$

Hence for $n \geq 1$ and $t := \log n$ we have

$$\max\{n^{-\frac{1}{p}}\|f_n\|_p, n^{-\frac{1}{q}}\|f_n\|_q\} = \|T(\log n)f_n\| \leq 3CM \, n^{\omega+\alpha}\|f\|^{1-\alpha}\|Af\|^\alpha.$$

Since this holds for all $n \geq 1$, it follows that $\omega + \alpha \geq -\frac{1}{q}$. This proves that

$$\omega_\alpha(\mathbf{T}) \geq -\frac{1}{q} - \alpha, \quad 0 < \alpha < 1. \tag{4.2.5}$$

Of course, by what is already known this also holds for $\alpha = 0$ and $\alpha = 1$. On the other hand, since $X$ has Fourier type $p$, by Theorem 4.2.4 we have

$$\omega_{\frac{1}{p} - \frac{1}{q}}(\mathbf{T}) \leq s_0(A).$$

For all $\beta > \frac{1}{p}$, the function $f_\beta(s) := s^{-\beta}$ is an eigenfunction of $A$ with eigenvalue $-\beta$. It follows that $e^{-\beta t}$ is an eigenvalue of $T(t)$ for all $\beta > \frac{1}{p}$, and therefore $\omega_\alpha(\mathbf{T}) \geq -\frac{1}{p}$ for all $\alpha \geq 0$. In particular, using Theorem 1.4.1 we have

$$-\frac{1}{p} \leq \omega_{\frac{1}{p} - \frac{1}{q}}(\mathbf{T}) \leq s_0(A) = s(A) = \omega_1(\mathbf{T}) = -\frac{1}{p}.$$

This proves that

$$\omega_\alpha(\mathbf{T}) = -\frac{1}{p}, \quad \forall \frac{1}{p} - \frac{1}{q} \leq \alpha \leq 1.$$

Let $0 \leq \theta \leq 1$. Then for $\alpha_\theta := \theta(\frac{1}{p} - \frac{1}{q})$, by the Weis-Wrobel convexity theorem and (4.2.5) we have

$$-\frac{1}{q} - \alpha_\theta \leq \omega_{\alpha_\theta}(\mathbf{T}) \leq (1 - \theta)\omega_0(\mathbf{T}) + \theta\omega_{\frac{1}{p} - \frac{1}{q}}(\mathbf{T})$$

$$= -(1 - \theta)\frac{1}{q} - \theta\frac{1}{p} = -\frac{1}{q} - \alpha_\theta.$$

This proves that

$$\omega_\alpha(\mathbf{T}) = -\frac{1}{q} - \alpha, \quad \forall 0 \leq \alpha \leq \frac{1}{p} - \frac{1}{q}$$

and the proof of (4.2.4) is complete.

By rescaling, for every $0 < \alpha < \frac{1}{p} - \frac{1}{q}$ we obtain a positive $C_0$-semigroup $\mathbf{T}$ for which $s_0(A) = \omega_1(\mathbf{T}) < 0$ and $\omega_\alpha(\mathbf{T}) > 0$. This shows that Theorem 4.2.4 is indeed the best possible.

## 4.3. Individual stability in $B$-convex Banach spaces

Let $A$ be the generator of a $C_0$-semigroup on a Banach space $X$ with Fourier type $p \in (1, 2]$. In this section we prove that

$$\lim_{t \to \infty} \|T(t)(\lambda_0 - A)^{-\alpha} x_0\| = 0$$

for all $\alpha \geq 0$ with $\alpha > \frac{1}{p}$ and all $\operatorname{Re}\lambda_0 > \omega_0(\mathbf{T})$, provided the local resolvent $\lambda \mapsto R(\lambda, A)x_0$ extends to a bounded holomorphic function in the open right half-plane. This result can be interpreted as an individual version of Theorem 4.2.4. The case $p = 1$ is more delicate and will be discussed separately in the next two sections.

Actually we shall address the following slightly more general problem: describe the asymptotic behaviour of $t \mapsto PT(t)(\lambda_0 - A)^{-\alpha}x_0$ assuming boundedness of a holomorphic extension of the map $\lambda \mapsto PR(\lambda, A)x_0$, where $P$ is an arbitrary bounded operator on $X$. Accordingly we assume that $\overline{PX}$ has Fourier type rather than $X$ itself. This enables us to treat the strong case and the weak case simultaneously: weak analogues are obtained by considering the rank one operators $P = x^* \otimes x$ (cf. Corollary 4.4.3 below). This extra generality may also be relevant to applications, for instance to matrix semigroups arising from higher order abstract Cauchy problems, in which case $P$ could be a coordinate projection.

For an non-negative real number $\alpha$ we let $[\alpha]$ denote the integer part of $\alpha$, i.e. the unique $n \in \mathbb{N}$ such that $n \leq \alpha < n+1$.

**Lemma 4.3.1.** *Let $A$ be the generator of a $C_0$-semigroup $\mathbf{T}$ on a Banach space $X$ and let $P$ be a bounded operator on $X$. Let $x_0 \in X$ be such that the map $\lambda \mapsto PR(\lambda, A)x_0$ admits a bounded holomorphic extension $F(\lambda)$ in the open right half-plane. Then for all $\alpha \geq 0$ and $\operatorname{Re}\lambda_0 > \max\{0, \omega_0(\mathbf{T})\}$, the map $\lambda \mapsto PR(\lambda, A)(\lambda_0 - A)^{-\alpha}x_0$ admits a holomorphic extension $G(\lambda)$ in the open right half-plane, and for all $\omega_0 \in (0, \operatorname{Re}\lambda_0)$ there exists an $M > 0$ such that*

$$\|G(\lambda)\| \leq M(1+|\lambda|)^{\max\{-\alpha, -1-\alpha+[\alpha]\}}, \quad 0 < \operatorname{Re}\lambda < \omega_0. \tag{4.3.1}$$

*Moreover, $G$ is $\overline{PX}$-valued.*

*Proof:* Fix $\operatorname{Re}\lambda_0 > \max\{0, \omega_0(\mathbf{T})\}$; by rescaling we may assume that $\lambda_0$ is real. Also fix $\omega_0 \in (0, \lambda_0)$.

Write $\alpha = n + \delta$ with $n \in \mathbb{N}$ and $0 \leq \delta < 1$ and put $y_0 := R(\lambda_0, A)^n x_0$. In view of the identity

$$R(\lambda, A)y_0 = \frac{R(\lambda, A)x_0}{(\lambda_0 - \lambda)^n} - \sum_{k=0}^{n-1} \frac{R(\lambda_0, A)^{k+1}x_0}{(\lambda_0 - \lambda)^{n-k}},$$

the map $\lambda \mapsto PR(\lambda, A)y_0$ admits a holomorphic extension $F_1(\lambda)$ to $\{\operatorname{Re}\lambda > 0\}$ which satisfies

$$\|F_1(\lambda)\| \leq M'(1+|\lambda|)^{\max\{-n, -1\}}, \quad 0 < \operatorname{Re}\lambda < \omega_0, \tag{4.3.2}$$

for some constant $M' > 0$. Consider the holomorphic function

$$G(\lambda) := PR(\lambda, A)(\lambda_0 - A)^{-\alpha}x_0 = PR(\lambda, A)(\lambda_0 - A)^{-\delta}y_0, \quad \operatorname{Re}\lambda > \omega_0(\mathbf{T}).$$

By the resolvent identity and the real representation of fractional powers, cf. (A.1.3) in the Appendix, for $\operatorname{Re}\lambda > \omega_0(\mathbf{T})$ we have

$$G(\lambda) = \frac{\sin \pi \delta}{\pi} \int_0^\infty t^{-\delta} PR(\lambda, A) R(\lambda_0 + t, A) y_0 \, dt$$

$$= \frac{\sin \pi \delta}{\pi} \int_0^\infty t^{-\delta} \frac{PR(\lambda, A) y_0 - PR(\lambda_0 + t, A) y_0}{t + \lambda_0 - \lambda} \, dt.$$

Passing to the holomorphic extension, we see from (4.3.2) that

$$G(\lambda) := \frac{\sin \pi \delta}{\pi} \int_0^\infty t^{-\delta} \frac{F_1(\lambda) - F_1(\lambda_0 + t)}{t + \lambda_0 - \lambda} \, dt \qquad (4.3.3)$$

converges absolutely and defines a holomorphic extension of $G$ in the strip $\{0 < \operatorname{Re}\lambda < \lambda_0\}$.

We claim that $G$ takes values in $\overline{PX}$. To see this, it is enough to show that this is true for $F_1$. To this end let $q : X \to X/\overline{PX}$ be the quotient map. Then $qF_1$ is a holomorphic extension of $\lambda \mapsto qPR(\lambda, A)y_0$. Since the latter is identically 0 in $\{\operatorname{Re}\lambda > \omega_0(\mathbf{T})\}$, by analytic continuation we have $qF_1 \equiv 0$ on $\{\operatorname{Re}\lambda > 0\}$. This means that $F_1(\lambda) \in \overline{PX}$ for all $\operatorname{Re}\lambda > 0$.

For $\omega > 0$ consider the functions $G_\omega : \mathbb{R} \to X$ defined by

$$G_\omega(s) := G(\omega - is), \quad s \in \mathbb{R}.$$

Then $G_\omega(s) = PR(\omega - is, A)(\lambda_0 - A)^{-\alpha} x_0$ for $\omega > \omega_0(\mathbf{T})$. In view of the estimate $\|R(\lambda, A)\| \leq \text{const} \cdot (\operatorname{Re}\lambda - \omega_0)^{-1}$ for all $\operatorname{Re}\lambda > \lambda_0$, we see that $c := \sup_{\tau > 0} \tau \|F_1(\tau)\| < \infty$. Hence by (4.3.2) and (4.3.3), for all $0 < \omega < \omega_0$ and $s \in \mathbb{R}$ we have

$$\|G_\omega(s)\| \leq \frac{\sin \pi \delta}{\pi} \int_0^\infty t^{-\delta} \frac{M'(1 + (\omega^2 + s^2)^{\frac{1}{2}})^{\max\{-n, -1\}} + c(\lambda_0 + t)^{-1}}{((t + \lambda_0 - \omega)^2 + s^2)^{\frac{1}{2}}} \, dt$$

$$\leq \text{const} \cdot (1 + s^2)^{\frac{1}{2} \cdot \max\{-n-\delta, -1-\delta\}}$$

$$= \text{const} \cdot (1 + s^2)^{\frac{1}{2} \cdot \max\{-\alpha, -1-\delta\}},$$

where the constant is independent of $s \in \mathbb{R}$ and $\omega \in (0, \omega_0)$. From this estimate the lemma follows. ////

The main result of this section imposes no restriction on the Fourier type.

**Theorem 4.3.2.** *Let $P$ be a bounded operator on a Banach space $X$ and assume that $\overline{PX}$ has Fourier type $p \in [1, 2]$. Let $A$ be the generator of a $C_0$–semigroup $\mathbf{T}$ on $X$ and let $x_0 \in X$ be such that the map $\lambda \mapsto PR(\lambda, A)x_0$ admits a bounded holomorphic extension $F(\lambda)$ in the open right half-plane. Then for all $\alpha > \frac{1}{p}$ and all $\operatorname{Re}\lambda_0 > \omega_0(\mathbf{T})$ we have*

$$PT(\cdot)(\lambda_0 - A)^{-\alpha} x_0 \in L^q(\mathbb{R}_+, X), \quad \frac{1}{p} + \frac{1}{q} = 1.$$

*Proof:* Without loss of generality we may assume that $\omega_0(\mathbf{T}) \geq 0$. By rescaling we may also assume that $\lambda_0$ is real. Fix $\omega_0 \in (\omega_0(\mathbf{T}), \lambda_0)$.

Lemma 4.3.1 shows that $\lambda \mapsto PR(\lambda, A)(\lambda_0 - A)^{-\alpha}x_0$ has a $\overline{PX}$-valued holomorphic extension $G(\lambda)$ in the open right half-plane which satisfies (4.3.1). Let $G_\omega(s) := G(\omega - is)$, $s \in \mathbb{R}$. Then by (4.3.1) and the assumption $\alpha > \frac{1}{p}$, $G_\omega \in L^p(\mathbb{R}, X)$ for all $\omega \in (0, \omega_0)$ and

$$\sup_{0 < \omega < \omega_0} \|G_\omega\|_p \leq C := \mathrm{const} \cdot \left( \int_{-\infty}^{\infty} (1 + |s|)^{p \cdot \max\{-\alpha, -1-\delta\}} ds \right)^{\frac{1}{p}},$$

where $\delta = \alpha - [\alpha]$. Since $\overline{PX}$ has Fourier type $p$, the Fourier transform $g_\omega := \frac{1}{2\pi} \mathcal{F} G_\omega$ of $G_\omega$ defines an element of $L^q(\mathbb{R}, X)$, $\frac{1}{p} + \frac{1}{q} = 1$.

Let $\omega \in (0, \omega_0)$ be fixed. We claim that

$$g_\omega(t) = e^{-\omega t} PT(t)(\lambda_0 - A)^{-\alpha} x_0 \quad \text{for a.a. } t > 0.$$

Because of the operator $P$ we cannot proceed as in Lemma 4.2.2 and have to rely more closely on Laplace transform techniques. Define, for each $r > 0$,

$$G_{\omega,r} := G_\omega \cdot \chi_{[-r,r]}.$$

Then $\lim_{r \to \infty} G_{\omega,r} = G_\omega$ in the norm of $L^p(\mathbb{R}, X)$, so for the Fourier transforms $g_{\omega,r} = \frac{1}{2\pi} \mathcal{F} G_{\omega,r}$ we have $\lim_{r \to \infty} g_{\omega,r} = g_\omega$ in $L^q(\mathbb{R}, X)$. Let $\Gamma$ be the rectangle spanned by the points $\omega - ir$, $\omega + ir$, $\omega_0 + ir$, and $\omega_0 - ir$. By Cauchy's theorem, for all $t > 0$ we have

$$\begin{aligned} \frac{1}{2\pi i} \int_{\omega - ir}^{\omega + ir} e^{zt} G(z)\, dz &= \frac{1}{2\pi i} \int_{\omega_0 - ir}^{\omega_0 + ir} e^{zt} G(z)\, dz + R_r(t) \\ &= \frac{1}{2\pi i} \int_{\omega_0 - ir}^{\omega_0 + ir} e^{zt} PR(z, A)(\lambda_0 - A)^{-\alpha} x_0\, dz + R_r(t), \end{aligned} \quad (4.3.4)$$

where $R_r(t)$ represents the integrals over the two horizontal parts of $\Gamma$. From (4.3.1) we see that $\lim_{r \to \infty} \|R_r(t)\| = 0$ for all $t > 0$. By the complex inversion theorem for the Laplace transform, the Cesàro means of the integral on the right hand side in (4.3.4) converge to $PT(t)(\lambda_0 - A)^{-\alpha} x_0$ as $r \to \infty$; here we use that $\omega_0 > \omega_0(\mathbf{T})$. It follows that for all $t > 0$,

$$\lim_{m \to \infty} \frac{1}{m} \int_0^m \frac{1}{2\pi i} \int_{\omega - ir}^{\omega + ir} e^{zt} G(z)\, dz\, dr = PT(t)(\lambda_0 - A)^{-\alpha} x_0. \quad (4.3.5)$$

On the other hand, for $t > 0$ we have

$$g_{\omega,r}(t) = \frac{1}{2\pi} \int_{-r}^{r} e^{-ist} G(\omega - is)\, ds = \frac{1}{2\pi i} e^{-\omega t} \int_{\omega - ir}^{\omega + ir} e^{zt} G(z)\, dz \quad (4.3.6)$$

It follows from (4.3.5) and (4.3.6) that

$$\lim_{m\to\infty}\left(\left(\frac{1}{m}\int_0^m g_{\omega,r}\,dr\right)(t)\right) = \lim_{m\to\infty}\frac{1}{m}\int_0^m g_{\omega,r}(t)\,dr$$
$$= e^{-\omega t}PT(t)(\lambda_0 - A)^{-\alpha}x_0$$

for all $t > 0$. In the first identity we used the fact that the map $r \mapsto g_{\omega,r}$ is continuous as a map into $C_0(\mathbb{R}, X)$ by the Riemann-Lebesgue lemma. Therefore the integrals with respect to $r$ can be regarded as Bochner integrals in $C_0(\mathbb{R}, X)$ and we may use the continuity of point evaluations.

We also have

$$\lim_{m\to\infty}\left(\frac{1}{m}\int_0^m g_{\omega,r}\,dr\right) = \lim_{r\to\infty} g_{\omega,r} = g_\omega$$

in the norm of $L^q(\mathbb{R}, X)$. Since norm convergent sequences have pointwise a.e. convergent subsequences, we see that $g_\omega(t) = e^{-\omega t}PT(t)(\lambda_0 - A)^{-\alpha}x_0$ for almost all $t > 0$ and the claim is proved.

It follows that $t \mapsto e^{-\omega t}PT(t)(\lambda_0 - A)^{-\alpha}x_0 \in L^q(\mathbb{R}_+, X)$ and

$$\|e^{-\omega(\cdot)q}PT(\cdot)(\lambda_0 - A)^{-\alpha}x_0\|_q \le \|g_\omega\|_q \le \frac{c_p}{2\pi}\|G_\omega\|_p \le \frac{c_p C}{2\pi},$$

where $c_p$ is the norm of the Fourier transform from $L^p(\mathbb{R}, \overline{PX})$ into $L^q(\mathbb{R}, X)$. By the monotone convergence theorem, upon letting $\omega \downarrow 0$ we obtain

$$\|PT(\cdot)(\lambda_0 - A)^{-\alpha}x_0\|_q \le \frac{c_p C}{2\pi}.$$

////

For $\alpha = n \in \mathbb{N}$ the use of fractional powers in the above proof can be avoided by considering the function $G(\lambda) := PR(\lambda, A)R(\lambda_0, A)^n x_0$. This simplifies the proof.

In Section 4.6 we will show that, at least for $p = 2$, Theorem 4.3.2 is the best possible in the sense that it fails for all $\alpha \in [0, \frac{1}{2})$.

For $P = I$ and $p \in (1, 2]$, Theorem 4.3.2 has the following consequence:

**Corollary 4.3.3.** *Let $X$ be a Banach space with Fourier type $p \in (1, 2]$, let $A$ be the generator of a $C_0$-semigroup $\mathbf{T}$ on $X$. Let $x_0 \in X$ be such that the local resolvent $\lambda \mapsto R(\lambda, A)x_0$ admits a bounded holomorphic extension $F(\lambda)$ in the open right half-plane. Then for all $\alpha > \frac{1}{p}$ and all $\operatorname{Re}\lambda_0 > \omega_0(\mathbf{T})$ we have*

$$\lim_{t\to\infty}\|T(t)(\lambda_0 - A)^{-\alpha}x_0\| = 0.$$

*Proof:* By Theorem 4.3.2 applied to the case $P = I$ we find that the function $f(t) := T(t)(\lambda_0 - A)^{-\alpha}x_0$ defines an element of $L^q(\mathbb{R}_+, X)$, $\frac{1}{p} + \frac{1}{q} = 1$. Hence by Lemma 3.1.9, $\lim_{t\to\infty}\|f(t)\| = 0$.    ////

Although we do not know whether the bound $\alpha > \frac{1}{p}$ in this corollary is optimal, the following example shows that the assertion is wrong for $\alpha = 0$, even if **T** is a positive $C_0$-semigroup on $L^p$, $1 < p < \infty$. In other words, there is *no* individual version of Weis's theorem (Theorem 3.5.3).

**Example 4.3.4.** Let $1 < p < \infty$. Define the $C_0$-semigroup **T** on $X := L^p(1, \infty)$ by $T(t)f(s) = f(se^t)$. Let $A$ be its generator. Let $Y := X \cap L^1(1, \infty)$ and let $\mathbf{T}_Y$ denote the restriction of **T** to $Y$. By Example 1.4.4, $s(A_Y) = -1$. Hence for all $y \in Y$, the map $\lambda \mapsto R(\lambda, A)y = R(\lambda, A_Y)y$ has a holomorphic $Y$-valued extension to $\{\operatorname{Re} \lambda > -1\}$. In particular, these extensions are holomorphic as $X$-valued maps.

On the other hand, $\omega_0(\mathbf{T}_Y) = -\frac{1}{p}$. By the uniform boundedness theorem, for each $\omega \in (\frac{1}{p}, 1)$ there exists a $y_\omega \in Y$ such that

$$\limsup_{t \to \infty} \|e^{\omega t} T(t) y_\omega\|_Y = \infty.$$

Since $\omega_0(\mathbf{T}_{L^1(1,\infty)}) = -1$ this implies that

$$\limsup_{t \to \infty} \|e^{\omega t} T(t) y_\omega\|_X = \infty.$$

This shows that for the semigroup **T**, Corollary 4.3.3 fails for $\alpha = 0$.

This example also shows that there is no individual Gearhart's theorem (Corollary 2.2.5). Indeed, take $p = 2$ and let $\omega \in (\frac{1}{2}, 1)$ and $y_\omega \in L^2(1, \infty)$ be as above. We have $y_\omega \in L^1(1, \infty) \cap L^2(1, \infty)$, and hence the same is true for the real and imaginary parts of $y_\omega$, and for the positive and negative parts of these. Since $\limsup_{t \to \infty} \|e^{\omega t} T(t) y_\omega\| = \infty$, the same must be true for at least one of these four positive vectors. This shows that without loss of generality we may assume that $y_\omega \geq 0$.

Consider the local resolvent $\lambda \mapsto R(\lambda, A)y_\omega$. This map is holomorphically extendable to $\{\operatorname{Re} \lambda > -1\}$. We claim that its extension $F$ is bounded in each half-plane $\{\operatorname{Re} \lambda > -1 + \epsilon\}$. To see this, note that by the Pringsheim-Landau theorem the holomorphic map

$$\lambda \mapsto \tilde{F}(\lambda) := \lim_{t \to \infty} \int_0^t e^{-\lambda s} T(s) y_\omega \, ds$$

extends holomorphically to $\{\operatorname{Re} \lambda > -1\}$; it agrees with $R(\lambda, A)y_\omega$ for $\operatorname{Re} \lambda$ large enough. It follows that $F(\lambda) = \tilde{F}(\lambda)$ on $\{\operatorname{Re} \lambda > -1\}$. Fix $\lambda \in \mathbb{C}$ with $\operatorname{Re} \lambda \geq -1 + \epsilon$. Then,

$$|F(\lambda)| = \lim_{t \to \infty} \left| \int_0^t e^{-\lambda s} T(s) y_\omega \, ds \right| \leq \lim_{t \to \infty} \int_0^t e^{(1-\epsilon)s} T(s) y_\omega \, ds = F(-1+\epsilon).$$

Therefore, $\sup_{\operatorname{Re} \lambda \geq -1+\epsilon} \|F(\lambda)\| \leq \|F(-1+\epsilon)\|$. We now obtain a counterexample to the individual Gearhart's theorem by taking $\epsilon$ so small that $-1 + \epsilon < -\omega$

and rescaling: this results in a (positive) $C_0$-semigroup $\mathbf{S}$ with generator $B$ on $L^2(1, \infty)$, along with a (positive) element $y_\omega \in L^2(1, \infty)$ such that the map $\lambda \mapsto R(\lambda, B)y_\omega$ extends to a bounded holomorphic map in the half-plane $\{\operatorname{Re} \lambda > 0\}$, although the orbit $t \mapsto S(t)y_\omega$ is unbounded.

In Section 4.5 we will show that, in spite of these counterexamples, the identity $s(A) = \omega_1(\mathbf{T})$ for positive $C_0$-semigroups *does* admit an individual version.

A Banach space is said to be $B$–*convex* if it has non-trivial type $p$, i.e. type $p \in (1, 2]$. By a result of J. Bourgain [Bo], every $B$–convex Banach space has non-trivial Fourier type, and conversely it is not difficult to show that every Banach space with non-trivial Fourier type is $B$–convex (cf. [BP, p. 354]).

**Corollary 4.3.5.** *Let $\mathbf{T}$ be a $C_0$–semigroup on a $B$–convex Banach space $X$, with generator $A$. If $x_0 \in X$ is such that the local resolvent $\lambda \mapsto R(\lambda, A)x_0$ admits a bounded holomorphic extension to the open right half-plane, then for all $\operatorname{Re} \lambda_0 > \omega_0(\mathbf{T})$ we have*
$$\lim_{t \to \infty} \|T(t)R(\lambda_0, A)x_0\| = 0.$$

We next discuss the analogue of Corollary 4.3.3 for $P \ne I$. Although the proof of Corollary 4.3.3 breaks down, for slightly larger values of $\alpha$ we can prove:

**Theorem 4.3.6.** *Let $P$ be a bounded operator on a Banach space $X$ and assume that $\overline{PX}$ is $B$–convex. Let $A$ be the generator of a $C_0$–semigroup $\mathbf{T}$ on $X$ and let $x_0 \in X$ be such that the map $\lambda \mapsto PR(\lambda, A)x_0$ extends to a bounded holomorphic function $F(\lambda)$ in the open right half-plane. Then for all $\alpha > 1$ and $\operatorname{Re} \lambda_0 > \omega_0(\mathbf{T})$ we have*
$$\lim_{t \to \infty} \|PT(t)(\lambda_0 - A)^{-\alpha} x_0\| = 0.$$

*Proof:* Let $p \in (1, 2]$ be the Fourier type of $\overline{PX}$. Choose $\delta \ge 0$ with $\delta > \frac{1}{p}$ in such a way that $\frac{1}{q} < \beta := \alpha - \delta < 1$. Consider the functions
$$f(t) := PT(t)(\lambda_0 - A)^{-\delta} x_0, \quad t \ge 0,$$
and
$$g(t) := PT(t)(\lambda_0 - A)^{-\alpha} x_0, \quad t \ge 0.$$
By Theorem 4.3.2, $f \in L^q(\mathbb{R}_+, X)$. For $t \ge 0$ we have
$$\begin{aligned} g(t) &= PT(t)(\lambda_0 - A)^{-\delta - \beta} x_0 \\ &= PT(t)(\lambda_0 - A)^{-\delta} \left( \frac{\sin \pi \beta}{\pi} \int_0^\infty s^{-\beta} R(\lambda_0 + s, A)x_0 \, ds \right) \\ &= \frac{\sin \pi \beta}{\pi} P(\lambda_0 - A)^{-\delta} \int_0^\infty s^{-\beta} \int_0^\infty e^{-(\lambda_0 + s)r} T(t+r) x_0 \, dr \, ds \\ &= \frac{\sin \pi \beta}{\pi} \int_0^\infty s^{-\beta} \int_0^\infty e^{-(\lambda_0 + s)r} f(t+r) x_0 \, dr \, ds. \end{aligned} \tag{4.3.7}$$

Now,

$$\left\| \int_0^\infty e^{-(\lambda_0+s)r} f(t+r) x_0 \, dr \right\|$$

$$\leq \left( \int_0^\infty e^{-(\lambda_0+s)rp} \, dr \right)^{\frac{1}{p}} \cdot \left( \int_0^\infty \|f(t+r)\|^q \, dr \right)^{\frac{1}{q}}$$

$$= \frac{1}{(p(\lambda_0+s))^{\frac{1}{p}}} \left( \int_t^\infty \|f(r)\|^q \, dr \right)^{\frac{1}{q}}.$$

Combining this estimate with (4.3.7) yields

$$\|g(t)\| \leq \frac{\sin \pi \beta}{\pi p^{\frac{1}{p}}} \int_0^\infty s^{-\beta} (\lambda_0+s)^{-\frac{1}{p}} \, ds \cdot \left( \int_t^\infty \|f(r)\|^q \, dr \right)^{\frac{1}{q}}.$$

Since $\frac{1}{q} < \beta < 1$, the first integral in the above expression is absolutely convergent, and the second tends to 0 as $t \to \infty$. This proves that $\lim_{t \to \infty} \|g(t)\| = 0$. ////

Theorem 4.3.6 is optimal in the sense that it fails for every $0 \leq \alpha < 1$. Indeed, consider the case that the resolvent $R(\lambda, A)$ itself is uniformly bounded in $\{\mathrm{Re}\, \lambda > 0\}$. Then the assumptions of Theorem 4.3.6 are satisfied for all $x_0 \in X$ and all operators $P = x^* \otimes x$. Hence if the theorem holds for some $\alpha \geq 0$, then from the uniform boundedness principle we conclude that

$$\sup_{t \geq 0} \|T(t)(\lambda_0 - A)^{-\alpha}\| < \infty.$$

For $0 \leq \alpha < 1$, Example 4.2.9 shows that this need not be true.

In the next section we will give an example which shows that the results of this section break down if no restrictions on the underlying Banach space are imposed.

## 4.4. Individual stability in spaces with the analytic RNP

The case $p = 1$ of Theorem 4.3.2 implies that for each $\alpha > 1$ the orbit $T(\cdot)(\lambda_0 - A)^{-\alpha} x_0$ is bounded provided the local resolvent at $x_0$ extends to a bounded holomorphic function in the open right half-plane. This can be improved in two ways: if $X$ has the analytic Radon-Nikodym property, we actually have strong convergence to 0, and if $X$ is an arbitrary Banach space we have weak convergence to 0. These results will be proved in the present section.

We start by recalling some facts concerning vector-valued Hardy spaces over the disc $D = \{z \in \mathbb{C} : |z| < 1\}$.

For $p \in [1, \infty]$ we let $H^p(D, X)$ denote the set of all holomorphic functions $f : D \to X$ for which

$$\|f\|_{H^p} := \sup_{0<r<1} \left( \int_0^{2\pi} \|f(re^{i\theta})\|^p \, d\theta \right)^{\frac{1}{p}} < \infty.$$

In case $p = \infty$ we interpret the above integral in terms of the supremum norm in the obvious way. It is not difficult to see that $H^p(D, X)$ is a Banach space with respect to the norm $\|\cdot\|_{H^p}$. We let $H_0^p(D, X)$ denote the subspace of $H^p(D, X)$ consisting of all functions $f$ for which the radial limits

$$\tilde{f}(e^{i\theta}) := \lim_{r \uparrow 1} f(re^{i\theta})$$

exist for almost all $\theta$. By Fatou's lemma,

$$\int_0^{2\pi} \|\tilde{f}(e^{i\theta})\|^p \, d\theta \leq \liminf_{r \uparrow 1} \int_0^{2\pi} \|f(re^{i\theta})\|^p \, d\theta.$$

This shows that the boundary function $\tilde{f}$, if it exists a.e., belongs to $L^p(\Gamma)$, where $\Gamma = \{z \in \mathbb{C} : |z| = 1\}$. In this case, $f$ can be recovered from $\tilde{f}$ by the Poisson integral

$$f(re^{i\theta}) = \frac{1}{2\pi} \int_0^{2\pi} \tilde{f}(e^{i\eta}) \frac{1 - r^2}{1 - 2r\cos(\theta - \eta) + r^2} \, d\eta.$$

Defining $f_r(e^{i\theta}) := f(re^{i\theta})$, as in the scalar case it follows from this representation that

$$\lim_{r \uparrow 1} \|\tilde{f} - f_r\|_{L^p(\Gamma)} = 0.$$

A Banach space $X$ has the *analytic Radon-Nikodym property* if $H_0^p(D, X) = H^p(D, X)$. Equivalently, $X$ has the analytic Radon-Nikodym property if for all $f \in H^p(D, X)$ the radial limits $\tilde{f}(e^{i\theta}) := \lim_{r \uparrow 1} f(re^{i\theta})$ exist for almost all $\theta$, and in this case we have $f_r \to \tilde{f}$ in the $L^p$-norm.

The role of the exponent $p$ needs some clarification: it can be shown that if $H_0^p(D, X) = H^p(D, X)$ holds for some $p \in [1, \infty]$, then it holds for all $p \in [1, \infty]$.

The following facts are well-known:

(i) If $X$ has the Radon-Nikodym property, then $X$ has the analytic Radon-Nikodym property;
(ii) If $X$ has the analytic Radon-Nikodym property, then $X$ contains no closed subspace isomorphic to $c_0$;
(iii) A Banach lattice $X$ has the analytic Radon-Nikodym property if and only if $X$ contains no closed subspace isomorphic to $c_0$.

It follows from (i) that every reflexive Banach space and every separable dual Banach space has the analytic Radon-Nikodym property. By (iii), the spaces $L^1(\mu)$ have the analytic Radon-Nikodym property. The proofs can be found in [Bu] and [BD].

By mapping a rectangle conformally onto the unit disc it is not difficult to prove the following.

**Proposition 4.4.1.** *Let $\Gamma$ and $\Gamma_r$, $0 < r < 1$, be the rectangles in $\mathbb{C}$ spanned by the points $\pm a \pm ib$ and $\pm ra \pm irb$, respectively. Let $f$ be a holomorphic $X$-valued function in the interior of $\Gamma$. Assume that $X$ has the analytic Radon-Nikodym property and that*

$$\sup_{0<r<1} \int_{\Gamma_r} \|f(z)\| \, |dz| < \infty.$$

*Then, the strong limits $\lim_{r \uparrow 1} f(rz)$ exist for almost all $z \in \Gamma$ and define a function $\tilde{f} \in L^1(\Gamma)$. Moreover,*

$$\lim_{r \uparrow 1} \int_{\Gamma} \|\tilde{f}(z) - f(rz)\| \, |dz| = 0.$$

////

**Theorem 4.4.2.** *Let $P$ be a bounded operator on a Banach space $X$ and assume that $\overline{PX}$ has the analytic Radon-Nikodym property. Let $A$ be the generator of a $C_0$-semigroup $\mathbf{T}$ on $X$. Assume that for some $x_0 \in X$, the map $\lambda \mapsto PR(\lambda, A)x_0$ admits a bounded holomorphic extension $F(\lambda)$ to the open right half-plane. Then for all $\operatorname{Re} \lambda_0 > \omega_0(\mathbf{T})$ and $\alpha > 1$ we have*

$$\lim_{t \to \infty} \|PT(t)(\lambda_0 - A)^{-\alpha} x_0\| = 0.$$

*Proof:* Without loss of generality we may assume that $\omega_0(\mathbf{T}) \geq 0$. Fix $\lambda_0 > \omega_0(\mathbf{T})$; by rescaling we may also assume that $\lambda_0$ is real. Fix $\omega_0 \in (\omega_0(\mathbf{T}), \lambda_0)$.

By Lemma 4.3.1, the $\overline{PX}$-valued functions $G_\omega(s) := PR(\omega - is, A)(\lambda_0 - A)^{-\alpha} x_0$ belong to $L^1(\mathbb{R}, X)$, uniformly for $\omega \in (0, \omega_0)$. By the complex inversion theorem for the Laplace transform and Cauchy's theorem (the use of which is justified by (4.3.1)), for all $\omega \in (0, \omega_0)$ we have

$$PT(t)(\lambda_0 - A)^{-\alpha} x_0 = \frac{1}{2\pi i} \int_{\omega + i\mathbb{R}} e^{\lambda t} G(\lambda) \, d\lambda = \frac{1}{2\pi} \int_{-\infty}^{\infty} e^{(\omega - is)t} G_\omega(s) \, ds, \quad (4.4.1)$$

where $G(\lambda) = PR(\lambda, A)(\lambda_0 - A)^{-\alpha} x_0$. Since $\overline{PX}$ has the analytic Radon-Nikodym property, we may apply Proposition 4.4.1 and conclude that the boundary function $\tilde{G}$ of $G$ exists a.e., defines an element in $L^1_{loc}(i\mathbb{R}, X)$, and that

$$\lim_{\omega \downarrow 0} \int_{-r}^{r} \|\tilde{G}(is) - G(\omega + is)\| \, ds = 0$$

for all $r > 0$. But then (4.3.1) easily implies that we actually have $\tilde{G} \in L^1(i\mathbb{R}, X)$ and

$$\lim_{\omega \downarrow 0} \int_{-\infty}^{\infty} \|\tilde{G}(is) - G(\omega + is)\| \, ds = 0.$$

Hence by passing to the limit $\omega \downarrow 0$ in (4.4.1), we obtain

$$PT(\cdot)(\lambda_0 - A)^{-\alpha}x_0 = \lim_{\omega \downarrow 0} e^{-\omega t} PT(\cdot)(\lambda_0 - A)^{-\alpha}x_0$$
$$= \frac{1}{2\pi} \lim_{\omega \downarrow 0} \mathcal{F}G(\omega - i(\cdot))(t)$$
$$= \frac{1}{2\pi} \mathcal{F}\tilde{G}(-i(\cdot))(t).$$

Therefore, $PT(\cdot)(\lambda_0 - A)^{-\alpha}x_0 \in C_0(\mathbb{R}_+, X)$ by the Riemann-Lebesgue lemma.
////

Taking $P := x_0^* \otimes x$ with $x \in X$ non-zero and $x_0^* \in X^*$, and noting that the one-dimensional space spanned by $x$ has the analytic Radon-Nikodym property, we obtain the following result.

**Corollary 4.4.3.** *Let $A$ be the generator of a $C_0$-semigroup $\mathbf{T}$ on a Banach space $X$. Assume that for some $x_0 \in X$ and $x_0^* \in X^*$, the map $\lambda \mapsto \langle x_0^*, R(\lambda, A)x_0 \rangle$ admits a bounded holomorphic extension $F(\lambda)$ to the open right half-plane. Then for all $\alpha > 1$ and $\operatorname{Re}\lambda_0 > \omega_0(\mathbf{T})$ we have*

$$\lim_{t\to\infty} \langle x_0^*, T(t)(\lambda_0 - A)^{-\alpha}x_0 \rangle = 0.$$

Of course, this can be proved without reference to the analytic Radon-Nikodym property: just take $P := x_0^* \otimes x$ in the proof of Theorem 4.4.2 and apply the scalar version of Proposition 4.4.1.

The following example shows that the main results of this section and the previous one fail in arbitrary Banach spaces.

**Example 4.4.4.** Let $X = C_0(\mathbb{R})$ and consider the left translation group $\mathbf{T}$ on $X$. Let $f \in X$ be any non-zero function with support in $[0, 1]$. Then for all $\operatorname{Re}\lambda > 0$ and $s \in \mathbb{R}$ we have

$$|(R(\lambda, A)f)(s)| = \left| \int_0^\infty e^{-\lambda t} f(t+s) \, dt \right| \leq \|f\|_\infty.$$

Consequently,

$$\sup_{\operatorname{Re}\lambda > 0} \|R(\lambda, A)f\|_\infty \leq \|f\|_\infty,$$

but since $\mathbf{T}$ is isometric and $(\lambda_0 - A)^{-\alpha}$ is injective (cf. Appendix A.1) we see that

$$\lim_{t\to\infty} \|T(t)(\lambda_0 - A)^{-\alpha}f\|_\infty = \|(\lambda_0 - A)^{-\alpha}f\|_\infty \neq 0; \quad \forall \alpha \geq 0, \lambda_0 > 0.$$

## 4.5. Individual stability in arbitrary Banach spaces

In the preceding sections we proved that

$$\lim_{t\to\infty} \|T(t)R(\lambda_0, A)x_0\| = 0$$

if the local resolvent $\lambda \mapsto R(\lambda, A)x_0$ extends to a bounded holomorphic function in the open right half-plane and the underlying space $X$ is either $B$−convex or has the analytic Radon-Nikodym property, and by means of an example we showed that this result fails if no restrictions on $X$ are imposed. Also, for arbitrary $X$ we proved that

$$\lim_{t\to\infty} \langle x^*, T(t)(\lambda_0 - A)^{-\alpha} x_0\rangle = 0$$

for all $x^* \in X^*$ and $\alpha > 1$. It thus remains open what can be said for $\alpha = 1$, i.e. about the behaviour of $T(t)R(\lambda_0, A)x_0$, in arbitrary Banach spaces. In this direction, we will prove next that there exists a constant $M > 0$ such that

$$\|T(t)R(\lambda_0, A)x_0\| \le M(1+t), \quad t \ge 0.$$

We start with a simple lemma.

**Lemma 4.5.1.** *For all $r > 0$ and $t > 0$,*

$$\left| \int_r^\infty \frac{e^{i\lambda t}}{\lambda} d\lambda \right| \le \frac{3\pi}{2rt}.$$

*Proof:* Integrate $z \mapsto z^{-1} e^{itz}$ along the closed contour consisting of the semi-circle $\Gamma_r$ of radius $r$ in the upper half-plane, the interval $[r, R]$, the semi-circle $\Gamma_R$, and the interval $[-R, -r]$. By letting $R \to \infty$, we find that

$$\left| \int_r^\infty \frac{\sin \lambda t}{\lambda} d\lambda \right| \le \left| \frac{1}{2i} \int_{\Gamma_r} \frac{e^{izt}}{z} dz \right| \le \frac{1}{2} \int_0^\pi e^{-rt \sin\theta} d\theta \le \frac{\pi}{2rt},$$

in the last estimate using the obvious facts that $\sin(\pi - \theta) = \sin\theta$ and $\sin\theta \ge \frac{2\theta}{\pi}$ for all $0 \le \theta \le \frac{\pi}{2}$. We also have

$$\left| \int_r^\infty \frac{\cos \lambda t}{\lambda} d\lambda \right| = \left| \int_{rt}^\infty \frac{\sin(\tau + \frac{\pi}{2})}{\tau} d\tau \right|$$
$$\le \frac{\pi}{2(rt + \frac{\pi}{2})} + \int_{rt}^\infty \left|\sin(\tau + \frac{\pi}{2})\right| \left(\frac{1}{\tau} - \frac{1}{\tau + \frac{\pi}{2}}\right) d\tau \le \frac{\pi}{rt}.$$

From these two estimates, the lemma follows. ////

**Theorem 4.5.2.** *Let $A$ be the generator of a $C_0$–semigroup $\mathbf{T}$ on a Banach space $X$ and let $P$ be a bounded operator on $X$. Let $x_0 \in X$ be such that the map $\lambda \mapsto PR(\lambda, A)x_0$ admits a bounded holomorphic extension $F(\lambda)$ to the open right half-plane. Let $\operatorname{Re}\lambda_0 > \omega_0(\mathbf{T})$. Then there exists a constant $M > 0$ such that*

$$\|PT(t)R(\lambda_0, A)x_0\| \leq M(1+t), \quad t \geq 0.$$

*Proof:* Choose $N > 0$ and $\omega > \omega_0(\mathbf{T})$ such that $\|T(t)\| \leq Ne^{\omega t}$ for all $t \geq 0$. Choose $K > 0$ in such a way that $\sup_{\operatorname{Re}\lambda > 0} \|F(\lambda)\| \leq K\|P\|\,\|x_0\|$.

First we assume that $\lambda_0 = \omega + 1$; the case of a general $\operatorname{Re}\lambda_0 > \omega_0(\mathbf{T})$ will be reduced to this special case by a rescaling procedure.

Put $A_{\omega+1} := A - \omega - 1$ and let $F_0(\lambda)$ denote the bounded holomorphic extension of $PR(\lambda, A_{\omega+1})x_0$ to $\{\operatorname{Re}\lambda > -\omega - 1\}$.

Fix $t > 0$. By Theorem 1.3.3 and the resolvent identity, for all $\xi > \omega_0(\mathbf{T}_{\omega+1})$ we have

$$PT_{\omega+1}(t)A_{\omega+1}^{-1}x_0 = \frac{1}{2\pi i}P\int_{\xi+i\mathbb{R}} e^{\lambda t}R(\lambda, A_{\omega+1})A_{\omega+1}^{-1}x_0\,d\lambda$$
$$= \frac{1}{2\pi i}\int_{\xi+i\mathbb{R}} e^{\lambda t}\lambda^{-1}(PA_{\omega+1}^{-1}x_0 + F_0(\lambda))\,d\lambda.$$

By Cauchy's theorem, we may shift the path of integration to $\Gamma = \Gamma_1 \cup \Gamma_2 \cup \Gamma_3 \cup \Gamma_4 \cup \Gamma_5$, where

$$\Gamma_1 = \{z = i\eta :\ \eta \leq -r\};$$
$$\Gamma_2 = \{z = \xi + i\eta :\ -\delta \leq \xi \leq 0,\ \eta = -r\};$$
$$\Gamma_3 = \{z = \xi + i\eta :\ \xi = -\delta,\ -r \leq \eta \leq r\};$$
$$\Gamma_4 = \{z = \xi + i\eta :\ -\delta \leq \xi \leq 0,\ \eta = r\};$$
$$\Gamma_5 = \{z = i\eta :\ \eta \geq r\}.$$

Here, $\delta \in (0, \omega+1)$ is arbitrary and fixed, and $r > 0$ is to be chosen later. We are going to estimate the integrals over $\Gamma_i$, $i = 1, ..., 5$, separately.

We start with the integral over $\Gamma_1$. From $\|T(t)\| \leq Ne^{\omega t}$, for all $x^* \in X^*$ we have

$$\left(\int_0^\infty |\langle x^*, PT_{\omega+1}(t)x_0\rangle|^2\,dt\right)^{\frac{1}{2}} \leq \frac{N}{\sqrt{2}}\|P\|\,\|x_0\|\,\|x^*\|.$$

Hence by the Plancherel theorem,

$$\left(\frac{1}{2\pi}\int_{-\infty}^\infty |\langle x^*, PR(i\eta, A_{\omega+1})x_0\rangle|^2\,d\eta\right)^{\frac{1}{2}} \leq \frac{N}{\sqrt{2}}\|P\|\,\|x_0\|\,\|x^*\|.$$

Therefore, by Lemma 4.5.1 and Hölder's inequality,

$$\left| \int_{\Gamma_1} e^{\lambda t} \lambda^{-1} (\langle x^*, PA_{\omega+1}^{-1} x_0 + F_0(\lambda) \rangle) \, d\lambda \right|$$

$$= \left| \int_{\Gamma_1} e^{\lambda t} \lambda^{-1} \langle x^*, PA_{\omega+1}^{-1} x_0 + PR(\lambda, A_{\omega+1}) x_0 \rangle \, d\lambda \right|$$

$$\leq \frac{3\pi}{2rt} \|P\| \|A_{\omega+1}^{-1} x_0\| \|x^*\| + N\sqrt{\pi} \, \|P\| \|x_0\| \|x^*\| \cdot \left( \int_r^\infty \frac{1}{\eta^2} \, d\eta \right)^{\frac{1}{2}}$$

$$\leq \frac{3\pi N}{2rt} \|P\| \|x_0\| \|x^*\| + \frac{N\sqrt{\pi}}{r^{\frac{1}{2}}} \|P\| \|x_0\| \|x^*\|,$$

noting that $\|A_{\omega+1}^{-1}\| = \|R(\omega+1, A)\| \leq \int_0^\infty Ne^{-t} \, dt = N$.

The same estimate holds for the integral over $\Gamma_5$. Also, we have

$$\left\| \int_{\Gamma_2} e^{\lambda t} \lambda^{-1} (PA_{\omega+1}^{-1} x_0 + F_0(\lambda)) \, dz \right\| \leq \delta \, r^{-1} (N+K) \|P\| \|x_0\|,$$

and the same estimate holds for the integral over $\Gamma_4$. Finally, for the integral over $\Gamma_3$ we have

$$\left\| \int_{\Gamma_3} e^{\lambda t} z^{-1} (PA_{\omega+1}^{-1} x_0 + F_0(\lambda)) \, d\lambda \right\|$$

$$\leq e^{-\delta t} (N+K) \|P\| \|x_0\| \cdot \int_{-r}^r \frac{1}{|-\delta + i\eta|} \, d\eta$$

$$\leq 2(N+K) e^{-\delta t} \log(1 + \frac{r}{\delta}) \|P\| \|x_0\|.$$

Putting everything together, we find

$$|\langle x^*, PT_{\omega+1}(t) A_{\omega+1}^{-1} x_0 \rangle|$$

$$\leq \left( \frac{3\pi N}{rt} + \frac{2N\sqrt{\pi}}{r^{\frac{1}{2}}} + \frac{2\delta(N+K)}{r} + \frac{2(N+K)}{e^{\delta t}} \log\left(1 + \frac{r}{\delta}\right) \right) \|P\| \|x_0\| \|x^*\|.$$

So far, $r > 0$ and $x^* \in X^*$ were arbitrary. For fixed $t > 0$ we now take $r = e^{2(\omega+1)t}$. This yields

$$\|PT_{\omega+1}(t) A_{\omega+1}^{-1} x_0\| \leq \Big( \frac{3\pi N}{te^{2(\omega+1)t}} + \frac{2N\sqrt{\pi}}{e^{(\omega+1)t}} + \frac{2\delta(N+K)}{e^{2(\omega+1)t}}$$

$$+ \frac{2(N+K)}{e^{\delta t}} \log\Big(1 + \frac{e^{2(\omega+1)t}}{\delta}\Big) \Big) \|P\| \|x_0\|.$$

Letting $\delta \to \omega + 1$, it follows that for all $t \geq 1$ we have

$$\|PT(t) R(\omega+1, A) x_0\| = e^{(\omega+1)t} \|PT_{\omega+1}(t) A_{\omega+1}^{-1} x_0\|$$

$$\leq \Big( 3\pi N + 2N\sqrt{\pi} + 2(\omega+1)(N+K)$$

$$+ 2(N+K) \log\Big(1 + \frac{e^{2(\omega+1)t}}{\omega+1}\Big) \Big) \|P\| \|x_0\|$$

$$\leq \big( 13N + 2(\omega+1)(N+K)(1+2t) \big) \|P\| \|x_0\|.$$

This proves the theorem for $\lambda_0 = \omega + 1$.

The general case follows by rescaling: first we may assume that $\lambda_0 \in \mathbb{R}$ and then we apply the special case proved above to the semigroup $\mathbf{T}^{(\epsilon)} = \{T(\epsilon t)\}_{t \geq 0}$, where $\epsilon = (\lambda_0 - \omega)^{-1}$. ////

Note that an alternative proof of Corollary 4.2.7 is obtained by combining Theorem 4.5.2 with the uniform boundedness theorem.

The following proposition gives a sufficient condition for boundedness of $PR(\lambda, A)x_0$ in the right half-plane.

**Proposition 4.5.3.** *Let $\mathbf{T}$ be a $C_0$-semigroup on a Banach space $X$, with generator $A$. Let $P$ be a bounded operator on $X$. If, for some $x_0 \in X$,*

$$\sup_{\omega \in \mathbb{R}} \sup_{s>0} \left\| \int_0^s e^{-i\omega t} PT(t)x_0 \, dt \right\| < \infty,$$

*then the map $\lambda \mapsto PR(\lambda, A)x_0$ admits a bounded holomorphic extension to $\{\operatorname{Re}\lambda > 0\}$.*

*Proof:* We use the technique of Theorem 3.3.1. For reasons of completeness we give the details. Choose a constant $K > 0$ such that

$$\sup_{\omega \in \mathbb{R}} \sup_{s>0} \left\| \int_0^s e^{-i\omega t} PT(t)x_0 \, dt \right\| \leq K \|P\| \|x_0\|.$$

Consider the entire functions $F_s(\lambda) = \int_0^s e^{-\lambda t} PT(t)x_0 \, dt$. By assumption, each $F_s$ is bounded on the imaginary axis, with bound $K\|P\|\|x_0\|$. Also, a simple estimate shows that each $F_s$ is bounded on vertical lines. Choose constants $N > 0$ and $\omega_0 \geq 0$ such that $\|T(t)\| \leq Ne^{\omega_0 t}$ for all $t \geq 0$, and let $\xi = \omega_0 + 1$. Then,

$$\|F_s(\xi + i\eta)\| \leq \int_0^s e^{-\xi t} \|PT(t)x_0\| \, dt \leq \int_0^s Ne^{-t} \|P\| \|x_0\| \, dt \leq N\|P\|\|x_0\|.$$

Therefore, by the Phragmen-Lindelöf theorem, each $F_s$ is uniformly bounded in the strip $S_\xi := \{0 \leq \operatorname{Re}\lambda \leq \xi\}$, with bound $\max\{K, N\}\|P\|\|x_0\|$. Moreover, for $\operatorname{Re}\lambda > \omega_0$ we have

$$\lim_{s \to \infty} F_s(\lambda) = PR(\lambda, A)x_0.$$

By Vitali's theorem, the limit $\lim_{s \to \infty} F_s(\lambda)$ exists for all $\lambda \in S_\xi$, the convergence being uniformly on compacta. The limit function $F$ is analytic in the interior of $S_\xi$ and coincides with $PR(\lambda, A)x_0$ for $\omega_0 < \operatorname{Re}\lambda \leq \xi$. Moreover, $F$ is uniformly bounded in $S_\xi$, with bound $\max\{K, N\}\|P\|\|x_0\|$. This proves that $PR(\lambda, A)x_0$ admits a bounded holomorphic extension $F$ to the interior of $S_\xi$. By the Hille-Yosida Theorem, $PR(\lambda, A)x_0$ is also uniformly bounded in $\{\operatorname{Re}\lambda \geq \xi\}$. Therefore, $F$ is uniformly bounded in $\{\operatorname{Re}\lambda > 0\}$. ////

We recall from Theorem 1.2.3 that the growth bound $\omega_1(\mathbf{T})$ and the abscissa of improper convergence of the Laplace transform of $\mathbf{T}$ coincide. The following result is a quantitative version of this. The reader should compare this result with Corollary 3.4.3 and the remarks following it.

**Corollary 4.5.4.** *Let $\mathbf{T}$ be a $C_0$–semigroup on a Banach space $X$, with generator $A$. If, for all $x \in X$ and $x^* \in X^*$,*

$$\sup_{\omega \in \mathbb{R}} \sup_{s>0} \left| \int_0^s e^{-i\omega t} \langle x^*, T(t)x \rangle \, dt \right| < \infty,$$

*then $\omega_1(\mathbf{T}) < 0$.*

Finally we give and individual version of the identity $s(A) = \omega_1(\mathbf{T})$ for positive $C_0$–semigroups. In order to treat the strong- and the weak case simultaneously, we formulate it in terms of a positive operator $P$.

**Corollary 4.5.5.** *Let $\mathbf{T}$ be a positive $C_0$–semigroup on a Banach lattice $X$, with generator $A$. Let $P \geq 0$ be a positive operator on $X$. If, for some $0 \leq x_0 \in X$, the map $\lambda \mapsto PR(\lambda, A)x_0$ has a holomorphic extension to $\{\operatorname{Re} \lambda > 0\}$, then for all $\operatorname{Re} \lambda_0 > \omega_0(\mathbf{T})$ the map $t \mapsto PT(t)R(\lambda_0, A)x_0$ has exponential type less than or equal to 0.*

*Proof:* Let $\omega_1(P, x_0)$ denote the abscissa of simple convergence of the Laplace transform of the positive function $t \mapsto PT(t)x_0$. By the Pringsheim-Landau theorem (Theorem 1.3.4), $\omega_1(P, x_0)$ is a singular point for the holomorphic function

$$\lambda \mapsto \lim_{s \to \infty} \int_0^s e^{-\lambda t} PT(t)x_0 \, dt.$$

Therefore, the fact that $PR(\lambda, A)x_0$ has an extension $F(\lambda)$ to $\{\operatorname{Re} \lambda > 0\}$ implies that $\omega_1(P, x_0) \leq 0$. Then it is evident that for all $\lambda_0, \lambda_1 \in \mathbb{C}$ with $\operatorname{Re} \lambda_1 \geq \operatorname{Re} \lambda_0 > 0$,

$$|F(\lambda_1)| = \left| \lim_{s \to \infty} \int_0^s e^{-\lambda_1 t} PT(t)x_0 \, dt \right|$$
$$\leq \lim_{s \to \infty} \int_0^s e^{-\operatorname{Re} \lambda_1 t} PT(t)x_0 \, dt$$
$$\leq \lim_{s \to \infty} \int_0^s e^{-\operatorname{Re} \lambda_0 t} PT(t)x_0 \, dt = F(\operatorname{Re} \lambda_0).$$

This implies that $F(\lambda)$ is uniformly bounded in each half-plane $\{\operatorname{Re} \lambda > \epsilon\}$. It then follows from Theorem 4.5.2 that $t \mapsto PT(t)R(\lambda_0, A)x_0$ has exponential type $\leq \epsilon$ for each $\epsilon > 0$. ////

## 4.6. Scalarly integrable semigroups

In this section we study the stability properties of scalarly integrable semigroups.

We start by showing that the resolvent of the generator of a $C_0$-semigroup, all of whose weak orbits belong to certain rearrangement invariant Banach function spaces, is uniformly bounded in the right half-plane. The proof is based on the following simple estimate for the Laplace transform in rearrangement invariant Banach function spaces. We will use freely the results about this class of spaces that are collected in Appendix A.4.

**Lemma 4.6.1.** *Let $E$ be a rearrangement invariant Banach function space over $\mathbb{R}_+$. Then the Laplace transform of elements of $E$ converges absolutely in the right half-plane $\{\operatorname{Re}\lambda > 0\}$ and*

$$\left|\int_0^\infty e^{-\lambda t} f(t)\, dt\right| \le \frac{1}{1-e^{-1}} \varphi_{E'}\left(\frac{1}{\operatorname{Re}\lambda}\right) \|f\|_E, \qquad \forall f \in E,\ \operatorname{Re}\lambda > 0.$$

*Proof:* The absolute convergence of the Laplace transform in the open right half-plane follows from the fact that $E \subset L^1 + L^\infty$. For $\mu > 0$, define the function $e_\mu : \mathbb{R}_+ \to \mathbb{R}_+$ by $e_\mu(t) = e^{-\mu t}$. Then for all $s > 0$ we have

$$0 \le e_\mu \le \chi_{[0,s]} + e_\mu|_{(s,\infty)}.$$

Let $E'$ denote the associate space of $E$. By the rearrangement invariance of $E'$ we have

$$\|e_\mu|_{(s,\infty)}\|_{E'} = e^{-\mu s}\|e_\mu\|_{E'}.$$

Therefore,

$$\|e_\mu\|_{E'} \le \varphi_{E'}(s) + e^{-\mu s}\|e_\mu\|_{E'}.$$

By taking $s = \mu^{-1}$ we obtain $\|e_\mu\|_{E'} \le \varphi_{E'}(\mu^{-1}) + e^{-1}\|e_\mu\|_{E'}$ and hence

$$\|e_\mu\|_{E'} \le \frac{1}{1-e^{-1}}\varphi_{E'}(\mu^{-1}).$$

Now fix $f \in E$ and $\lambda \in \mathbb{C}$ with $\operatorname{Re}\lambda > 0$ arbitrary. By Hölder's inequality,

$$\left|\int_0^\infty e^{-\lambda t} f(t)\, dt\right| \le \|e_{\operatorname{Re}\lambda}\|_{E'}\|f\|_E \le \frac{1}{1-e^{-1}} \varphi_{E'}((\operatorname{Re}\lambda)^{-1})\|f\|_E.$$

////

**Theorem 4.6.2.** *Let $\mathbf{T}$ be a $C_0$-semigroup on a Banach space $X$, with generator $A$. Let $E$ be a rearrangement invariant Banach function space over $\mathbb{R}_+$ with $\lim_{t\to\infty} \varphi_E(t) = \infty$. If, for all $x \in X$ and $x^* \in X^*$, the map $t \mapsto \langle x^*, T(t)x\rangle$ defines an element of $E$, then $R(\lambda, A)$ is uniformly bounded in the right half-plane. In particular, $\mathbf{T}$ is exponentially stable.*

*Proof:* Since the Laplace transforms of $\langle x^*, T(\cdot)x\rangle$ are holomorphic in $\{\operatorname{Re}\lambda > 0\}$, it follows from Lemma 1.1.6 that $\{\operatorname{Re}\lambda > 0\} \subset \varrho(A)$ and
$$\langle x^*, R(\lambda, A)x\rangle = \int_0^\infty e^{-\lambda t}\langle x^*, T(t)x\rangle\, dt, \quad x \in X,\ x^* \in X^*,\ \operatorname{Re}\lambda > 0.$$
By Lemma 4.6.1 this implies
$$\|R(\lambda, A)\| \leq \frac{C}{1-e^{-1}}\varphi_{E'}((\operatorname{Re}\lambda)^{-1}), \quad \operatorname{Re}\lambda > 0, \tag{4.6.1}$$
where $C > 0$ is such that $\|\langle x^*, T(\cdot)x\rangle\|_E \leq C\|x\|\,\|x^*\|$ for all $x \in X$ and $x^* \in X^*$. Note that such $C$ exists by virtue of Lemma 3.1.2. Since $\lim_{t\to\infty}\varphi_E(t) = \infty$, there is a $t_0 > 0$ such that $\varphi_E(t_0^{-1}) \geq 2C(1-e^{-1})^{-1}$. Then by (4.6.1) and the identity $\varphi_E(t)\varphi_{E'}(t) = t$ (A.4.1), for all $s \in \mathbb{R}$ we have $\|R(t_0+is, A)\| \leq (2t_0)^{-1}$. Moreover, by the resolvent identity, for all $0 < t \leq t_0$ and $s \in \mathbb{R}$ we have
$$\|R(t+is, A)\| = \left\|\sum_{k=0}^\infty (t_0-t)^k R(t_0+is, A)^{k+1}\right\| \leq \sum_{k=0}^\infty t_0^k \cdot \frac{1}{(2t_0)^{k+1}} = \frac{1}{t_0}. \tag{4.6.2}$$
By (4.6.1) and (4.6.2), the resolvent is uniformly bounded in $\{\operatorname{Re}\lambda > 0\}$.  ////

By going through the above argument for $\phi(t) = t^p$ and then applying Lemma 2.3.4 (to $\Omega = \{\operatorname{Re}\lambda = t_0\}$, using the estimate (4.6.2), and observing that the constant 2 in the lemma can be improved to $1 + \epsilon$) one can prove that the condition
$$\left(\int_0^\infty |\langle x^*, T(t)x\rangle|^p\, dt\right)^{\frac{1}{p}} \leq C\|x\|\,\|x^*\|, \quad \forall x \in X,\ x^* \in X^*,$$
implies $s_0(A) \leq -1/(pC)$. This is due to G. Weiss and can be considered as a weak analogue of Theorem 3.1.8.

As an application of Theorem 4.6.2 we prove a weak analogue of Rolewicz's theorem. In contrast to the situation of integrability of $\|T(\cdot)x\|$ with respect to some function $\phi$, integrability with respect to $\phi$ of the weak orbits does not imply their boundedness. Because of this we restrict ourselves to uniformly bounded semigroups.

**Theorem 4.6.3.** *Let $\phi : \mathbb{R}_+ \to \mathbb{R}_+$ be non-decreasing with $\phi(t) > 0$ for all $t > 0$.*
*(i) If* **T** *is a uniformly bounded $C_0$–semigroup on a Banach space $X$ such that*
$$\int_0^\infty \phi(|\langle x^*, T(t)x\rangle|)\, dt < \infty$$
*for all norm one vectors $x \in X$ and $x^* \in X^*$, then* **T** *is exponentially stable.*
*(ii) If* **T** *is a uniformly $C_0$–semigroup on a Hilbert space $H$ such that*
$$\int_0^\infty \phi(|\langle x^*, T(t)x\rangle|)\, dt < \infty$$
*for all norm one vectors $x \in H$ and $x^* \in H$, then* **T** *is uniformly exponentially stable.*

*Proof:* By Lemma 3.2.1, the maps $t \mapsto \langle x^*, T(t)x \rangle$ define elements of an Orlicz space $E$ satisfying $\lim_{t\to\infty} \varphi_E(t) = \infty$. By linearity, the same holds for arbitrary $x \in X$ and $x^* \in X^*$, resp. $x \in H$ and $x^* \in H$. By Theorem 4.6.2, the resolvent of $A$ exists and is uniformly bounded in the right half-plane. Therefore (i) and (ii) follow from Corollaries 4.2.7 and 2.2.5, respectively.    ////

Condition in (i) does not imply uniform exponential stability, even if $\phi(t) = t$, **T** is uniformly bounded, and $X$ is uniformly convex. A counterexample is obtained by rescaling the semigroup of Example 1.4.4 with $e^{\frac{t}{q}}$; the resulting semigroup **S** is uniformly bounded and $\omega_0(\mathbf{S}) = 0$. On the other hand, since for its generator $B$ we have $s(B) = \frac{1}{q} - \frac{1}{p} < 0$, Theorem 4.6.5 below shows that **S** is scalarly integrable with respect to $\phi(t) = t$.

The content of the following theorem is that a uniformly bounded $C_0$-semigroup whose growth bound $\omega_1(\mathbf{T})$ is non-negative has weak orbits that decay arbitrarily slowly. It can be regarded as a weak analogue of Lemma 3.1.7.

**Theorem 4.6.4.** *Let* **T** *be a uniformly bounded $C_0$-semigroup on a Banach space $X$ with $\omega_1(\mathbf{T}) \geq 0$. Then there exists an $\epsilon > 0$ with the following property. For each $t > 0$, there exist norm one vectors $x \in X$ and $x^* \in X^*$ such that*

$$\mathrm{meas}\,\{|\langle x^*, T(\cdot)x \rangle| \geq \epsilon\} \geq t.$$

*The same conclusion holds for uniformly bounded $C_0$-semigroups* **T** *on Hilbert spaces with $\omega_0(\mathbf{T}) \geq 0$.*

*Proof:* Assume the contrary. Then, for each $n = 1, 2, ...$ we have

$$\mu_n := \sup\left\{\mathrm{meas}\,\left\{|\langle x^*, T(\cdot)x \rangle| \geq \frac{1}{n+1}\right\}\right\} < \infty,$$

where the supremum is taken over all norm one vectors $x \in X$ and $x^* \in X^*$. Note that $0 < \mu_1 \leq \mu_2 \leq ...$ . Define the function $\phi: \mathbb{R}_+ \to \mathbb{R}_+$ by

$$\phi(t) = \begin{cases} \mu_1^{-1}, & t \geq \tfrac{1}{2}; \\ (n+1)^{-2}\mu_{n+1}^{-1}, & (n+2)^{-1} \leq t < (n+1)^{-1},\ n = 1, 2, ...; \\ 0, & t = 0. \end{cases}$$

Then $\phi$ is non-decreasing and strictly positive away from 0. Fix $x \in X$ and $x^* \in X^*$ of norm one arbitrary and put $E_0 := \{|\langle x^*, T(t)x \rangle| \geq \tfrac{1}{2}\}$, $E_n := \{(n+2)^{-1} \leq |\langle x^*, T(t)x \rangle| < (n+1)^{-1}\}$, $n = 1, 2, ...$ Then $\mathrm{meas}\, E_n \leq \mu_{n+1}$, $n = 0, 1, 2, ...$, and

$$\int_0^\infty \phi(|\langle x^*, T(t)x \rangle|)\, dt = \sum_{n=0}^\infty \int_{E_n} \phi(|\langle x^*, T(t)x \rangle|)\, dt$$

$$\leq \sum_{n=0}^\infty \mu_{n+1} \cdot \frac{1}{(n+1)^2 \mu_{n+1}} < \infty.$$

Therefore we can apply Theorem 4.6.3.    ////

For positive semigroups we have the following simple result.

**Theorem 4.6.5.** *For a positive $C_0$-semigroup on a Banach lattice $X$, with generator $A$, the following assertions are equivalent:*

(i) $s(A) < 0$;

(ii) *For all $x \in X$ and $x^* \in X^*$ we have $\int_0^\infty |\langle x^*, T(t)x\rangle|\, dt < \infty$.*

*Proof:* If $s(A) = \omega_1(\mathbf{T}) < 0$, then Theorem 1.2.3 shows that for all $x \in X$ and $x^* \in X^*$ we have
$$\int_0^\infty |\langle x^*, T(t)x\rangle|\, dt \leq \int_0^\infty \langle |x^*|, T(t)|x|\rangle\, dt = \langle |x^*|, R(0,A)|x|\rangle < \infty.$$

Conversely, (ii) implies (i) by Theorem 4.6.2. ////

We call a $C_0$-semigroup $\mathbf{T}$ on $X$ *scalarly $p$-integrable* if
$$\int_0^\infty |\langle x^*, T(t)x\rangle|^p\, dt < \infty, \quad \forall x \in X,\ x^* \in X^*.$$

Instead of 'scalarly 1-integrable', we shall simply say 'scalarly integrable'. Theorem 4.6.2 shows that $\omega_1(\mathbf{T}) < 0$ if $\mathbf{T}$ is scalarly $p$-integrable for some $1 \leq p < \infty$. In terms of the fractional growth bounds $\omega_\alpha(\mathbf{T})$, this result can be improved. To this we turn next.

**Lemma 4.6.6.** *Let $\mathbf{T}$ be $C_0$-semigroup on a Banach space $X$ with uniformly bounded resolvent in the right half-plane. Let $p \in [1,2]$, $\alpha \geq 0$, and $\mathrm{Re}\,\lambda_0 > \omega_0(\mathbf{T})$ be such that*
$$\int_0^\infty |\langle x^*, T(t)(\lambda_0 - A)^{-\alpha}x\rangle|^p\, dt < \infty, \quad \forall x \in X,\ x^* \in X^*.$$
*Then for all $\beta > \frac{1}{p}$ we have*
$$\sup_{t \geq 0} \|T(t)(\lambda_0 - A)^{-\alpha - \beta}\| < \infty.$$

*Proof:* For $x \in X$ and $x^* \in X^*$ put
$$f_{x,x^*}(t) := \langle x^*, T(t)(\lambda_0 - A)^{-\alpha}x\rangle, \quad t \geq 0.$$
Then $f_{x,x^*} \in L^p(\mathbb{R}_+)$ and by Lemma 3.1.2 there exists a constant $C > 0$ such that $\|f_{x,x^*}\|_p \leq C\|x\|\,\|x^*\|$ for all $x \in X$ and $x^* \in X^*$. By the Hausdorff-Young theorem, $s \mapsto \langle x^*, R(is,A)(\lambda_0 - A)^{-\alpha}x\rangle \in L^q(\mathbb{R})$, $\frac{1}{p} + \frac{1}{q} = 1$. Hence for all $\beta > \frac{1}{p}$ and $\omega > 0$, by Hölder's inequality the function
$$g_{\omega,x,x^*}(s) := (\omega + is)^{-\beta}\langle x^*, R(-is,A)(\lambda_0 - A)^{-\alpha}x\rangle$$

belongs to $L^1(\mathbb{R})$. In particular, the Fourier transforms $\mathcal{F}g_{\omega,x,x^*}$ are bounded. We claim that
$$\mathcal{F}g_{\omega,x,x^*}(t) = \langle x^*, T(t)(\lambda_0 - A)^{-\alpha-\beta}x\rangle, \quad \text{a.a. } t > 0.$$
Indeed, for $t > 0$ we have, with $A_\omega := A - \omega$,
$$\frac{1}{2\pi}\mathcal{F}g_{\omega,x,x^*}(t) = \frac{1}{2\pi}\int_{-\infty}^{\infty} e^{-ist}(\omega+is)^{-\beta}\langle x^*, R(-is, A)(\lambda_0 - A)^{-\alpha}x\rangle\, ds$$
$$= \frac{1}{2\pi i}e^{\omega t}\int_{\operatorname{Re}\lambda = -\omega} e^{\lambda t}(-\lambda)^{-\beta}\langle x^*, R(\lambda, A_\omega)(\lambda_0 - A)^{-\alpha}x\rangle\, d\lambda$$

If $x \in D(A^2) = D(A_\omega^2)$, then by Lemma 4.2.3 the right most hand equals
$$e^{\omega t}\langle x^*, T_\omega(t)(-A_\omega)^{-\beta}(\lambda_0 - A)^{-\alpha}x\rangle = \langle x^*, T(t)(\omega - A)^{-\beta}(\lambda_0 - A)^{-\alpha}x\rangle.$$

For general $x \in X$ we choose a sequence $x_n \to x$ with $x_n \in D(A^2)$ for all $n$. Then $f_{x_n,x^*} \to f_{x,x^*}$ in $L^p(\mathbb{R}_+)$ for all $x^* \in X^*$, hence $g_{\omega,x_n,x^*} \to g_{\omega,x,x^*}$ in $L^1(\mathbb{R})$, and so $\mathcal{F}g_{\omega,x_n,x^*} \to \mathcal{F}g_{\omega,x,x^*}$ in $L^\infty(\mathbb{R})$. Therefore, for almost all $t > 0$,
$$\frac{1}{2\pi}\mathcal{F}g_{\omega,x,x^*}(t) = \lim_{n\to\infty}\frac{1}{2\pi}\mathcal{F}g_{\omega,x_n,x^*}(t)$$
$$= \lim_{n\to\infty}\langle x^*, T(t)(\omega - A)^{-\beta}(\lambda_0 - A)^{-\alpha}x_n\rangle$$
$$= \langle x^*, T(t)(\omega - A)^{-\beta}(\lambda_0 - A)^{-\alpha}x\rangle.$$

This proves the claim. If follows that $t \mapsto \langle x^*, T(t)(\omega - A)^{-\beta}(\lambda_0 - A)^{-\alpha}x\rangle$ is bounded. The theorem follows from this by taking $\omega = \lambda_0$ and applying the uniform boundedness theorem. ////

This lemma implies that, at least for $p = 2$, Theorem 4.3.2 is optimal in the sense that it fails for all $\alpha \in [0, \frac{1}{2})$:

**Example 4.6.7.** Fix $\alpha \in [0, \frac{1}{2})$ and choose $\beta > \frac{1}{2}$ such that $\alpha + \beta < 1$. By Example 4.2.9, there exists a $C_0$-semigroup, with generator $A$, on a Banach space $X$, such that
(i) The resolvent $R(\lambda, A)$ is uniformly bounded in $\{\operatorname{Re}\lambda > 0\}$;
(ii) $\omega_{\alpha+\beta}(\mathbf{T}) > 0$.

Now suppose Theorem 4.3.2 holds for $p = 2$ and the above value of $\alpha$. Then by (i) for all $\lambda_0 > \omega_0(\mathbf{T})$ we have
$$\int_0^\infty |\langle x^*, T(t)(\lambda_0 - A)^{-\alpha}x\rangle|^2\, dt < \infty, \quad \forall x \in X,\, x^* \in X^*.$$
Hence from Lemma 4.6.6 we see that
$$\sup_{t\geq 0}\|T(t)(\lambda_0 - A)^{-\alpha-\beta}\| < \infty.$$
It follows that $\omega_{\alpha+\beta}(\mathbf{T}) \leq 0$, which contradicts (ii).

**Theorem 4.6.8.** Let $\mathbf{T}$ be a $C_0$-semigroup which is scalarly $p$-integrable for some $1 \leq p < \infty$. Then $\omega_\alpha(\mathbf{T}) < 0$ for all $\alpha > \max\{\frac{1}{p}, \frac{1}{2}\}$ and $\omega_{\max\{\frac{1}{p}, \frac{1}{2}\}}(\mathbf{T}) \leq 0$.

*Proof:* We start by observing that $s_0(A) < 0$ by Theorem 4.6.2 and Lemma 2.3.4. Hence, by the Weis-Wrobel convexity theorem and the fact that $\omega_1(\mathbf{T}) < 0$, it is enough to show that $\omega_\beta(\mathbf{T}) \leq 0$ for all $\beta > \max\{\frac{1}{p}, \frac{1}{2}\}$. For $1 \leq p \leq 2$ this follows from Lemma 4.6.6 by taking $\alpha = 0$. If $2 < p < \infty$, then for all $\epsilon > 0$ the rescaled semigroup $\mathbf{T}_\epsilon$ is scalarly square integrable and we can apply what we already know for $p = 2$ to see that $\omega_\beta(\mathbf{T}_\epsilon) \leq s_0(A_\epsilon)$ for all $\beta > \frac{1}{2}$. In view of Proposition A.1.2 this implies $\omega_\beta(\mathbf{T}) \leq s_0(A)$ for all $\beta > \frac{1}{2}$. ////

We saw in Theorem 4.6.5 that for a positive $C_0$-semigroup, uniform boundedness of the resolvent in the right half-plane implies scalar integrability; recall that $s(A) = s_0(A)$ for positive semigroups. Conversely, we have seen in Theorem 4.6.2 that scalar $p$-integrability for some $1 \leq p < \infty$ implies uniform boundedness of the resolvent in the right half-plane. We are going to apply Theorem 4.6.8 to construct a positive $C_0$-semigroup on a reflexive Banach lattice $X$ which has uniformly bounded resolvent in the right half-plane but fails to be scalarly $p$-integrable for *all* $p > 1$. This shows that the converse of Theorem 4.6.2 does not hold for $\phi(t) = t^p$, $1 < p < \infty$, and that scalar integrability in Theorem 4.6.5 cannot be replaced by scalar $p$-integrability, $1 < p < \infty$.

**Example 4.6.9.** For $n = 1, 2, \ldots$ choose $1 < r_n < 2$ close enough to 1 in order that $\frac{1}{r_n} - \frac{1}{r'_n} > \frac{n}{n+1}$, where $r'_n$ is the conjugate exponent of $r_n$. Let $\mathbf{T}^{(n)}$ be the semigroup of Example 1.4.4 on $X_n := L^{r_n}(1, \infty) \cap L^{r'_n}(1, \infty)$. Since by Example 4.2.9 the map $\alpha \mapsto \omega_\alpha(\mathbf{T}^{(n)})$ is linear and decreasing on $[0, \frac{1}{r_n} - \frac{1}{r'_n}]$ and constant on $[\frac{1}{r_n} - \frac{1}{r'_n}, 1]$, by rescaling with some $\lambda_n > 0$ we obtain a $C_0$-semigroup $\mathbf{T}^{(n)}_{-\lambda_n} = \{e^{\lambda_n t} T^{(n)}(t)\}_{t \geq 0}$ with the properties that $s(A^{(n)}_{-\lambda_n}) = \omega_1(\mathbf{T}^{(n)}_{-\lambda_n}) < 0$ and $\omega_{n/(n+1)}(\mathbf{T}^{(n)}_{-\lambda_n}) > 0$. By Theorem 4.6.8, $\mathbf{T}^{(n)}_{-\lambda_n}$ is not scalarly $(1 + \frac{1}{n})$-integrable. Choosing $\epsilon_n > 0$ so small that still $s(A^{(n)}_{-\lambda_n - \epsilon_n}) = \omega_1(\mathbf{T}^{(n)}_{-\lambda_n - \epsilon_n}) < 0$, Hölder's inequality implies that $\mathbf{T}^{(n)}_{-\lambda_n - \epsilon_n}$ fails to be scalarly $p$-integrable for all $p \geq 1 + \frac{1}{n}$.

Let $\mathbf{T}^{[n]}$ be the positive $C_0$-semigroup on $X_n$ defined by

$$T^{[n]}(t) := T^{(n)}_{-\lambda_n - \epsilon_n}(\alpha_n t),$$

where $\alpha_n := \|(A^{(n)}_{-\lambda_n - \epsilon_n})^{-1}\|_{X_n}$. Then $\mathbf{T}^{[n]}$ fails to be scalarly $p$-integrable for all $p \geq 1 + \frac{1}{n}$, $0 \in \varrho(A^{[n]})$, and

$$\|(A^{[n]})^{-1}\|_{X_n} = 1 \qquad (4.6.3)$$

by the choice of $\alpha_n$, using the Laplace transform representation of the the resolvent of Theorem 1.4.1.

Now let $X$ be the Banach lattice of all sequences $x = (x_n)_{n\geq 1}$ with $x_n \in X_n$ and

$$\|x\| := \left(\sum_{n=1}^{\infty} \|x_n\|_{X_n}^2\right)^{\frac{1}{2}} < \infty.$$

Define the positive $C_0$–semigroup $\mathbf{T}$ on $X$ coordinatewise by $\mathbf{T} = (\mathbf{T}^{[n]})_{n\geq 1}$. The generator $A$ of $\mathbf{T}$ is given coordinatewise by $A = (A^{[n]})_{n\geq 1}$ with maximal domain. By (4.6.3), the positive operator $((A^{[n]})^{-1})_{n\geq 1}$ is bounded on $X$ and agrees coordinatewise with the resolvent of $\mathbf{T}$ at the origin. It follows that $0 \in \varrho(A)$ and $A^{-1} = ((A^{[n]})^{-1})_{n\geq 1}$, and Theorem 1.4.1 implies that $s_0(A) = s(A) < 0$. Thus, $A$ has uniformly bounded resolvent in the right half-plane. On the other hand, it is obvious from the construction that $\mathbf{T}$ fails to be scalarly $p$-integrable for all $p > 1$. Finally, $X$ is reflexive, being the $l^2$-sum of reflexive spaces.

If we assume certain integrability properties of the resolvent *and its adjoint* along vertical lines, then uniform boundedness of the resolvent does imply scalar $p$-integrability of the semigroup for certain $p$. This is the content of the next theorem.

**Theorem 4.6.10.** *Let $\mathbf{T}$ be a $C_0$–semigroup on a Banach space $X$ whose resolvent is uniformly bounded in the right half-plane. Let $1 \leq r_0, r_1 \leq \infty$ be such that $\frac{1}{2} \leq \frac{1}{p} := \frac{1}{r_0} + \frac{1}{r_1} \leq 1$. If there exists an $\omega_0 > \omega_0(\mathbf{T})$ such that*

$$\int_{-\infty}^{\infty} \|R(\omega_0 + is, A)x\|^{r_0}\, ds < \infty, \quad \forall x \in X,$$

*and*

$$\int_{-\infty}^{\infty} \|R(\omega_0 + is, A^*)x^*\|^{r_1}\, ds < \infty, \quad \forall x^* \in X^*,$$

*then $\mathbf{T}$ is scalarly $r$-integrable for all $2 \leq r \leq q$, $\frac{1}{p} + \frac{1}{q} = 1$. In particular, $\omega_{\frac{1}{2}}(\mathbf{T}) < 0$.*

*Proof:* Fix $r \in [2, q]$ arbitrary.

Since $\omega_0 > \omega_0(\mathbf{T})$, the functions $s \mapsto R(\omega_0+is, A)x$ and $s \mapsto R(\omega_0+is, A^*)x^*$ are bounded. Since they also belong to $L^{r_0}(\mathbb{R})$ and $L^{r_1}(\mathbb{R})$ respectively, it follows that in the assumptions of the theorem we may replace $r_0$ and $r_1$ by larger values in such a way that $\frac{1}{r_0} + \frac{1}{r_1} = \frac{1}{r'}$, where $\frac{1}{r} + \frac{1}{r'} = 1$. In other words, without loss of generality we may assume that $r = q$.

For $x \in X$ and $x^* \in X^*$ we define

$$g_{x,x^*}(s) := \langle x^*, R(-is, A)R(\omega_0 - is, A)x\rangle, \quad s \in \mathbb{R}.$$

Since the resolvent is uniformly bounded in the right half-plane and

$$\|R(-is, A)x\| \leq \|(I + \omega_0 R(-is, A))\|\, \|R(\omega_0 - is, A)x\|,$$

Hölder's inequality implies that $g_{x,x^*}$ belongs to $L^p(\mathbb{R})$. By the Hausdorff-Young theorem, there exists a constant $c_p > 0$ such that

$$\left\| \left\langle x^*, (C,1) \int_{-\infty}^{\infty} e^{is(\cdot)} R(is, A) R(\omega_0 + is, A) x \, ds \right\rangle \right\|_q = \|\mathcal{F} g_{x,x^*}\|_q \leq c_p \|g_{x,x^*}\|_p. \tag{4.6.4}$$

Also, by Theorem 1.3.3, for all $t > 0$ we have

$$\frac{1}{2\pi}(C,1)\int_{-\infty}^{\infty} e^{ist} R(\omega_0 + is, A) x \, ds = e^{-\omega_0 t} T(t) x.$$

Since $\omega_0 > \omega_0(\mathbf{T})$, it follows that there is a constant $C > 0$ such that

$$\left\|(C,1)\int_{-\infty}^{\infty} e^{is(\cdot)} R(\omega_0 + is, A) x \, ds\right\|_q = 2\pi \|e^{-\omega_0(\cdot)} T(\cdot) x\|_q \leq C\|x\|. \tag{4.6.5}$$

Multiply the resolvent identity

$$R(is, A) = R(\omega_0 + is, A) + \omega_0 R(is, A) R(\omega_0 + is, A)$$

on both sides with $e^{ist}$ and integrate. Using Theorem 1.3.3, (4.6.4), and (4.6.5), we obtain

$$2\pi \|\langle x^*, T(\cdot)x\rangle\|_q = \left\|(C,1)\int_{-\infty}^{\infty} e^{is(\cdot)} \langle x^*, R(is, A)x\rangle \, ds\right\|_q$$

$$\leq C\|x\|\|x^*\| + c_p|\omega_0|\|g_{x,x^*}\|_p < \infty.$$

This proves that $\mathbf{T}$ is scalarly $q$-integrable. Moreover, Theorem 4.6.8 implies that $\omega_{\frac{1}{2}}(\mathbf{T}) \leq 0$. Applying this to the semigroup $\mathbf{T}_\epsilon = \{e^{-\epsilon t} T(t)\}_{t \geq 0}$, where $\epsilon > 0$ is so small that the resolvent of $\mathbf{T}_\epsilon$ is still uniformly bounded in the right half-plane, it follows that $\omega_{\frac{1}{2}}(\mathbf{T}) \leq -\epsilon$. ////

Notice that the function $R : s \mapsto \|R(\omega_0 + is, A^*)x^*\|$ is always measurable, even though the adjoint semigroup $\mathbf{T}^*$ need not be strongly continuous. Indeed, let $(s_n)$ be a countable dense set in $\mathbb{R}$. For each $n$, choose a countable set $(x_{nm}) \subset X$ with $\|x_{nm}\| \leq 1$ such that

$$\|R(\omega_0 + is_n, A^*)x^*\| = \sup_{m \in \mathbb{N}} |\langle R(\omega_0 + is_n, A^*)x^*, x_{nm}\rangle|$$

for all $n$. Then by the weak*-continuity of $s \mapsto R(\omega_0 + is, A^*)x^*$, for all $s \in \mathbb{R}$ we have

$$\|R(\omega_0 + is, A^*)x^*\| = \sup_{n,m \in \mathbb{N}} |\langle R(\omega_0 + is, A^*)x^*, x_{nm}\rangle|.$$

This shows that $R$ is measurable, being the supremum of countably many continuous functions.

If both $X$ and $X^*$ have Fourier type $\frac{4}{3} \leq p \leq 2$, the assumption of the corollary is satisfied and together with Theorem 4.6.2 we obtain:

**Corollary 4.6.11.** *Let* **T** *be a $C_0$-semigroup on a Banach space $X$. Assume that both $X$ and $X^*$ have Fourier type $\frac{4}{3} \leq p \leq 2$ and let $\frac{1}{p} + \frac{1}{q} = 1$. Then the following assertions are equivalent:*

(i) *The resolvent is uniformly bounded in the right half-plane;*
(ii) **T** *is scalarly $r$-integrable for some $2 \leq r \leq q$;*
(iii) **T** *is scalarly $r$-integrable for all $2 \leq r \leq q$.*

In the situation of this corollary, we have $\omega_{\frac{1}{2}}(\mathbf{T}) < 0$ by Theorem 4.6.10. This, however, follows already if we only assume that $X$ has Fourier type $\frac{4}{3} \leq p \leq 2$. Indeed, since $\frac{1}{p} - \frac{1}{q} \leq \frac{1}{2}$ an application of Theorem 4.2.4 shows that $\omega_{\frac{1}{2}}(\mathbf{T}) \leq s_0(A) < 0$.

Theorem 4.6.10 applied to $r_0 = r_1 = 2$ gives the following continuous analogue of Corollary 2.2.3.

**Corollary 4.6.12.** *Let $A$ be the generator of a $C_0$-semigroup* **T** *on a Banach space $X$. Assume that there is an $\omega_0 > \omega_0(\mathbf{T})$ such that*

$$\int_{-\infty}^{\infty} \|R(\omega_0 + is, A)x\|^2 \, ds < \infty, \quad \forall x \in X,$$

*and*

$$\int_{-\infty}^{\infty} \|R(\omega_0 + is, A^*)x^*\|^2 \, ds < \infty, \quad \forall x^* \in X^*.$$

*Then, $\omega_0(\mathbf{T}) = s_0(A)$. In particular, if in addition the resolvent of $A$ is uniformly bounded in the right half-plane, then $\omega_0(\mathbf{T}) < 0$.*

**Notes.** Theorems 4.1.1 and 4.2.4 are due to Weis and Wrobel [WW]. Theorem 4.2.4 solved the problem whether $\omega_1(\mathbf{T}) \leq s_0(A)$ for arbitrary $C_0$-semigroups. This had been open for quite some time. The following partial results into this direction were known:

- M. Slemrod [Sl] showed that $\omega_{n+2}(\mathbf{T}) \leq s_n(A)$ for all $n = 0, 1, ...$;
- V. Wrobel [Wr] showed that $\omega_{n+1}(\mathbf{T}) \leq s_n(A)$ in uniformly convex Banach spaces; $n = 0, 1, ...$;
- G. Weiss [Ws3] showed that $\omega_n(\mathbf{T}) = s_n(A)$ for $C_0$-semigroup on a Hilbert space; $n = 0, 1, ...$;
- It was shown in [NSW] that $\omega_\alpha(\mathbf{T}) \leq s_0(A)$ for all $\alpha > \frac{1}{p} - \frac{1}{q}$. At least for $p = 1$, it is shown in [NS] that this inequality generalizes to a wider class of operators $A$.

In these results, the the quantities $s_n(A)$ are defined as the infimum of all $\omega \in \mathbb{R}$ for which there exists a constant $M = M_\omega > 0$ such that $\{\operatorname{Re} \lambda > \omega\} \subset \varrho(A)$ and $\|R(\lambda, A)\| \leq M(1 + |\lambda|)^n$ for all $\operatorname{Re} \lambda > \omega$.

Example 4.2.9 is taken from [WW].

The material of Sections 4.3 and 4.4 is taken from [HN], except Example 4.3.4, which is due to W. Arendt [Ar2]. For Hilbert spaces, the case $\alpha = 1$ of Corollary 4.3.3 is due to S.Z. Huang [Hu4]. Proposition 4.4.1 is a special case of a result in [Ch]. For more information concerning $B$−convex spaces we refer to the monograph [Pi].

The results of Section 4.5 slightly extend those of [Ne5].

The results of Sections 4.2, 4.3, 4.4, and 4.5 can be extended to resolvents with polynomial growth in a right half-plane. Defining for real $\beta \geq 0$ the abscissae $s_\beta(A)$ in the natural way, in spaces with Fourier type $p$ we have the inequality [WW]

$$\omega_{\beta+\frac{1}{p}-\frac{1}{q}}(\mathbf{T}) \leq s_\beta(A), \quad \beta \geq 0,$$

if $X$ has Fourier type $p$. Similarly, if in Theorems 4.3.2 and 4.3.6 we assume that the function $F$ is polynomially bounded of order $\beta$ in a strip $\{0 < \operatorname{Re}\lambda < \omega_0\}$ for some $\beta \geq -1$ and $\omega_0 > \max\{0, \omega_0(\mathbf{T})\}$, then the conclusions of the theorems hold for all $\alpha \geq 0$ with $\alpha > \beta + \frac{1}{p}$ and all $\alpha > \max\{0, \beta+1\}$, respectively [HN]. In both cases, the proofs for the case $\beta = 0$ need only minor modifications. Finally, from Theorem 4.5.2 and Lemma 4.3.1 it follows immediately that

$$\|PT(t)(\lambda_0 - A)^{-\beta-1}x_0\| \leq M(1+t), \quad t \geq 0,$$

whenever $x_0 \in X$ and $\beta \geq 0$ are such that $\lambda \mapsto PR(\lambda, A)x_0$ admits a holomorphic extension to the open right half-plane which is polynomially bounded of order $\beta$ in a strip $\{0 < \operatorname{Re}\lambda < \omega_0\}$ for some $\omega_0 > \max\{0, \omega_0(\mathbf{T})\}$.

That scalarly $p$-integrable semigroups have uniformly bounded resolvent is proved by G. Weiss in [Ws1] (he actually proves the quantitative result mentioned after Theorem 4.6.2); a slightly simpler proof is given in [NSW]. Whether scalar $p$-integrability is equivalent to uniform exponential stability in Hilbert spaces was asked by A.J. Pritchard and J. Zabczyk [PZ]. The affirmative answer was obtained independently by Falun Huang and Kangsheng Liu [HL] and G. Weiss [Ws1]; their proofs are essentially equivalent. Our approach generalizes their arguments to the setting of rearrangement invariant Banach function spaces. Results about (scalar) integrability have applications to control theory; cf. [Zb1], [PZ], [Pl], [Ws1].

Most of Theorem 4.6.8 is proved in [NSW]: there it is shown that if $\mathbf{T}$ is scalarly $p$-integrable for some $1 \leq p \leq 2$, then $\omega_\alpha(\mathbf{T}) \leq 0$ for all $\alpha > \frac{1}{p}$.

Example 4.6.9 provides a counterexample to Lemma 4 in [HF3]. In the same paper, this lemma is used to prove that uniform boundedness of the resolvent implies scalar integrability of the semigroup. Whether this is indeed the case is left open by our results.

Corollary 4.6.12 is due to M.A. Kaashoek and S.M. Verduyn Lunel [KV]. The present proof simplifies their arguments somewhat.

We close with some open questions:

(i) Do some of the results of Sections 4.3 and 4.4 still hold in the limit case (Theorem 4.3.2 for $\alpha = \frac{1}{p}$, etc.)?

(ii) In the situation of Theorem 4.5.2, is $T(\cdot)R(\lambda_0, A)x_0$ necessarily bounded?

(iii) Does there exist a $C_0$-semigroup with uniformly bounded resolvent in the right half-plane which is not scalarly integrable?

(iv) If $\mathbf{T}$ is an arbitrary $C_0$-semigroup on a space $C_0(\Omega)$ or $L^p(\mu)$, $1 \leq p < \infty$, does it follow that $s_0(A) = \omega_0(\mathbf{T})$?

Question (iv) is motivated by the observation that on these spaces we always have $s(A) = \omega_0(\mathbf{T})$ for *positive* $C_0$-semigroups. The results of this chapter seem to indicate that for general semigroups one should consider $s_0(A)$ rather than $s(A)$; furthermore in Hilbert space the identity $s_0(A) = \omega_0(\mathbf{T})$ indeed holds.

# Chapter 5

*Countability of the unitary spectrum*

In this chapter we study sufficient conditions for certain orbits $T(\cdot)x$ of a $C_0$–semigroup $\mathbf{T}$ to be stable, i.e. $\lim_{t\to\infty} \|T(t)x\| = 0$.

We start in Section 5.1 with the theorem of Arendt, Batty, Lyubich, and Vũ: a uniformly bounded $C_0$–semigroup is uniformly stable if the unitary spectrum $\sigma(A) \cap i\mathbb{R}$ of its generator $A$ is countable and contains no residual spectrum.

In Section 5.2 we apply this result to derive a theorem of Katznelson-Tzafriri type for $C_0$–semigroups: if $\mathbf{T}$ is uniformly bounded and $f$ is of spectral synthesis with respect to $i\sigma(A) \cap \mathbb{R}$, then $\lim_{t\to\infty} \|T(t)\hat{f}(\mathbf{T})\| = 0$. In both sections, we actually prove versions of these theorems for an individual orbit of a uniformly bounded semigroup. Instead of making global spectral assumptions on the semigroups, we consider holomorphic extension properties of the local resolvent $\lambda \mapsto R(\lambda, A)x_0$.

In Section 5.3 we extend the results to individual bounded uniformly continuous orbits of a (possibly unbounded) semigroup.

In Section 5.4 we give an elementary semigroup proof of the fact that closed subsets of $\mathbb{R}$ with countable boundary are sets of spectral synthesis. This result is needed in Section 5.5, where we derive an explicit expression for $\lim_{t\to\infty} \|T(t)x\|$ for contraction semigroups with $\sigma(A) \cap i\mathbb{R}$ countable in terms of the distance of $x$ to the subspace $X_0 = \{x \in X : \lim_{t\to\infty} \|T(t)x\| = 0\}$.

In Section 5.6 the stability problem for semigroups is approached from the Laplace transform point of view. An abstract Tauberian theorem for Laplace transforms of bounded strongly measurable functions is proved and then applied to bounded orbits of semigroups to yield further individual stability theorems.

In the final Section 5.7 we prove the Glicksberg-DeLeeuw splitting theorem: an almost periodic $C_0$–semigroup can be decomposed into a uniformly bounded group and a uniformly stable semigroup. Moreover, necessary and sufficient conditions for almost periodicity are given for a $C_0$–semigroup whose generator has countable spectrum on the imaginary axis.

## 5.1. The stability theorem of Arendt, Batty, Lyubich, and Vũ

In this section we prove the following stability theorem: if $\mathbf{T}$ is a uniformly bounded $C_0$–semigroup such that the unitary spectrum $\sigma(A) \cap i\mathbb{R}$ is at most countable and

$\sigma_p(A^*) \cap i\mathbb{R} = \emptyset$, then **T** is uniformly stable, i.e. $\lim_{t\to\infty} \|T(t)x\| = 0$ for all $x \in X$. The proof relies of a reduction to isometric semigroups and then to isometric groups, which allows us to apply the spectral theory of uniformly bounded $C_0$–groups of Section 2.4.

A bounded operator $T$ on a Banach space $X$ is called an *isometry* if $\|Tx\| = \|x\|$ for all $x \in X$; we do not suppose $T$ to be invertible.

**Lemma 5.1.1.** *Let* **T** *be a $C_0$–semigroup of isometries on a Banach space $X$, and let $A$ be its generator.*

(i) *For all $x \in D(A)$ and $\lambda \in \mathbb{C}$ we have $\|(\lambda - A)x\| \geq |\text{Re }\lambda|\,\|x\|$;*

(ii) *If $E \subset \mathbb{R}$ is closed and $x \in X$ is such that the map $\lambda \mapsto R(\lambda, A)x$ has a holomorphic extension $F$ to a connected neighbourhood $V$ of $\{\text{Re }\lambda \geq 0\} \backslash iE$, then for all $\lambda \in V \backslash i\mathbb{R}$ we have $\|F(\lambda)\| \leq |\text{Re }\lambda|^{-1}\|x\|$.*

*Proof:* (i): We may assume that $\text{Re }\lambda \neq 0$. From the identity

$$e^{-\lambda t}T(t)x = x + \int_0^t e^{-\lambda s}(A - \lambda)T(s)x\,ds$$

we have

$$e^{-\text{Re }\lambda t}\|x\| = e^{-\text{Re }\lambda t}\|T(t)x\|$$
$$\leq \|x\| + \int_0^t e^{-\text{Re }\lambda s}\|T(s)(\lambda - A)x\|\,ds$$
$$= \|x\| + \left(\int_0^t e^{-\text{Re }\lambda s}\,ds\right)\|(\lambda - A)x\|$$
$$= \|x\| + \frac{e^{-\text{Re }\lambda t} - 1}{-\text{Re }\lambda}\|(\lambda - A)x\|.$$

This proves the lemma for $\text{Re }\lambda < 0$. For $\text{Re }\lambda > 0$ the inequality follows from the Laplace transform representation of the resolvent.

(ii): This is proved in the same way, after first substituting $R(\lambda, A)x$ for $x$ in the first formula and passing to the holomorphic extension. ////

**Theorem 5.1.2.** *Let* **T** *be $C_0$–semigroup of contractions on a Banach space $X$, with generator $A$. Then there exists a Banach space $Y$, a bounded operator $\pi : X \to Y$ with dense range, and a $C_0$–semigroup* **U** *of isometries on $Y$ with generator $B$ such that:*

(i) $U(t)\pi = \pi T(t)$ *for all $t \geq 0$. Moreover, $\pi D(A) \subset D(B)$ and $B\pi x = \pi A x$ for all $x \in D(A)$;*

(ii) $\lim_{t\to\infty} \|T(t)x\| = \|\pi x\|$ *for all $x \in X$;*

(iii) $\sigma(B) \subset \sigma(A)$.

*If $\sigma(A) \cap i\mathbb{R}$ is a proper subset of $i\mathbb{R}$, then* **U** *extends to a $C_0$–group of isometries.*

*Proof:* On $X$ we define the semi-norm $l$ by $l(x) := \lim_{t\to\infty} \|T(t)x\|$. Since **T** is contractive, this limit indeed exists. Let $\pi : X \to Y_0 := X/\ker l$ be the quotient mapping. The semi-norm $l$ induces a norm $l_0$ on $Y_0$ in the natural way by $l_0(\pi x) := l(x)$, and we have

$$l_0(\pi x) = l(x) = \lim_{t\to\infty} \|T(t)x\|.$$

For $t \geq 0$, we define $U_0(t) : Y_0 \to Y_0$ by $U_0(t)\pi x := \pi T(t)x$. We have

$$l_0(U_0(t)\pi x) = l_0(\pi T(t)x) = l(T(t)x) = l(x) = l_0(\pi x).$$

This shows that $U_0(t)$ is isometric with respect to the norm $l_0$. Let $Y$ be the completion of $Y_0$ with respect to $l_0$. Then each operator $U_0(t)$ extends to an isometry $U(t)$ on $Y$. Strong continuity of the family $\mathbf{U} = \{U(t)\}_{t\geq 0}$ follows from the density of $\pi X$ in $Y$, the contractivity of the operators $U(t)$, and the estimate

$$\limsup_{t\downarrow 0} \|U(t)\pi x - \pi x\| = \limsup_{t\downarrow 0} \left(\lim_{s\to\infty} \|T(t+s)x - T(s)x\|\right)$$
$$\leq \limsup_{t\downarrow 0} \|T(t)x - x\| = 0, \quad x \in X.$$

If $x \in D(A)$, then

$$\lim_{t\downarrow 0} \frac{1}{t}(U(t)\pi x - \pi x) = \pi \lim_{t\downarrow 0} \frac{1}{t}(T(t)x - x) = \pi A x,$$

proving that $\pi x \in D(B)$ and $B\pi x = \pi A x$.

We have proved (i) and (ii). Next we prove (iii). Let $\lambda \in \varrho(A)$ be given. We define the linear operator $R_\lambda$ on $Y_0$ by

$$R_\lambda \pi x := \pi R(\lambda, A)x.$$

This operator is well-defined and

$$l_0(R_\lambda \pi x) = \lim_{t\to\infty} \|T(t)R(\lambda, A)x\|$$
$$\leq \|R(\lambda, A)\| \lim_{t\to\infty} \|T(t)x\| = \|R(\lambda, A)\| l_0(\pi x).$$

Therefore, $R_\lambda$ extends to a bounded operator on $Y$ and $\|R_\lambda\| \leq \|R(\lambda, A)\|$. For all $x \in X$ we have $R_\lambda \pi x = \pi R(\lambda, A)x \in \pi D(A) \subset D(B)$ and $(\lambda - B)R_\lambda \pi x = (\lambda - B)\pi R(\lambda, A)x = \pi(\lambda - A)R(\lambda, A)x = x$. Similarly, for all $x \in D(A)$ we have $\pi x \in D(B)$ and $R_\lambda(\lambda - B)\pi x = R_\lambda \pi(\lambda - A)x = \pi R(\lambda, A)(\lambda - A)x = x$. Therefore, to prove that $R_\lambda$ is a two-sided inverse of $\lambda - B$, in view of the closedness of $B$ it remains to prove that $\pi D(A)$ is dense in $D(B)$ with respect to the graph norm. So let $y \in D(B)$ be arbitrary. Fix $\mu > 0$ arbitrary and choose $z \in Y$ such that $y = R(\mu, B)z$. Pick a sequence $(x_n) \subset X$ such that $\pi x_n \to z$ in $Y$ and put $y_n := \pi R(\mu, A)x_n$. Then $y_n \in \pi D(A)$, $y_n = R(\mu, B)\pi x_n \to R(\mu, B)z = y$, and $By_n = BR(\mu, B)\pi x_n \to BR(\mu, B)z = By$. Here we used that $\pi R(\mu, A) = R(\mu, B)\pi$ by the Laplace transform representation of the resolvents and that $BR(\mu, B) = \mu R(\mu, B) - I$ is a bounded operator on $Y$. Thus, $y_n \to y$ in $D(B)$ with respect to the graph norm. This concludes the proof of (iii).

Suppose $\sigma(A) \cap i\mathbb{R}$ is properly contained in $i\mathbb{R}$. We have to prove that $\mathbf{U}$ extends to a $C_0$-group of isometries. By Lemma 5.1.1 (i), for all $y \in D(B)$ and $\mathrm{Re}\,\lambda < 0$ we have
$$\|(\lambda - B)y\| \geq |\mathrm{Re}\,\lambda|\,\|y\|.$$
It follows from this that the open left half-plane $\mathbb{C}_-$ contains no approximate eigenvalues for $B$. In particular, $\mathbb{C}_-$ contains no elements of the boundary of $\sigma(B)$. Hence, either $\mathbb{C}_- \subset \sigma(B)$ or $\mathbb{C}_- \cap \sigma(B) = \emptyset$. But in the first case, also $i\mathbb{R} \subset \sigma(B)$ since $\sigma(B)$ is closed. This contradicts the assumption, so we must have the second alternative.

It follows that $\sigma(B) \subset i\mathbb{R}$. For $\mathrm{Re}\,\lambda > 0$ we have
$$\|R(\lambda, -B)\| = \|R(-\lambda, B)\| \leq \frac{1}{|\mathrm{Re}\,(-\lambda)|} = \frac{1}{\mathrm{Re}\,\lambda}.$$
By the Hille-Yosida theorem, $-B$ is the generator of a $C_0$-semigroup $\mathbf{V}$ of contractions on $Y$. We check that $U(t)$ is invertible for all $t \geq 0$ with inverse $V(t)$. For all $x \in D(B) = D(-B)$, the maps $t \mapsto U(t)V(t)x$ and $t \mapsto V(t)U(t)x$ are differentiable with derivative identically zero. It follows that the maps are constant, and by letting $t \downarrow 0$ it follows that $U(t)V(t) = V(t)U(t) = I$ on the dense set $D(B)$, hence on all of $Y$. Finally, each operator $V(t)$ is an isometry, being the inverse of an isometry.    ////

The condition $\sigma(A) \cap i\mathbb{R} \neq i\mathbb{R}$ cannot be omitted from the last statement. Indeed, let $X = L^2(\mathbb{R}_+)$ and consider the right translation semigroup on $X$:
$$(T(t)f)(s) = \begin{cases} f(s-t), & s \geq t; \\ f(0), & \text{else.} \end{cases}$$
Then $\mathbf{T}$ is isometric, $\sigma(A) \cap i\mathbb{R} = i\mathbb{R}$, and $\mathbf{T}$ does not extend to a $C_0$-group.

The triple $(Y, \pi, \mathbf{U})$ will be called the *isometric limit (semi)group* associated to $\mathbf{T}$.

**Corollary 5.1.3.** *If $\mathbf{T}$ is an isometric $C_0$-semigroup on $X$ with $\sigma(A) \cap i\mathbb{R} \neq i\mathbb{R}$, then $\mathbf{T}$ extends to an isometric $C_0$-group.*

*Proof:* By the isometric nature of $\mathbf{T}$ we have $l(x) = x$ for all $x \in X$, so $Y = X$ and $\mathbf{T} = \mathbf{U}$. But $\mathbf{U}$ extends to a $C_0$-group since $\sigma(A) \cap i\mathbb{R} \neq i\mathbb{R}$.    ////

An invertible operator is called *doubly power bounded* if
$$\sup_{k \in \mathbb{Z}} \|T^k\| < \infty.$$

**Lemma 5.1.4.** *Let $\mathbf{T}$ a doubly power bounded operator on a Banach space $X$ with $\sigma(T) = \{1\}$. Then $T = I$.*

*Proof:* Since $\log z$ is holomorphic in a neighbourhood of $z = 1$, by Dunford calculus we may define the bounded operator $S := -i \log T$. Then $T = e^{iS}$ and the spectral mapping theorem implies that $\sigma(mS) = \{0\}$ for all $m \in \mathbb{N}$. Also, for all $m \in \mathbb{N}$ we have $\sigma(\sin(mS)) = \sin(\sigma(mS)) = \{\sin 0\} = \{0\}$, and

$$\|(\sin(mS))^n\| = \left\|\left(\frac{T^m - T^{-m}}{2i}\right)^n\right\| \leq \sup_{k \in \mathbb{Z}} \|T^k\|.$$

Let $\sum_{n=0}^{\infty} c_n z^n$ be Taylor series of the principle branch of $\arcsin z$ at $z = 0$. As is well-known, $c_n \geq 0$ for all $n$ and $\sum_{n=0}^{\infty} c_n = \arcsin(1) = \frac{\pi}{2}$. Consequently,

$$\|mS\| = \|\arcsin(\sin(mS))\| \leq \sum_{n=0}^{\infty} c_n \|(\sin(mS))^n\| \leq \frac{\pi}{2} \sup_{k \in \mathbb{Z}} \|T^k\|.$$

Since this holds for all $m \in \mathbb{N}$, it follows that $S = 0$ and $T = e^{iS} = I$. ////

Now we are in a position to prove the theorem of Arendt, Batty, Lyubich, and Vũ.

**Theorem 5.1.5.** *Let $\mathbf{T}$ be a uniformly bounded $C_0$-semigroup on a Banach space $X$, with generator $A$. If*

(i) $\sigma(A) \cap i\mathbb{R}$ *is countable, and*
(ii) $\sigma_p(A^*) \cap i\mathbb{R} = \emptyset$,

*then $\mathbf{T}$ is uniformly stable, i.e. $\lim_{t \to \infty} \|T(t)x\| = 0$ for all $x \in X$.*

*Proof:* By renorming with the equivalent norm $\|x\| := \sup_{t \geq 0} \|T(t)x\|$, we may assume that $\mathbf{T}$ is contractive. Let $(Y, \pi, \mathbf{U})$ be the isometric limit semigroup associated to $\mathbf{T}$, and let $B$ be the generator of $\mathbf{U}$. By (i), $\sigma(A) \cap i\mathbb{R}$ cannot be all of $i\mathbb{R}$, and therefore $\mathbf{U}$ extends to an isometric group on $Y$. Assuming that $\mathbf{T}$ is not uniformly stable, we shall prove that (i) implies $\sigma_p(A^*) \cap i\mathbb{R} \neq \emptyset$.

Since $\mathbf{T}$ is not uniformly stable, the definition of $Y$ implies that $Y \neq \{0\}$. By Lemma 2.4.3, $\sigma(B) \neq \emptyset$. Also, since $\sigma(B) \subset \sigma(A)$, it follows that $\sigma(B)$ is countable. In particular, by a well-known result from point set topology, it contains an isolated point, say $i\omega$. Let $P_\omega$ be the associated spectral projection in $Y$, let $\mathbf{U}_\omega$ be the restriction of $\mathbf{U}$ to $P_\omega Y$ and let $B_\omega$ denote its generator. Since $\sigma(B_\omega) = \{i\omega\}$, Theorem 2.4.4 implies that $\sigma(U_\omega(t)) = \{e^{i\omega t}\}$ for all $t \in \mathbb{R}$. By Lemma 5.1.4, this implies that $U_\omega(t) = e^{i\omega t} I$. Hence, for all $y \in Y$ we have $U_\omega(t) P_\omega y = e^{i\omega t} P_\omega y$, so $P_\omega y \in D(B_\omega)$ and $B_\omega P_\omega y = i\omega P_\omega y$. Fix an arbitrary non-zero $y_\omega^* \in (P_\omega Y)^*$ and define $x^* \in X^*$ by

$$\langle x^*, x \rangle := \langle y_\omega^*, P_\omega \pi x \rangle, \quad x \in X.$$

Then $x^* \neq 0$. For all $x \in D(A)$ we have $\pi x \in D(B)$, $B\pi x = \pi A x$, and

$$\langle x^*, Ax \rangle = \langle y_\omega^*, P_\omega \pi A x \rangle = \langle y_\omega^*, P_\omega B \pi x \rangle$$
$$= \langle y_\omega^*, B_\omega P_\omega \pi x \rangle = i\omega \langle y_\omega^*, P_\omega \pi x \rangle = i\omega \langle x^*, x \rangle.$$

Hence, $x^* \in D(A^*)$ and $A^* x^* = i\omega x^*$. ////

The following example shows that the countability assumption in Theorem 5.1.5 cannot be relaxed.

**Example 5.1.6.** Let $E \subset \mathbb{R}$ be uncountable and let $\mu$ be a positive diffuse measure (i.e. a measure without atoms) supported by $E$. Let $X = L^2(E, \mu)$ and define
$$(T(t)f)(s) = e^{ist} f(s), \quad s \in E, \, t \geq 0.$$
This defines a $C_0$–semigroup $\mathbf{T}$ on $X$, each operator $T(t)$ is unitary, and hence $\|T(t)f\| = \|f\|$ for all $t \geq 0$. On the other hand, one checks that $\sigma(A) \subset iE$ and $\sigma_p(A^*) = \sigma_p(A) = \emptyset$.

Example 5.1.12 below shows that also the uniform boundedness assumption cannot be omitted from Theorem 5.1.5.

The condition $\sigma_p(A^*) \cap i\mathbb{R} = \emptyset$ is necessary for $\mathbf{T}$ to be uniformly stable. Indeed, if $A^* x^* = i\omega x^*$ for some non-zero $x^* \in D(A^*)$, then the identity
$$\langle e^{-i\omega t} T^*(t)x^* - x^*, x \rangle = \int_0^t \langle (A^* - i\omega)x^*, e^{-i\omega s} T(s)x \rangle \, ds = 0, \quad x \in X,$$
shows that $T^*(t)x^* = e^{i\omega t}x^*$ for all $t \geq 0$. Choosing $x_0 \in X$ such that $\langle x^*, x_0 \rangle \neq 0$, we have
$$\langle x^*, T(t)x_0 \rangle = \langle T^*(t)x^*, x_0 \rangle = e^{i\omega t} \langle x^*, x_0 \rangle,$$
proving that $\mathbf{T}$ is not even weakly uniformly stable.

Similarly, a necessary condition for uniform stability is $\sigma_p(A) \cap i\mathbb{R} = \emptyset$. Interestingly, the latter is actually implied by the first, and in certain situations the two conditions are actually equivalent. This is will be proved at the end of this section.

Countability of the unitary spectrum is by no means necessary for uniform stability: the $C_0$–semigroup $\mathbf{T}$ on $C_0[0, \infty)$ defined by $T(t)f(s) := f(s+t)$ is trivially seen to be uniformly stable, but $\sigma(A) = \{\operatorname{Re} \lambda \leq 0\}$. For this semigroup, however, it is easy to see that there is a dense subspace $Y$ with the property that for each $y \in Y$, the map $\lambda \mapsto R(\lambda, A)y$ extends holomorphically across the imaginary axis. Indeed, for $Y$ we may take the set of all $f \in C_0[0, \infty)$ such that $\sup_{t \geq 0} e^t |f(t)| < \infty$. We shall show below that a bounded semigroup is uniformly stable whenever such a dense subspace can be found. To prepare for these developments we start with a lemma about holomorpic extensions of the local resolvent.

In the rest of this chapter, $X_{x_0}$ denotes the closed linear span of the orbit $\{T(t)x_0 : t \geq 0\}$. The restriction of $\mathbf{T}$ to $X_{x_0}$ is denoted by $\mathbf{T}_{x_0}$ and its generator by $A_{x_0}$.

**Lemma 5.1.7.** *Let $\mathbf{T}$ be an isometric $C_0$–semigroup on a Banach space $X$. Let $E \subset \mathbb{R}$ be a closed subset and let $x_0 \in X$ be such that the map $\lambda \mapsto R(\lambda, A)x_0$ admits a holomorphic extension to a connected neighbourhood $V$ of $\{\operatorname{Re} \lambda \geq 0\} \setminus iE$. Then there exists a connected neighbourhood $W \subset V$ of $\{\operatorname{Re} \lambda \geq 0\} \setminus iE$ with the following property: for each $x \in X_{x_0}$ the map $\lambda \mapsto R(\lambda, A)x$ admits a holomorphic extension to $W$. Moreover, these extensions are $X_{x_0}$-valued.*

*Proof:* Let $Y$ denote the linear span of the orbit of $x_0$. Fix $y \in X_{x_0} = \overline{Y}$ and choose a sequence $(y_n) \subset Y$ with $y_n \to y$. The first thing we observe is that each $R(\lambda, A)y_n$ admits a holomorphic extension to $V$: this follows by linearity and the fact that for each $t \geq 0$, $T(t)F(\lambda)$ extends $R(\lambda, A)T(t)x_0$ if $F(\lambda)$ extends $R(\lambda, A)x_0$.

For each $\lambda_0 \in i\mathbb{R}\backslash iE$ we choose $r(\lambda_0) > 0$ such that the closure of the open ball $B(\lambda_0, r(\lambda_0))$ of radius $r$ and centre $\lambda_0$ is contained in $V$. Let $B$ denote the union of all balls $B(\lambda_0, \frac{1}{2}r(\lambda_0))$ and put $W = \{\text{Re}\,\lambda > 0\} \cup B$. Then $W \subset V$ and $W$ is a connected neighbourhood of $\{\text{Re}\,\lambda \geq 0\}\backslash iE$. In order to prove that $R(\lambda, A)y$ extends holomorphically to $W$, it suffices to show that it extends to each $B(\lambda_0, \frac{1}{2}r)$.

So let $\lambda_0 \in i\mathbb{R}\backslash iE$ and $r = r(\lambda_0)$ as above be fixed. By rescaling **T**, we may assume that $\lambda_0 = 0$. For each $n$, let $F_n(\lambda)$ denote the holomorphic extension of $R(\lambda, A)y_n$ to $V$. We claim that the functions $F_n$ are bounded on $B(0, \frac{1}{2}r)$, uniformly in $n$. Once this has been shown, from the fact that $R(\lambda, A)y_n \to R(\lambda, A)y$ for $\text{Re}\,\lambda > 0$ it follows by Vitali's theorem that the functions $F_n$ converge, uniformly on compacta, to a holomorphic function $F$ on $B(0, \frac{1}{2}r)$. This $F$ is the desired extension of $R(\lambda, A)y$.

Define the continuous functions $f_n : B(0, r) \to X$ by

$$f_n(\lambda) := \left(1 + \frac{\lambda^2}{r^2}\right) F_n(\lambda).$$

Each $f_n$ is holomorphic on $B(0, r)$. By virtue of Lemma 5.1.1 (ii), for $\lambda = re^{i\theta} \in \partial B(0, r)\backslash i\mathbb{R}$ we have

$$\|f_n(\lambda)\| \leq \left|1 + \frac{\lambda^2}{r^2}\right| \cdot \frac{1}{|\text{Re}\,\lambda|} \|y_n\|$$
$$= |1 + e^{2i\theta}| \cdot \frac{1}{r|\cos\theta|}\|y_n\| = \frac{2}{r}\|y_n\|.$$

Therefore, by the maximum modulus theorem,

$$\sup_{\lambda \in B(0,r)} \|f_n(\lambda)\| \leq \frac{2}{r}\|y_n\|.$$

It follows that

$$\sup_n \sup_{\lambda \in B(0, \frac{r}{2})} \|F_n(\lambda)\| = \sup_n \sup_{\lambda \in B(0, \frac{r}{2})} \frac{r^2}{|r^2 + \lambda^2|}\|f_n(\lambda)\|$$
$$\leq \sup_n \sup_{\lambda \in B(0, \frac{r}{2})} \frac{2r}{|r^2 + \lambda^2|}\|y_n\| \leq \frac{8}{3r} \sup_n \|y_n\|.$$

This proves the claim.

It remains to prove that $F$ takes values in $X_{x_0}$. To see this, we consider the quotient space $X/X_{x_0}$. If $q : X \to X/X_{x_0}$ is the quotient mapping, then $qF(\lambda)$ is a holomorpic extension of $qR(\lambda, A)y$ to $W$. Moreover, $qF(\lambda) = 0$ for all $\text{Re}\,\lambda > 0$ since $R(\lambda, A)y \in X_{x_0}$ for these $\lambda$. Therefore, $qF \equiv 0$ on $W$ by uniqueness of analytic continuation. ////

As a consequence of this lemma we have

$$\sigma(A_{x_0}) \cap i\mathbb{R} \subset iE, \tag{5.1.1}$$

where $A_{x_0}$ denotes the generator of the restriction of $\mathbf{T}$ to $X_{x_0}$. Indeed, if $i\omega \in \sigma(A_{x_0}) \cap i\mathbb{R}$, then $\lim_{\epsilon \downarrow 0} \|R(i\omega + \epsilon, A_{x_0})\| = \infty$ by Proposition 1.1.5, and by the uniform boundedness theorem there exists $x \in X_{x_0}$ such that $\lim_{\epsilon \downarrow 0} \|R(i\omega + \epsilon, A_{x_0})x\| = 0$. Then Lemma 5.1.7 shows that $i\omega \in iE$.

In Section 5.3 we will give an example which shows that Lemma 5.1.7 is false for contraction semigroups. Thus, the assumption that $\mathbf{T}$ be isometric is essential.

Our next objective is to prove an individual version of the Arendt-Batty-Lyubich-Vũ theorem. The countability of $\sigma(A) \cap i\mathbb{R}$ will be replaced by countability of the set of singular points of the local resolvent $R(\lambda, A)x_0$ on the imaginary axis. In the next three lemmas we show how the other global spectral assumption, viz. $\sigma_p(A^*) \cap i\mathbb{R}$, can be localized to an individual $x_0 \in X$.

**Lemma 5.1.8.** *Let $\mathbf{T}$ be a uniformly bounded $C_0$-semigroup on $X$, with generator $A$. Let $x_0 \in X$ and let $i\omega \in \sigma_p(A_{x_0}^*) \cap i\mathbb{R}$. Then the map $\lambda \mapsto R(\lambda, A)x_0$ cannot be holomorphically extended to a neighbourhood of $i\omega$.*

*Proof:* Without loss of generality we may assume $\omega = 0$. Let $0 \neq y^* \in D(A_{x_0}^*)$ be such that $A_{x_0}^* y^* = 0$. Then $T_{x_0}^*(t) y^* = y^*$ for all $t \geq 0$. Hence if $\langle y^*, x_0 \rangle = 0$, then also $\langle y^*, T(t)x_0 \rangle = 0$ for all $t \geq 0$, so $y^* = 0$. Since the linear span of the set $\{T(t)x_0 : t \geq 0\}$ is dense in $X_{x_0}$, this leads to a contradiction. It follows that $\langle y^*, x_0 \rangle =: \alpha \neq 0$. Then for all $\mathrm{Re}\,\lambda > 0$ we have

$$0 = \langle A_{x_0}^* y^*, R(\lambda, A)x_0 \rangle = \langle y^*, AR(\lambda, A)x_0 \rangle = \langle y^*, \lambda R(\lambda, A)x_0 - x_0 \rangle.$$

Hence,

$$\langle y^*, R(\lambda, A)x_0 \rangle = \frac{\alpha}{\lambda}, \quad \forall \mathrm{Re}\,\lambda > 0.$$

This shows that $\lim_{\lambda \downarrow 0} \|R(\lambda, A)x_0\| = \infty$. ////

**Lemma 5.1.9.** *Let $\mathbf{T}$ be a uniformly bounded $C_0$-semigroup on $X$, with generator $A$. Then the following two assertions are equivalent:*

(i) $0 \notin \sigma_p(A^*)$;

(ii) *For all $x \in X$,* $\displaystyle\lim_{t \to \infty} \frac{1}{t} \left\| \int_0^t T(s)x\,ds \right\| = 0.$

*Proof:* Assume (i). By the Hahn-Banach theorem, the set $\{Ay : y \in D(A)\}$ is dense in $X$. For all $x = Ay$ in this set we have

$$\lim_{t \to \infty} \frac{1}{t} \left\| \int_0^t T(s)x\,ds \right\| = \lim_{t \to \infty} \frac{1}{t} \|T(t)y - y\| = 0$$

using that **T** is uniformly bounded. Since the operators $S_t x := t^{-1} \int_0^t T(s)x\, dt$, are uniformly bounded for $t > 0$, (ii) follows from this by density.

Conversely, assume (ii) and let $x^* \in D(A^*)$ be such that $A^* x^* = 0$. Then $T^*(t) x^* = x^*$ for all $t \geq 0$ by the remark following Proposition 2.1.6. Hence for all $x \in X$,

$$\langle x^*, x \rangle = \lim_{t \to \infty} \frac{1}{t} \int_0^t \langle T^*(s)x^*, x \rangle \, ds = \lim_{t \to \infty} \frac{1}{t} \langle x^*, \int_0^t T(s)x\, ds \rangle = 0.$$

Therefore, $x^* = 0$ and $0 \notin \sigma_p(A^*)$.  ////

If **T** is uniformly bounded and for some $x_0 \in X$ we have

$$\lim_{t \to \infty} \frac{1}{t} \left\| \int_0^t T(s)x_0\, ds \right\| = 0,$$

then by a density argument

$$\lim_{t \to \infty} \frac{1}{t} \left\| \int_0^t T(s)x\, ds \right\| = 0, \quad \forall x \in X_{x_0}.$$

Hence by the previous two lemmas applied to $X_{x_0}$ we obtain:

**Lemma 5.1.10.** *Let **T** be a uniformly bounded $C_0$-semigroup on $X$, with generator $A$. For a given $x_0 \in X$ the following are equivalent:*

(i) $\sigma_p(A^*_{x_0}) = \emptyset$;

(ii) $\lim_{t \to \infty} \frac{1}{t} \left\| \int_0^t e^{-i\omega s} T(s)x_0\, ds \right\| = 0$ *for all $i\omega \in i\mathbb{R}$ to which $R(\lambda, A)x_0$ cannot be holomorphically extended.*

**Theorem 5.1.11.** *Let **T** be a uniformly bounded $C_0$-semigroup on a Banach space $X$, with generator $A$. Let $x_0 \subset X$ and a countable closed set $E \subset \mathbb{R}$ be given such that:*

(i) *The map $\lambda \mapsto R(\lambda, A)x_0$ admits a holomorphic extension to a neighbourhood $V$ of $\{\mathrm{Re}\,\lambda \geq 0\} \setminus iE$;*

(ii) *For all $\omega \in E$, $\lim_{t \to \infty} \frac{1}{t} \left\| \int_0^t e^{-i\omega s} T(s)x_0\, ds \right\| = 0.$*

*Then, $\lim_{t \to \infty} \|T(t)x_0\| = 0$.*

*Proof:* Consider the isometric limit semigroup $(Y, \pi, \mathbf{U})$ associated to the restriction $\mathbf{T}_{x_0}$ of **T** to $X_{x_0}$; let $B$ denote the generator of **U**. We shall prove that $Y = \{0\}$.

By shrinking $V$, we may assume $V$ to be connected. Since $\pi R(\lambda, A)x_0 = R(\lambda, B)\pi x_0$ for $\mathrm{Re}\,\lambda > 0$, we see that $R(\lambda, B)\pi x_0$ can be holomorphically extended to $V$. Since the orbit of $\pi x_0$ spans a dense subspace of $Y$, (5.1.1) implies that $\sigma(B) \subset iE$. Hence $\sigma(B)$ is countable. Furthermore by (ii), for all $\omega \in E$ we have

$$\lim_{t \to \infty} \frac{1}{t} \left\| \int_0^t e^{-i\omega s} U(s) \pi x_0 \, ds \right\| = 0,$$

so $\sigma_p(B^*) = \emptyset$ by Lemma 5.1.10.

It follows that **U** is uniformly stable by virtue of the Arendt-Batty-Lyubich-Vũ theorem. Since **U** is also isometric, we conclude that $Y = \{0\}$.

By Lemma 5.1.9, the theorem remains true if we replace condition (ii) by the global assumption $\sigma_p(A^*) \cap i\mathbb{R} = \emptyset$. In this form, the theorem can also proved alternatively by mimicking the proof of Theorem 5.1.5.

It turns out that Theorem 5.1.11 can be further generalized: the uniform boundedness assumption may be replaced by boundedness and uniform continuity of the orbit $T(\cdot)x_0$. This will be proved in Section 5.3.

Another generalization of Theorem 5.1.11 is obtained by developing the assumption (ii). This assumption is an ergodic one, and a version of Theorem 5.1.11 can be formulated in which the limit in (ii) is assumed to exist but allowed to be different from zero. This point is taken up at the end of Section 5.7.

The next example shows that the uniform boundedness assumption in Theorem 5.1.5 cannot be omitted and that in Theorem 5.1.11 it is not enough to assume only boundedness of the orbit $T(\cdot)x_0$, even if $X$ is a Hilbert space and $x_0 \in D(A)$.

**Example 5.1.12.** Let $X = l^2$ and define the $C_0$-semigroup **T** on $X$ by

$$(T(t)x)_{2n-1} = e^{int - t/n^2}(x_{2n-1} + tx_{2n}),$$
$$(T(t)x)_{2n} = e^{int - t/n^2} x_{2n}.$$

Then, by the triangle inequality, for all $x \in X$ we have

$$\|T(t)x\| \leq \left( \sum_{n=1}^{\infty} |x_{2n-1} + tx_{2n}|^2 \right)^{\frac{1}{2}} + \left( \sum_{n=1}^{\infty} |x_{2n}|^2 \right)^{\frac{1}{2}}$$
$$\leq \left( \sum_{n=1}^{\infty} |x_{2n-1}|^2 \right)^{\frac{1}{2}} + (t+1) \left( \sum_{n=1}^{\infty} |x_{2n}|^2 \right)^{\frac{1}{2}}$$
$$\leq (t+2)\|x\|.$$

It follows that $\|T(t)\| \leq t+2$ for all $t \geq 0$. The generator $A$ is given by

$$D(A) = \{x \in X : (nx_n) \in X\};$$
$$(Ax)_{2n-1} = x_{2n} + (in - 1/n^2)x_{2n-1},$$
$$(Ax)_{2n} = (in - 1/n^2)x_{2n}.$$

Hence, $\sigma(A) = \{in - 1/n^2 : n = 1, 2, ...\}$ and $\sigma(A) \cap i\mathbb{R} = \emptyset$. Define $y \in X$ by
$$y_{2n-1} = 0,$$
$$y_{2n} = n^{-\frac{5}{2}}.$$
A simple calculation shows that $y \in D(A)$. Moreover,
$$(T(t)y)_{2n-1} = tn^{-\frac{5}{2}} e^{int - t/n^2},$$
$$(T(t)y)_{2n} = n^{-\frac{5}{2}} e^{int - t/n^2}.$$
By taking the Riemann sums of $s \mapsto s^{-5} e^{-2/s^2}$ in the points $nt^{-\frac{1}{2}}$, $n = 1, 2, ...$, applying the dominated convergence theorem, and noting that
$$\lim_{t \to \infty} \sum_{n=1}^{\infty} \frac{1}{n^5} e^{-2t/n^2} = 0,$$
we obtain
$$\lim_{t \to \infty} \|T(t)y\|^2 = \lim_{t \to \infty} \sum_{n=1}^{\infty} \frac{1 + t^2}{n^5} e^{-2t/n^2}$$
$$= \lim_{t \to \infty} \sum_{n=1}^{\infty} \frac{t^2}{n^5} e^{-2t/n^2}$$
$$= \int_0^{\infty} s^{-5} e^{-2/s^2} \, ds$$
$$= \frac{1}{2} \int_0^{\infty} u e^{-2u} \, du = \frac{1}{8}.$$
Thus, $\|T(\cdot)y\|$ is bounded but fails to converge to zero.

A sufficient condition for uniform stability is the existence of a dense subset of elements with countable local spectrum on the imaginary axis:

**Corollary 5.1.13.** *Let* **T** *be a uniformly bounded $C_0$-semigroup on a Banach space $X$, with generator $A$. Assume*
(i) $\sigma_p(A^*) \cap i\mathbb{R} = \emptyset$;
(ii) *There exist a dense subspace $Y \subset X$ with the following property: for each $y \in Y$ there is a countable closed subset $E \subset \mathbb{R}$ such that for all $y \in Y$ the map $\lambda \mapsto R(\lambda, A)y$ admits a holomorphic extension to a neighbourhood of $\{\text{Re}\,\lambda \geq 0\} \backslash iE$.*

*Then* **T** *is uniformly stable.*

Indeed, we apply Theorem 5.1.11 and the remark following it.

One may wonder whether the conditions of Theorem 5.1.11 and its corollaries are necessary for uniform stability. Unfortunately this is note the case: the following is an example of a uniformly stable $C_0$-semigroup with the property that whenever $\lambda \mapsto R(\lambda, A)x$ admits a holomorphic extension to a neighbourhood of some $iw \in i\mathbb{R}$, then $x = 0$.

**Example 5.1.14.** Let $X = L^1(\mathbb{R}_+, w(t)\,dt)$, where $w : \mathbb{R}_+ \to \mathbb{R}_+$ satisfies
   (i) $w$ is non-increasing;
   (ii) $\lim_{t\to\infty} w(t) = 0$;
   (iii) For each $a > 0$ there exists a constant $c > 0$ such that $w(t) \geq ce^{-at}$ for all $t \geq 0$.

Let $\mathbf{T}$ be the $C_0$–semigroup on $X$ defined by

$$T(t)f(s) := \begin{cases} f(s-t), & 0 \leq t \leq s; \\ 0, & \text{else}, \end{cases}$$

and let $A$ denote its generator. By (i) and (ii), for all $f \in X$ we have

$$\lim_{t\to\infty} \|T(t)f\| = \lim_{t\to\infty} \int_t^\infty |f(s-t)|\, w(s)\, ds = \lim_{t\to\infty} \int_0^\infty |f(s)|\, w(s+t)\, ds = 0,$$

so $\mathbf{T}$ is uniformly stable. We will prove that $0$ is the only element in $X$ whose local resolvent can be extended across some point of the imaginary axis.

By (ii), for $\operatorname{Re}\lambda < 0$ the function

$$g_\lambda(s) := \frac{e^{\lambda s}}{w(s)}, \quad s \geq 0$$

is bounded, so $h_\lambda(s) := e^{\lambda s}$ defines an element of $X^*$. For all $f \in X$ and $t \geq 0$ we have

$$\langle h_\lambda, T(t)f \rangle = \int_t^\infty f(s-t)e^{\lambda s}\, ds$$
$$= \int_0^\infty f(s)e^{\lambda(s+t)}\, ds$$
$$= e^{\lambda t} \int_0^\infty f(s)e^{\lambda s}\, ds$$
$$= e^{\lambda t} \langle h_\lambda, f \rangle.$$

Thus,
$$T^*(t)h_\lambda = e^{\lambda t} h_\lambda, \quad t \geq 0,$$

so $h_\lambda \in D(A^*)$ and $A^* h_\lambda = \lambda h_\lambda$.

Now suppose $f \in X$ is such that the map $\lambda \mapsto R(\lambda, A)f$ has a holomorphic extension $F$ to a connected neighbourhood $V$ of some point $i\omega \in i\mathbb{R}$. From the identity $R(1, A)f = (I + (\lambda - 1)R(1, A))R(\lambda, A)f$, by analytic continuation we obtain

$$R(1, A)f = (I + (\lambda - 1)R(1, A))F(\lambda) = (\lambda - A)R(1, A)F(\lambda)$$

for all $\lambda \in V$. Hence, for all $\lambda \in V$ with $\operatorname{Re}\lambda < 0$,

$$\int_0^\infty e^{\lambda s}(R(1, A)f)(s)\, ds = \langle h_\lambda, R(1, A)f \rangle = \langle (\lambda - A^*)h_\lambda, R(1, A)F(\lambda) \rangle = 0.$$

As a function of $\lambda$, the first of these expressions is holomorphic on $\{\operatorname{Re}\lambda < 0\}$ and vanishes in $\{\operatorname{Re}\lambda < 0\} \cap V$. Therefore,

$$\int_0^\infty e^{\lambda s}(R(1,A)f)(s)\,ds = 0, \qquad \forall \operatorname{Re}\lambda < 0.$$

By the uniqueness of the Laplace transform, this implies that $R(1,A)f = 0$ a.e. Hence $f = 0$ by the injectivity of $R(1,A)$.

We close this section with some propositions about the point spectra of $A$ and $A^*$. We use the existence of *left invariant means* on $BUC(\mathbb{R}_+)$, that is, positive linear functionals $0 \leq \phi \in (BUC(\mathbb{R}_+))^*$ such that

(i) $\langle \phi, \mathbf{1} \rangle = 1$,
(ii) $\langle \phi, f(\cdot) \rangle = \langle \phi, f(\cdot + s) \rangle$ for all $s \geq 0$.

In (i), $\mathbf{1}$ denotes the constant one function. The existence of such $\phi$ is easily proved as follows. Fix an arbitrary $0 \leq \phi_0 \in (BUC(\mathbb{R}_+))^*$ such that $\phi_0(\mathbf{1}) = 1$. For $n = 1, 2, \dots$ define $0 \leq \phi_n \in (BUC(\mathbb{R}_+))^*$ by

$$\langle \phi_n, f \rangle := \frac{1}{n} \int_0^n \langle \phi_0, f(\cdot + s) \rangle\, ds.$$

Clearly, $\|\phi_n\| \leq \|\phi_0\|$ and $\langle \phi_n, \mathbf{1} \rangle = \langle \phi_0, \mathbf{1} \rangle = 1$ for all $n$. By the Banach-Alaoglu theorem, the unit ball of a dual Banach space is weak*-compact. Therefore, the sequence $(\phi_n)$ has a weak*-cluster point $\phi$. This is the left invariant mean we are looking for: clearly, $\phi \geq 0$ and $\langle \phi, \mathbf{1} \rangle = 1$. For an arbitrary $f \in BUC(\mathbb{R}_+)$ and $s \geq 0$, let $(n_k)$ be a subsequence such that

$$\lim_{k\to\infty} \langle \phi_{n_k}, f \rangle = \langle \phi, f \rangle$$

and

$$\lim_{k\to\infty} \langle \phi_{n_k} f(\cdot + s) \rangle = \langle \phi, f(\cdot + s) \rangle.$$

Then,

$$\langle \phi, f(\cdot + s) \rangle = \lim_{k\to\infty} \frac{1}{n_k} \int_0^{n_k} \langle \phi_0, f(\cdot + s + \sigma) \rangle\, d\sigma$$
$$= \lim_{k\to\infty} \frac{1}{n_k} \int_0^{n_k} \langle \phi_0, f(\cdot + \sigma) \rangle\, d\sigma = \langle \phi, f \rangle.$$

This proves that $\phi$ is a left invariant mean.

A $C_0$-semigroup $\mathbf{T}$ is *weakly almost periodic* if for all $x \in X$ the set $\{T(t)x : t \geq 0\}$ is relatively weakly compact in $X$.

**Proposition 5.1.15.** *Let $\mathbf{T}$ be a uniformly bounded $C_0$-semigroup on a Banach space $X$, with generator $A$. Then $\sigma_p(A) \cap i\mathbb{R} \subset \sigma_p(A^*) \cap i\mathbb{R}$. If in addition $\mathbf{T}$ is weakly almost periodic, then $\sigma_p(A) \cap i\mathbb{R} = \sigma_p(A^*) \cap i\mathbb{R}$.*

*Proof:* Let $i\omega \in \sigma_p(A) \cap i\mathbb{R}$. By rescaling we may assume that $\omega = 0$. Let $x_0 \in D(A)$ be an eigenvector with eigenvalue 0. Then $T(t)x_0 = x_0$ for all $t \geq 0$ by Proposition 2.1.6. Let $y_0^* \in X^*$ be such that $\langle y_0^*, x_0 \rangle = 1$. Let $\phi \in (BUC(\mathbb{R}_+))^*$ be a left invariant mean and define $x_0^* \in X^*$ by

$$\langle x_0^*, x \rangle := \phi(\langle y_0^*, T(\cdot)x \rangle), \quad x \in X,$$

noting that $\langle x^*, T(\cdot)x \rangle \in BUC(\mathbb{R}_+)$ for all $x \in X$ and $x^* \in X^*$. Then, for all $x \in X$ we have

$$\langle T^*(t)x_0^*, x \rangle = \phi(\langle y_0^*, T(t+\cdot)x \rangle) = \phi(\langle y_0^*, T(\cdot)x \rangle) = \langle x_0^*, x \rangle.$$

Hence, $T^*(t)x_0^* = x_0^*$ for all $t \geq 0$, so $x_0^* \in D(A^*)$ and $A^*x_0^* = 0$. Also,

$$\langle x_0^*, x_0 \rangle = \phi(\langle y_0^*, T(\cdot)x_0 \rangle) = \phi(\langle y_0^*, x_0 \rangle \mathbf{1}) = \phi(\mathbf{1}) = 1,$$

which shows that $x_0^* \neq 0$. This proves that $x_0^*$ is an eigenvector of $A^*$ with eigenvalue 0.

Next assume that **T** is weakly almost periodic. Let $0 \in \sigma_p(A^*) \cap i\mathbb{R}$ and let $x_0^* \in D(A^*)$ be an eigenvector and let $y_0 \in X$ be such that $\langle x_0^*, y_0 \rangle = 1$. Define $x_0^{**} \in X^{**}$ by

$$\langle x_0^{**}, x^* \rangle := \phi(\langle x^*, T(\cdot)y_0 \rangle), \quad x^* \in X^*.$$

We claim that $x_0^{**} \in X$. To see this, let $y^* \in X^*$ and $\alpha \in \mathbb{R}$ be such that $\text{Re}\,\langle y^*, T(t)x_0 \rangle \geq \alpha$ for all $t \geq 0$. Then also $\text{Re}\,\langle x_0^{**}, y^* \rangle = \text{Re}\,\phi(\langle y^*, T(\cdot)x_0 \rangle) = \phi(\text{Re}\,\langle y^*, T(\cdot)x_0 \rangle) \geq \phi(\alpha) = \alpha$. By the Hahn-Banach separation theorem this implies that $x_0^{**}$ lies in the weak*-closure of the convex hull of $jH$, where $H := \{T(t)x_0 : t \geq 0\}$ and $j : X \to X^{**}$ is the canonical embedding. But the weak*-closure of $j(\operatorname{co} H)$ in $X^{**}$ is $j\left(\overline{\operatorname{co} H}^{weak}\right)$: since $H$ is relatively weakly compact in $X$, so is its convex hull $\operatorname{co} H$ by the Krein-Shmulyan theorem, and since $j : (X, \text{weak}) \to (X^{**}, \text{weak}^*)$ is continuous, it maps the weakly compact set $\overline{\operatorname{co} H}^{weak}$ onto the weak*-compact (hence weak*-closed) set $j\left(\overline{\operatorname{co} H}^{weak}\right)$.

We have proved that $x_0^{**} = jx_0$ for some $x_0$ in the weak closure of $\operatorname{co} H$. As in the first part of the proof, we check that $x_0 \neq 0$ and that $T(t)x_0 = x_0$ for all $t \geq 0$. Hence, $x_0$ is an eigenvector of $A$ with eigenvalue 0.    ////

As a consequence of the second part of this proposition and Theorem 5.1.11, we have:

**Corollary 5.1.16.** *Let* **T** *be a weakly almost periodic* $C_0$*-semigroup on a Banach space* $X$, *with generator* $A$, *and assume that* $\sigma_p(A) \cap i\mathbb{R} = \emptyset$. *Let* $E \subset \mathbb{R}$ *be closed and countable, and let* $x_0 \in X$ *be given such that the map* $\lambda \mapsto R(\lambda, A)x_0$ *admits a holomorphic extension to a neighbourhood of* $\{\text{Re}\,\lambda \geq 0\} \setminus iE$. *Then* $\lim_{t \to \infty} \|T(t)x_0\| = 0$.

This corollary will be developed further in Section 5.7.

Clearly, uniformly bounded semigroups in reflexive Banach spaces are weakly almost periodic. More generally, we shall prove now that uniformly bounded $\odot$-reflexive semigroups are weakly almost periodic. Recall that a $C_0$-semigroup $\mathbf{T}$ is $\odot$-*reflexive* if the canonical map $j : X \to X^{\odot\odot}$ defined by

$$\langle jx, x^\odot \rangle := \langle x^\odot, x \rangle, \quad x \in X,\ x^\odot \in X^\odot,$$

maps $X$ onto $X^{\odot\odot}$. By a theorem of de Pagter, $\mathbf{T}$ is $\odot$-reflexive if and only if $R(\lambda, A)$ is weakly compact for some (and hence for all) $\lambda \in \varrho(A)$.

In the next proposition we use Grothendieck's lemma: a set $H \subset X$ is relatively weakly compact if for all $\epsilon > 0$ there exists a weakly compact set $K$ such that $H \subset K + \epsilon B_X$, where $B_X$ is the unit ball of $X$.

**Proposition 5.1.17.** *Let $\mathbf{T}$ be a uniformly bounded $\odot$-reflexive $C_0$-semigroup. Then $\mathbf{T}$ is weakly almost periodic.*

*Proof:* Since $\mathbf{T}$ is uniformly bounded, for all $\lambda \in \varrho(A)$ and $x \in X$ the set $O_{\lambda,x} := \{T(t)(\lambda R(\lambda, A)x) : t \geq 0\}$ is relatively weakly compact by de Pagter's theorem. Let $x \in X$ be arbitrary, let $\epsilon > 0$, and choose $\lambda \in \mathbb{R}$ so large that $\|\lambda R(\lambda, A)x - x\| \leq \epsilon M^{-1}$, where $M$ is the boundedness constant of $\mathbf{T}$. Then, for all $t \geq 0$ we have $\|T(t)(\lambda R(\lambda, A)x) - T(t)x\| \leq \epsilon$. It follows that the orbit $O_x = \{T(t)x : t \geq 0\}$ is contained in $O_{\lambda,x} + \epsilon B_X$. Therefore, $O_x$ is relatively weakly compact by Grothendieck's lemma.  ////

## 5.2. The Katznelson-Tzafriri theorem

In this section we apply the isometric limit semigroup construction to prove a Katznelson-Tzafriri type theorem for $C_0$-semigroups.

A function $f \in L^1(\mathbb{R})$ is said to be of *spectral synthesis* for a closed set $E \subset \mathbb{R}$ if $f$ can be approximated in the norm of $L^1(\mathbb{R})$ by a sequence $(f_n)$ of functions, each of which has the property that its Fourier transform $\hat{f}_n$ vanishes in a neighbourhood of $E$. If $f$ is of spectral synthesis for $E$, then the Fourier transform of $f$ vanishes on $E$. Upon identifying $L^1(\mathbb{R}_+)$ in the natural way with a closed subspace of $L^1(\mathbb{R})$, we say that $f \in L^1(\mathbb{R}_+)$ is of *spectral synthesis* with respect to $E$ if $f$ is so when regarded as an element of $L^1(\mathbb{R})$. The set $E$ is called *spectral* if every $f \in L^1(\mathbb{R})$ whose Fourier transform vanishes on $E$ is of spectral synthesis for $E$.

For $f \in L^1(\mathbb{R}_+)$ and a uniformly bounded $C_0$-semigroup $\mathbf{T}$ we define the bounded operator $\hat{f}(\mathbf{T})$ by

$$\hat{f}(\mathbf{T})x := \int_0^\infty f(t) T(t) x\, dt, \quad x \in X.$$

If $\mathbf{T}$ is a $C_0$-group, this definition is consistent with that in Section 2.4.

**Theorem 5.2.1.** *Let* $\mathbf{T}$ *be a uniformly bounded $C_0$–semigroup on a Banach space $X$, with generator $A$. Let $E \subset \mathbb{R}$ be closed and let $f \in L^1(\mathbb{R}_+)$ be of spectral synthesis with respect to $-E$. If, for some $x_0 \in X$, the map $\lambda \mapsto R(\lambda, A)x_0$ extends holomorphically to a neighbourhood of $\{\operatorname{Re} \lambda \geq 0\}\setminus iE$, then*

$$\lim_{t\to\infty} \|T(t)\hat{f}(\mathbf{T})x_0\| = 0.$$

*Proof:* By renorming $X$ we may assume that $\mathbf{T}$ is a contraction semigroup.

For $\lambda > 0$, let $K_\lambda$ denote the Fejér kernel for the real line, $K_\lambda(t) := \lambda K(\lambda t)$, where

$$K(t) = \frac{1}{2\pi}\left(\frac{\sin(t/2)}{t/2}\right)^2 = \frac{1}{2\pi}\int_{-1}^{1}(1-|s|)e^{ist}\,ds.$$

As is well-known (see, e.g., [Ka]), $\hat{K}_\lambda(s) = \max\{1 - \lambda^{-1}|s|, 0\}$, so $\hat{K}_\lambda$ is compactly supported, and for all $g \in L^1(\mathbb{R})$ we have $\lim_{\lambda\to\infty}\|K_\lambda * g - g\|_1 = 0$.

If $E = \mathbb{R}$, then $\hat{f}$ is identically zero, hence $f = 0$ and there is nothing to prove. Therefore, we may assume that $E$ is a proper subset of $\mathbb{R}$. Choose a sequence $(f_n) \subset L^1(\mathbb{R})$ such that $\lim_{n\to\infty}\|f - f_n\|_1 = 0$ and each $\hat{f}_n$ vanishes in a neighbourhood of $-E$. By replacing $f_n$ by $K_{\lambda_n} * f_n$ for large enough $\lambda_n$, we may assume that the Fourier transform of $f_n$ is compactly supported.

Let $X_{x_0}$ denote the closed linear span of the orbit of $x_0$, let $(Y_0, \pi_0, \mathbf{U}_0)$ be the isometric limit semigroup associated to the restriction $\mathbf{T}_{x_0}$ of $\mathbf{T}$, and let $B_0$ be the generator of $\mathbf{U}_0$. The $\mathbf{U}_0$-orbit of $\pi_0 x_0$ is dense in $Y_0$. Since $R(\lambda, B)\pi_0 x_0 = \pi_0 R(\lambda, A)x_0$ extends holomorphically to a neighbourhood of $\{\operatorname{Re} \lambda \geq 0\}\setminus iE$, it follows that $\sigma(B_0) \cap i\mathbb{R} \subset iE$ by (5.1.1). By Corollary 5.1.3, $\mathbf{U}_0$ extends to an isometric $C_0$–group. Each $\hat{f}_n$ vanishes in a neighbourhood of $i\sigma(B_0) = i\sigma(B_0)\cap\mathbb{R}$, hence $\hat{f}_n(\mathbf{U}_0) = 0$ by Lemma 2.4.3. Since the map $g \mapsto \hat{g}(\mathbf{U}_0)$ is continuous from $L^1(\mathbb{R})$ into $\mathcal{L}(Y_0)$, it follows that also $\hat{f}(\mathbf{U}_0) = 0$. But then

$$\lim_{t\to\infty}\|T(t)\hat{f}(\mathbf{T})x_0\| = \lim_{t\to\infty}\left\|T(t)\int_0^\infty f(s)T(s)x_0\,ds\right\|$$

$$= \left\|\pi_0\int_0^\infty f(s)T(s)x_0\,ds\right\|$$

$$= \left\|\int_0^\infty f(s)U_0(s)\pi_0 x_0\,ds\right\|$$

$$= \|\hat{f}(\mathbf{U}_0)\pi_0 x_0\| = 0.$$

////

It will be shown in the next section that the uniform boundedness assumption on $\mathbf{T}$ is unnecessarily strong; it is enough to know that $T(\cdot)x_0$ is bounded.

If $f$ is of spectral synthesis with respect to $i\sigma(A) \cap \mathbb{R}$ a global version of Theorem 5.2.1 holds. In this case we even obtain convergence to 0 in the norm topology. The proof of this is based on the following construction.

Let **T** be a $C_0$–semigroup on a Banach space $X$ and let $\mathcal{L}_0(X)$ be the closed subspace of $\mathcal{L}(X)$ consisting of all operators $S$ such that $\lim_{t\downarrow 0}\|T(t)S-S\|=0$. On $\mathcal{L}_0(X)$, **T** induces a $C_0$–semigroup $\mathcal{T}$ by the formula $\mathcal{T}(t)S := T(t)S$, $S \in \mathcal{L}_0(X)$. Let $\mathcal{A}$ denote its generator.

**Lemma 5.2.2.** *Let* **T** *be a $C_0$–semigroup on a Banach space $X$, with generator $A$. Then $\sigma(\mathcal{A}) \subset \sigma(A)$ and their peripheral spectra agree.*

*Proof:* For $\lambda \in \varrho(A)$ we define the bounded operator $\mathcal{R}_\lambda$ on $\mathcal{L}_0(X)$ by $\mathcal{R}_\lambda(S) := R(\lambda, A)S$. Then $\mathcal{R}_\lambda$ defines a two-sided inverse for $\lambda - \mathcal{A}$: from the identity

$$\frac{1}{t}(T(t) - I)R(\lambda, A)S = \frac{\lambda}{t}R(\lambda, A)\int_0^t T(s)S - S\,ds,$$

valid for all $S \in \mathcal{L}_0(X)$, it follows upon letting $t \downarrow 0$ that $\mathcal{R}_\lambda \in D(\mathcal{A})$ and $(\lambda - \mathcal{A})\mathcal{R}_\lambda S = S$. Also, if $S \in D(\mathcal{A})$, then $Sx \in D(A)$ and $\mathcal{A}Sx = ASx$ for all $x \in X$, so $\mathcal{R}_\lambda(\lambda - \mathcal{A})Sx = R_\lambda(\lambda - A)Sx = Sx$. Hence $\lambda \in \varrho(\mathcal{A})$ and $R(\lambda, \mathcal{A}) = \mathcal{R}_\lambda$.

For the proof of the second assertion we may assume that $s(A) = 0$. Let $i\omega \in \sigma(A) \cap i\mathbb{R}$. By Proposition 1.1.5 and the uniform boundedness theorem there exists an $x \in X$ of norm one such that $\lim_{\epsilon\downarrow 0}\|R(i\omega+\epsilon, A)x\| = \infty$. By what we just proved, for all $\epsilon > 0$ we have $i\omega + \epsilon \in \varrho(\mathcal{A})$ and $R(i\omega+\epsilon, \mathcal{A})S = R(i\omega+\epsilon, A)S$. Let $x^* \in X^*$ be such that $\langle x^*, x\rangle \neq 0$ and consider the rank one operator $S := x^* \otimes x$. Then $S \in \mathcal{L}_0(X)$ and

$$\|R(i\omega + \epsilon, \mathcal{A})Sx\| = \|R(i\omega + \epsilon, A)x\| \cdot |\langle x^*, x\rangle|.$$

This proves that $\lim_{\epsilon\downarrow 0}\|R(i\omega + \epsilon, \mathcal{A})\| = \infty$. But this implies that $i\omega \in \sigma(\mathcal{A})$. Hence $\sigma(\mathcal{A}) \cap i\mathbb{R} \supset \sigma(A) \cap i\mathbb{R}$ and the lemma is proved. ////

**Theorem 5.2.3.** *Let* **T** *be a uniformly bounded $C_0$–semigroup on a Banach space $X$, with generator $A$. If $f \in L^1(\mathbb{R}_+)$ is of spectral synthesis with respect to $i\sigma(A) \cap \mathbb{R}$, then*

$$\lim_{t\to\infty} \|T(t)\hat{f}(\mathbf{T})\| = 0.$$

*Proof:* By renorming $X$ we may assume that **T** is a contraction semigroup.
By Theorem 5.2.1,

$$\lim_{t\to\infty}\|T(t)\hat{f}(\mathbf{T})x\| = 0, \quad \forall x \in X.$$

By Lemma 5.2.2, we can apply this to the semigroup $\mathcal{T}$ on $\mathcal{L}_0(X)$. It is easily checked that $\mathcal{L}_0(X)$ contains the closure with respect to the uniform operator topology of the set $\{\hat{g}(\mathbf{T}) : g \in L^1(\mathbb{R}_+)\}$: in fact, if $g \in L^1(\mathbb{R}_+)$, then

$$\|T(t)\hat{g}(\mathbf{T})x - \hat{g}(\mathbf{T})x\| = \left\|\int_0^\infty g(s)(T(t+s) - T(s))x\,ds\right\|$$

$$\leq \left\|\int_0^\infty (g(s) - g(s+t))T(s+t)x\,ds\right\| + \left\|\int_0^t g(s)T(s)x\,ds\right\|$$

$$\leq \left(\int_0^\infty |g(s) - g(s+t)|\,ds + \int_0^t |g(s)|\,ds\right)\|x\|.$$

Since translation on $L^1(\mathbb{R}_+)$ is strongly continuous, the assertion follows from this.

We conclude that
$$\lim_{t\to\infty} \|T(t)\hat{f}(\mathbf{T})\hat{g}(\mathbf{T})\| = \lim_{t\to\infty} \|\mathcal{T}(t)\hat{f}(\mathcal{T})\hat{g}(\mathbf{T})\| = 0, \quad \forall g \in L^1(\mathbb{R}_+). \quad (5.2.1)$$

Finally, let $(h_n) \subset L^1(\mathbb{R}_+)$ an approximate identity, i.e. $\|h_n\|_1 = 1$ for all $n$ and $\lim_{n\to\infty} \|g * h_n - g\|_1 = 0$ for all $g \in L^1(\mathbb{R}_+)$. For all $n$ we have

$$\|T(t)\hat{f}(\mathbf{T})\| \leq \|T(t)(f * h_n)\hat{\ }(\mathbf{T})\| + \|T(t)(f - f * h_n)\hat{\ }(\mathbf{T})\|$$
$$\leq \|T(t)(f * h_n)\hat{\ }(\mathbf{T})\| + \|f - f * h_n\|_1.$$

Hence by applying (5.2.1) to $g = h_n$ and noting that $(f * h_n)\hat{\ }(\mathbf{T}) = \hat{f}(\mathbf{T})\hat{h}_n(\mathbf{T})$, it follows that

$$\limsup_{t\to\infty} \|T(t)\hat{f}(\mathbf{T})\| \leq \lim_{t\to\infty} \|T(t)(f * h_n)\hat{\ }(\mathbf{T})\| + \|f - f * h_n\|_1 = \|f - f * h_n\|_1.$$

Since this holds for all $n$, we conclude that $\lim_{t\to\infty} \|T(t)\hat{f}(\mathbf{T})\| = 0$. ////

Since we derived Theorem 5.2.3 from 5.2.1 it depends implicitly on Theorem 5.1.11 and hence on Lemma 5.1.7. If $f$ is of spectral synthesis with respect to $i\sigma(A) \cap \mathbb{R}$, this can be avoided by invoking the Arendt-Batty-Lyubich-Vũ theorem 5.1.5 instead.

Our next result is a partial converse of Theorem 5.2.3. We start with a lemma.

**Lemma 5.2.4.** *Let $\mathbf{T}$ be a uniformly bounded $C_0$−semigroup on a Banach space $X$, with generator $A$. Then, for all $\lambda \in \sigma_a(A)$ and all $f \in L^1(\mathbb{R}_+)$ we have*

$$\left| \int_0^\infty e^{\lambda t} f(t)\, dt \right| \leq \|\hat{f}(\mathbf{T})\|.$$

*Proof:* Choose a sequence $(x_n)$ of norm one vectors in $X$, $x_n \in D(A)$ for all $n$, such that $\lim_{n\to\infty} \|Ax_n - \lambda x_n\| \to 0$. By Proposition 2.1.6, $(x_n)$ is an approximate eigenvector of $T(t)$ with approximate eigenvalue $e^{\lambda t}$.

Let $f \in L^1(\mathbb{R}_+)$. By the dominated convergence theorem,
$$\lim_{n\to\infty} \left\| \int_0^\infty f(t)(T(t)x_n - e^{\lambda t}x_n)\, dt \right\| = 0.$$

Thus, using that $\|x_n\| = 1$,
$$\|\hat{f}(\mathbf{T})\| \geq \lim_{n\to\infty} \|\hat{f}(\mathbf{T})x_n\| = \lim_{n\to\infty} \left\| \int_0^\infty f(t)T(t)x_n\, dt \right\| = \left| \int_0^\infty e^{\lambda t} f(t)\, dt \right|.$$

////

**Theorem 5.2.5.** *Let* **T** *be a uniformly bounded $C_0$-semigroup on a Banach space $X$, with generator $A$, and let $f \in L^1(\mathbb{R}_+)$. If $\lim_{t \to \infty} \|T(t)\hat{f}(\mathbf{T})\| = 0$, then $\hat{f}$ vanishes on $i\sigma(A) \cap \mathbb{R}$.*

*Proof:* Let $i\omega \in \sigma(A) \cap i\mathbb{R}$ be such that $\hat{f}(-\omega) \neq 0$. Without loss of generality we assume that $\hat{f}(-\omega) = 1$. For $s \geq 0$, define $f_s \in L^1(\mathbb{R}_+)$ by

$$f_s(t) = \begin{cases} f(t-s), & t \geq s; \\ 0, & \text{else.} \end{cases}$$

Then,

$$\int_0^\infty e^{i\omega t} f_s(t)\, dt = \int_s^\infty e^{i\omega t} f(t-s)\, dt$$
$$= \int_0^\infty e^{i\omega(t+s)} f(t)\, dt = e^{i\omega s} \hat{f}(-\omega).$$

Similarly,

$$\hat{f}_s(\mathbf{T}) = T(s)\hat{f}(\mathbf{T}).$$

Therefore, by Lemma 5.2.4,

$$\|T(s)\hat{f}(\mathbf{T})\| = \|\hat{f}_s(\mathbf{T})\| \geq \left|\int_0^\infty e^{i\omega t} f_s(t)\, dt\right| = |e^{i\omega s}\hat{f}(-\omega)| = 1.$$

////

An interesting consequence is the following result.

**Corollary 5.2.6.** *Let* **T** *be a uniformly bounded $C_0$-semigroup on a Banach space $X$, with generator $A$. Then the following assertions are equivalent:*

(i) $\lim_{t \to \infty} \|T(t)R(\lambda, A)\| = 0$ for some $\lambda \in \varrho(A)$;
(ii) $\lim_{t \to \infty} \|T(t)R(\lambda, A)\| = 0$ for all $\lambda \in \varrho(A)$;
(iii) $\sigma(A) \cap i\mathbb{R} = \emptyset$.

*If one of the equivalent conditions holds, then* **T** *is uniformly stable.*

*Proof:* Assume (iii). Since all $f \in L^1(\mathbb{R}_+)$ are of spectral synthesis with respect to the empty set, this is in particular true for $f_\lambda(t) := e^{-\lambda t}$, $t \geq 0$, Re $\lambda > 0$. Since $\hat{f}_\lambda(\mathbf{T}) = R(\lambda, A)$, for Re $\lambda > 0$ (ii) follows from Theorem 5.2.3. For arbitrary $\lambda \in \varrho(A)$, (ii) follows from this special case via the resolvent identity.

(ii) Trivially implies (i).

If (i) holds with Re $\lambda > 0$, then

$$\lim_{t \to \infty} \|T(t)\hat{f}_\lambda(\mathbf{T})\| = \lim_{t \to \infty} \|T(t)R(\lambda, A)\| = 0$$

and therefore $\hat{f}_\lambda$ vanishes on $i\sigma(A) \cap \mathbb{R}$ by Theorem 5.2.5. But,

$$\hat{f}_\lambda(s) = \int_0^\infty e^{-ist} e^{-\lambda t}\, dt = \frac{1}{\lambda + is}$$

which does not vanish for any $s \in \mathbb{R}$. Therefore, $\sigma(A) \cap i\mathbb{R} = \emptyset$. The case of a general $\lambda \in \varrho(A)$ follows from this via the resolvent identity.

The last statement is an obvious consequence of (i) and the density of $D(A)$.
////

## 5.3. The unbounded case

In this section we shall generalize the main results of Sections 5.1 and 5.2 to individual bounded uniformly continuous orbits of a possibly unbounded semigroup. The idea is to consider a bounded uniformly continuous orbit $g_0 := T(\cdot)x_0$ as an element of $BUC(\mathbb{R}_+, X)$, the space of bounded, uniformly continuous $X$-valued functions on $\mathbb{R}_+$. Letting $\mathbf{U}$ denote the left translation semigroup on $BUC(\mathbb{R}_+, X)$, $(U(t)h)(s) := h(s+t)$, $s, t \geq 0$, it is obvious that $\lim_{t\to\infty} \|T(t)x_0\| = 0$ if and only if $\lim_{t\to\infty} \|U(t)g_0\| = 0$. Thus, in order to obtain results on the asymptotic behaviour of $T(t)x_0$, we will try to apply the individual stability results of the preceding sections to the (uniformly bounded) semigroup $\mathbf{U}$.

There is an obvious obstruction to this programme: if the map $\lambda \mapsto R(\lambda, A)x_0$ has a holomorphic extension to certain parts of the left half-plane, it is not a priori clear that the same should be true for the map $\lambda \mapsto R(\lambda, B)g_0$, where $B$ denotes the generator of $\mathbf{U}$. This problem will be addressed in a series of lemmas.

**Lemma 5.3.1.** *Let $g \in BUC(\mathbb{R}_+, X)$. For each $s \geq 0$, denote by $G_s$ the Laplace transform of the translated function $g_s(t) := g(s+t)$, $t \geq 0$. If $E \subset \mathbb{R}$ is such that $G_0$ admits a holomorphic extension to a connected neighbourhood $V$ of $\{\text{Re}\,\lambda \geq 0\} \setminus iE$, then so does each $G_s$. Moreover, for all $\lambda \in V$ and $r > 0$ such that the closure of the ball $B(\lambda, r)$ with centre $\lambda$ and radius $r$ is contained in $V$ we have*

$$\sup_{s \geq 0} \sup_{z \in B(\lambda, r)} \|G_s(\lambda)\| < \infty. \tag{5.3.1}$$

*Proof:* For all $\text{Re}\,z > 0$ we have

$$G_s(z) = \int_0^\infty e^{-zt} g(s+t)\, dt. \tag{5.3.2}$$

From this, a simple calculation leads to

$$G_s(z) = e^{zs} G_0(z) - e^{zs} \int_0^s e^{-zt} g(t)\, dt. \tag{5.3.3}$$

This identity defines a holomorphic extension of $G_s$ to $V$. It remains to prove (5.3.1). To this end fix $\lambda \in V$ arbitrary.

*Case 1*: If $\operatorname{Re}\lambda < 0$ and $0 < r < |\operatorname{Re}\lambda|$, (5.3.1) is an immediate consequence of (5.3.3): for all $s \geq 0$ and $z \in B(\lambda, r)$ we have

$$\|G_s(z)\| \leq e^{\operatorname{Re} zs}\|G_0(z)\| + e^{\operatorname{Re} zs} \int_0^s e^{-\operatorname{Re} zt}\|g\|_\infty \, dt$$

$$\leq \|G_0(z)\| + \frac{1 - e^{\operatorname{Re} zs}}{|\operatorname{Re} z|}\|g\|_\infty$$

$$\leq \sup_{\zeta \in B(\lambda,r)} \|G_0(\zeta)\| + (|\operatorname{Re}\lambda| - r)^{-1}\|g\|_\infty.$$

*Case 2*: Assume that $\operatorname{Re}\lambda = 0$. By rescaling we may assume that $\lambda = 0$. Let $r > 0$ be such that $B(0, 2r) \subset V$ and define, for $s \geq 0$, the functions $H_s : B(0, 2r) \to X$ by

$$H_s(z) := \left(1 + \frac{z^2}{4r^2}\right) G_s(z).$$

Let $z \in \partial B(0, 2r) \setminus \{2ir, -2ir\}$ be arbitrary; say $z = 2re^{i\theta}$. If $-\frac{\pi}{2} < \theta < \frac{\pi}{2}$, we have, using (5.3.2),

$$\|H_s(z)\| \leq |1 + e^{2i\theta}| \int_0^\infty e^{-2tr\cos\theta}\|g\|_\infty \, dt = 2\cos\theta \cdot \frac{1}{2r\cos\theta}\|g\|_\infty = \frac{1}{r}\|g\|_\infty.$$

If $\frac{\pi}{2} < \theta < \frac{3\pi}{2}$ we have, using (5.3.3),

$$\|H_s(z)\| \leq |1 + e^{2i\theta}| \left(e^{2sr\cos\theta}\|G_0(z)\| + \int_0^s e^{2(s-t)r\cos\theta}\|g\|_\infty \, dt\right)$$

$$\leq 2|\cos\theta| \left(\|G_0(z)\| + \frac{1}{2r\cos\theta}\|g\|_\infty\right)$$

$$\leq 2\|G_0(z)\| + \frac{1}{r}\|g\|_\infty.$$

Hence, by the maximum modulus principle,

$$\|H_s(z)\| \leq 2 \sup_{\zeta \in B(\lambda,r)} \|G_0(\zeta)\| + \frac{1}{r}\|g\|_\infty, \quad \forall z \in B(0, 2r).$$

For all $s \geq 0$ and all $z \in B(0, r)$, this implies

$$\|G_s(z)\| = \left|\frac{4r^2}{z^2 + 4r^2}\right| \|H_s(z)\| \leq \frac{4}{3}\left(2 \sup_{\zeta \in B(\lambda,r)} \|G_0(\zeta)\| + \frac{1}{r}\|g\|_\infty\right).$$

This proves (5.3.1) for $\operatorname{Re}\lambda = 0$.

*Case 3*: If $\operatorname{Re}\lambda > 0$ and $0 < r < \operatorname{Re}\lambda$, (5.3.1) follows immediately from (5.3.2).

*Case 4*: If $\lambda \in V$ and $r > 0$ with $B(\lambda, r) \subset V$ is arbitrary, (5.3.1) follows from the previous three cases.  ////

**Lemma 5.3.2.** *In the situation of Lemma 5.3.1, for each $\lambda \in V$ the map $s \mapsto G_s(\lambda)$ is uniformly continuous.*

*Proof:* Fix $\lambda \in V$.

*Case 1.* First let $\operatorname{Re} \lambda < 0$. In the right hand side of (5.3.3), the first term $e^{\lambda s} G_0(\lambda)$ is uniformly continuous as a function of $s$, so in order to prove uniform continuity of $s \mapsto G_s(\lambda)$, it suffices to prove that the second term is uniformly continuous. Let $0 \leq s_0 \leq s_1$. Then,

$$\left\| e^{\lambda s_1} \int_0^{s_1} e^{-\lambda t} g(t)\, dt - e^{\lambda s_0} \int_0^{s_0} e^{-\lambda t} g(t)\, dt \right\|$$

$$= \left\| (e^{\lambda s_1} - e^{\lambda s_0}) \int_0^{s_0} e^{-\lambda t} g(t)\, dt + \int_{s_0}^{s_1} e^{\lambda(s_1-t)} g(t)\, dt \right\|$$

$$\leq |e^{\lambda s_1} - e^{\lambda s_0}| \int_0^{s_0} e^{-\operatorname{Re}\lambda t} \|g\|_\infty\, dt + \int_{s_0}^{s_1} \|g\|_\infty\, dt$$

$$\leq e^{\operatorname{Re}\lambda s_0} |e^{\lambda(s_1-s_0)} - 1| \frac{e^{-\operatorname{Re}\lambda s_0} - 1}{|\operatorname{Re}\lambda|} \|g\|_\infty + (s_1 - s_0)\|g\|_\infty$$

$$\leq \frac{|e^{\lambda(s_1-s_0)} - 1|}{|\operatorname{Re}\lambda|} \|g\|_\infty + (s_1 - s_0)\|g\|_\infty.$$

This proves uniform continuity of $s \mapsto G_s(\lambda)$ for $\operatorname{Re}\lambda < 0$.

*Case 2.* If $\operatorname{Re}\lambda = 0$, then both terms in the right hand side of (5.3.3) are clearly uniformly continuous, and hence so is the map $s \mapsto G_s(\lambda)$.

*Case 3.* If $\operatorname{Re}\lambda > 0$, then from (5.3.2) we have

$$G_s(\lambda) = \int_s^\infty e^{-\lambda(t-s)} g(t)\, dt.$$

Let $0 \leq s_0 \leq s_1$. Then,

$$\|G_{s_1}(\lambda) - G_{s_0}(\lambda)\|$$

$$= \left\| \int_{s_1}^\infty \left( e^{-\lambda(t-s_1)} - e^{-\lambda(t-s_0)} \right) g(t)\, dt - \int_{s_0}^{s_1} e^{-\lambda(t-s_0)} g(t)\, dt \right\|$$

$$\leq |1 - e^{-\lambda(s_1-s_0)}| \int_{s_1}^\infty e^{-\operatorname{Re}\lambda(t-s_1)} \|g\|_\infty\, dt + (s_1 - s_0)\|g\|_\infty$$

$$= \frac{|1 - e^{-\lambda(s_1-s_0)}|}{\operatorname{Re}\lambda} \|g\|_\infty + (s_1 - s_0)\|g\|_\infty.$$

This proves uniform continuity for $\operatorname{Re}\lambda > 0$.  ////

**Lemma 5.3.3.** *In the situation of Lemmas 5.3.1 and 5.3.2, the map $\lambda \mapsto G_{(\cdot)}(\lambda)$ is a holomorpic $BUC(\mathbb{R}_+, X)$-valued function on $V$.*

*Proof:* Let $\lambda \in V$ and $r > 0$ be fixed such that (5.3.1) holds. By Cauchy's theorem, for any two $z_0, z_1$ in (the open set) $B(\lambda, r)$ we have

$$\|G_s(z_0) - G_s(z_1)\| = \left\| \frac{1}{2\pi i} \int_{|z-\lambda|=r} \left( \frac{G_s(z)}{z - z_0} - \frac{G_s(z)}{z - z_1} \right) dz \right\|$$

$$\leq r \cdot \sup_{|z-\lambda|=r} \|G_s(z)\| \cdot \sup_{|z-\lambda|=r} \left| \frac{1}{z - z_0} - \frac{1}{z - z_1} \right|.$$

By Lemma 5.3.1, the first of these suprema is finite, uniformly in $s \geq 0$, say $\leq K$. It follows that

$$\sup_{s \geq 0} \|G_s(z_0) - G_s(z_1)\| \leq rK \sup_{|z-\lambda|=r} \left| \frac{1}{z - z_0} - \frac{1}{z - z_1} \right|.$$

This proves that the map $z \mapsto G_{(\cdot)}(z)$ is continuous from $V$ to $BUC(\mathbb{R}_+, X)$.

For each $s \geq 0$, the map $z \mapsto G_s(z)$ is holomorphic as a map $V \to X$. Hence, if $\Gamma$ is any closed simple Jordan curve in $V$, then

$$\left( \int_\Gamma G_{(\cdot)}(z) \, dz \right)(s) = \int_\Gamma G_s(z) \, dz = 0$$

by Cauchy's theorem. It follows that

$$\int_\Gamma G_{(\cdot)}(z) \, dz = 0$$

for all such $\Gamma$. Therefore $z \mapsto G_{(\cdot)}(z)$ is holomorphic by Morera's theorem.  ////

In terms of the left translation semigroup $\mathbf{U}$ on $BUC(\mathbb{R}_+, X)$ and its generator $B$, the preceding lemmas can be summarized as follows.

**Theorem 5.3.4.** *For $\omega \in \mathbb{R}$ and $g \in BUC(\mathbb{R}_+, X)$ the following assertions are equivalent:*

(i) *The Laplace transform of $g$ admits a holomorphic extension to a neighbourhood of $i\omega$;*
(ii) *The local resolvent $\lambda \mapsto R(\lambda, B)g$ admits a holomorphic extension to a neighbourhood of $i\omega$.*

We obtain the following stability result for $BUC(\mathbb{R}_+, X)$-functions:

**Theorem 5.3.5.** *Let $g \in BUC(\mathbb{R}_+, X)$ and let $E \subset \mathbb{R}$ closed and countable be given, and assume that the Laplace transform of $g$ admits a holomorphic extension to a neighbourhood of $\{\operatorname{Re} \lambda \geq 0\} \setminus iE$. If*

$$\lim_{t \to \infty} \left( \sup_{s \geq 0} \left\| \frac{1}{t} \int_0^t e^{-i\omega t} g(\tau + s) \, d\tau \right\| \right) = 0, \quad \forall \omega \in E,$$

*then* $\lim_{t \to \infty} \|g(t)\| = 0.$

*Proof:* By Theorem 5.3.4, we can apply Theorem 5.1.11 to the left translation semigroup $\mathbf{U}$ on $BUC(\mathbb{R}_+)$. It follows that $\lim_{t \to \infty} \|U(t)g\| = 0$ and hence also $\lim_{t \to \infty} \|g(t)\| = 0$.  ////

By taking for $g$ the orbit of a semigroup, we have proved the following generalization of Theorem 5.1.11.

**Theorem 5.3.6.** *Let $\mathbf{T}$ be a $C_0$–semigroup on a Banach space $X$, with generator $A$. Let $x_0 \in X$ and $E \subset \mathbb{R}$ closed and countable be given, and assume that:*
  (i) *$t \mapsto T(t)x_0$ is bounded and uniformly continuous;*
  (ii) *$\lambda \mapsto R(\lambda, A)x_0$ admits a holomorphic extension $F(\lambda)$ to a neighbourhood of $\{\mathrm{Re}\,\lambda \geq 0\}\setminus iE$.*
  (iii) *For all $\omega \in E$, $\displaystyle\lim_{t \to \infty}\left(\sup_{s \geq 0}\left\|\frac{1}{t}\int_0^t e^{-i\omega\tau}T(\tau+s)x_0\,d\tau\right\|\right) = 0$.*

*Then $\lim_{t \to \infty}\|T(t)x_0\| = 0$.*

**Corollary 5.3.7.** *Let $\mathbf{T}$ be a $C_0$–semigroup on a Banach space $X$ and let $x_0 \in X$ be such that $T(\cdot)x_0$ is bounded and uniformly continuous. If the map $\lambda \mapsto R(\lambda, A)x_0$ admits a holomorphic extension across the imaginary axis, then $\lim_{t \to \infty}\|T(t)x_0\| = 0$.*

This corollary admits a considerably simpler proof which will be presented in Section 5.6.

Also in the Katznelson-Tzafriri theorem 5.2.1 the uniform boundedness assumption can be relaxed to a local boundedness assumption. For the proof we need the following lemma.

**Lemma 5.3.8.** *Let $\mathbf{T}$ be a $C_0$–semigroup on a Banach space $X$. Let $x_0 \in X$ be such that $T(\cdot)x_0$ is bounded and let $E \subset \mathbb{R}$ be such that $\lambda \mapsto R(\lambda, A)x_0$ admits a holomorphic extension $F(\lambda)$ to a connected neighbourhood $V$ of $\{\mathrm{Re}\,\lambda \geq 0\}\setminus iE$. Then, for all $f \in L^1(\mathbb{R}_+)$ the map $\lambda \mapsto R(\lambda, A)\hat{f}(\mathbf{T})x_0$ admits a holomorphic extension to $V$.*

*Proof:* For $\mathrm{Re}\,\lambda$ large enough we have

$$\begin{aligned}
R(\lambda, A)\hat{f}(\mathbf{T})x_0 &= \int_0^\infty e^{-\lambda t}T(t)\int_0^\infty f(s)T(s)x_0\,ds\,dt \\
&= \int_0^\infty f(s)T(s)\int_0^\infty e^{-\lambda t}T(t)x_0\,dt\,ds \\
&= \int_0^\infty f(s)T(s)R(\lambda, A)x_0\,ds \\
&= \int_0^\infty f(s)T(s)F(\lambda)\,ds.
\end{aligned} \qquad (5.3.4)$$

By Theorem 5.3.4, for all $\lambda \in V$ the map $s \mapsto T(s)F(\lambda)$ is bounded; moreover, for all $s \geq 0$ the map $\lambda \mapsto T(s)F(\lambda)$ is holomorphic. Therefore, the last term in (5.3.4) defines a continuous extension of $\lambda \mapsto R(\lambda, A)\hat{f}(\mathbf{T})x_0$ to $V$, which is holomorphic by Morera's theorem. ////

**Theorem 5.3.9.** Let $\mathbf{T}$ be a $C_0$-semigroup on a Banach space $X$. Let $x_0 \in X$ and let $E \subset \mathbb{R}$ be such that

(i) The orbit $t \mapsto T(t)x_0$ is bounded;
(ii) The map $\lambda \mapsto R(\lambda, A)x_0$ admits a holomorphic extension to a neighbourhood $V$ of $\{\operatorname{Re}\lambda \geq 0\}\backslash iE$.

If $f \in L^1(\mathbb{R}_+)$ is such that $\hat{f}$ vanishes on $-E$, then

$$\lim_{t\to\infty} \|T(t)\hat{f}(\mathbf{T})x_0\| = 0.$$

*Proof:* **Step 1.** First we prove the theorem under the additional assumption that $T(\cdot)x_0$ is uniformly continuous. Then $g_0 := T(\cdot)x_0 \in BUC(\mathbb{R}_+, X)$ and $\hat{f}(\mathbf{T})x_0 = \hat{f}(\mathbf{U})g_0$, where $\mathbf{U}$ is the left translation semigroup on $BUC(\mathbb{R}_+, X)$. By Theorem 5.3.4, we can apply Theorem 5.2.1 to $\mathbf{U}$ and obtain that

$$\lim_{t\to\infty} \|U(t)\hat{f}(\mathbf{U})g_0\| = 0.$$

For this the desired result immediately follows.

**Step 2.** The general case is deduced from this as follows. If $T(\cdot)x_0$ is bounded, then for all $g \in L^1(\mathbb{R}_+)$, the map $t \mapsto T(t)\hat{g}(\mathbf{T})x_0$ is bounded and uniformly continuous. Indeed, for all $t \geq 0$ we have

$$\|T(t)\hat{g}(\mathbf{T})x_0\| = \left\|\int_0^\infty g(s)T(t+s)x_0\,ds\right\| \leq M\|g\|_1,$$

where $M := \sup_{s\geq 0} \|T(s)x_0\|$. Let $\epsilon > 0$ be arbitrary and choose $\delta > 0$ so small that

$$\int_0^\infty |g(\tau+s) - g(\tau)|\,d\tau \leq \epsilon, \quad \forall 0 \leq s \leq \delta,$$

and

$$\int_0^\delta |g(\tau)|\,d\tau \leq \epsilon.$$

Then for all $t \geq 0$ and $0 \leq s \leq \delta$ we have

$$\|T(t+s)\hat{g}(\mathbf{T})x_0 - T(t)\hat{g}(\mathbf{T})x_0\| = \left\|\int_0^\infty g(\tau)(T(t+s+\tau) - T(t+\tau))x_0\,d\tau\right\|$$

$$\leq \left\|\int_0^\infty (g(\tau) - g(\tau+s))T(\tau+t+s)x_0\,d\tau\right\|$$

$$+ \left\|\int_0^s g(\tau)T(\tau+t)x_0\,d\tau\right\|$$

$$\leq 2\epsilon M.$$

This proves the boundedness and uniform continuity of $T(\cdot)\hat{g}(\mathbf{T})x_0$.

The local resolvent $R(\lambda, A)\hat{g}(\mathbf{T})x_0$ admits a holomorphic extension to a neighbourhood of $\{\operatorname{Re}\lambda \geq 0\}\setminus iE$ by Lemma 5.3.8. Hence by the first step,

$$\lim_{t\to\infty} \|T(t)\hat{f}(\mathbf{T})\hat{g}(\mathbf{T})x_0\| = 0.$$

Choosing an approximate identity $(g_n) \subset L^1(\mathbb{R}_+)$, we observe that

$$\|T(t)\hat{f}(\mathbf{T})(\hat{g}_n(\mathbf{T})x_0 - x_0)\| = \|T(t)(g_n * f - f)\hat{\,}(\mathbf{T})x_0\|$$
$$= \left\|\int_0^\infty (g_n * f - f)(s) T(t+s)x_0\, ds\right\|$$
$$\leq M\|g_n * f - f\|_1.$$

Hence by choosing $n$ large enough we have

$$\sup_{t\geq 0} \|T(t)\hat{f}(\mathbf{T})(\hat{g}_n(\mathbf{T})x_0 - x_0)\| \leq \epsilon$$

and therefore

$$\limsup_{t\to\infty} \|T(t)\hat{f}(\mathbf{T})x_0\| \leq \limsup_{t\to\infty} \|T(t)\hat{f}(\mathbf{T})\hat{g}(\mathbf{T})x_0\| + \epsilon = \epsilon.$$

////

Theorem 5.3.9 trivially implies the following result.

**Corollary 5.3.10.** *Let $\mathbf{T}$ be a $C_0$-semigroup on a Banach space $X$ and let $x_0 \in X$ be such that $T(\cdot)x_0$ is bounded. If the map $\lambda \mapsto R(\lambda, A)x_0$ admits a holomorphic extension across the imaginary axis, then for all $f \in L^1(\mathbb{R}_+)$ we have*

$$\lim_{t\to\infty} \|T(t)\hat{f}(\mathbf{T})x_0\| = 0.$$

*In particular, we may take $f(t) = e^{-\mu t}$ with $\operatorname{Re}\mu > \omega_0(\mathbf{T})$, and obtain that*

$$\lim_{t\to\infty} \|T(t)R(\mu, A)x_0\| = 0. \tag{5.3.5}$$

This also follows from Corollary 5.3.7. In fact, from the identity

$$T(t_1)R(\mu, A)x_0 - T(t_0)R(\mu, A)x_0 = \int_{t_0}^{t_1} T(s)AR(\mu, A)x_0\, ds$$

it follows that the map $t \mapsto T(t)R(\mu, A)x_0$ is uniformly continuous. Moreover, along with $R(\lambda, A)x_0$ the map $R(\lambda, A)R(\mu, A)x_0 = R(\mu, A)R(\lambda, A)x_0$ admits a holomorphic extension across the imaginary axis, so that Corollary 5.3.7 applies.

We note the following simple consequence of (5.3.5):

**Corollary 5.3.11.** Let **T** be a $C_0$-semigroup on a Banach space $X$, let $x_0 \in X$ be such that $T(\cdot)x_0$ is bounded. If the map $\lambda \mapsto R(\lambda, A)x_0$ admits a holomorphic extension across the imaginary axis, then for all $x^\odot \in X^\odot$ we have

$$\lim_{t\to\infty} |\langle x^\odot, T(t)x_0\rangle| = 0.$$

*Proof:* Fix $\operatorname{Re}\mu > \omega_0(\mathbf{T})$ arbitrary. For all $x^* \in X^*$ we have

$$\lim_{t\to\infty} |\langle R(\mu, A)^* x^*, T(t)x_0\rangle| = \lim_{t\to\infty} |\langle x^*, T(t)R(\mu, A)x_0\rangle| = 0$$

by (5.3.5). Since the range of $R(\mu, A)^*$ is dense in $X^\odot$, the result follows from this and the boundedness of $T(\cdot)x_0$. ////

We conclude this section with an example which shows that Theorem 5.1.11 is a genuine extension of Theorem 5.1.5 and that in Lemma 5.1.7 the assumption that **T** be isometric cannot be omitted.

**Example 5.3.12.** Let $X = C_{00}(\mathbb{R}_+)$, the subspace of all $f \in C_0(\mathbb{R}_+)$ with $f(0) = 0$, and let **U** be the left translation semigroup on $X$ with generator $B$. Choose a sequence $(f_k)$ of functions in the unit ball of $X$ whose linear span is dense in $X$ in such a way that $f_k$ has compact support contained in $[0, 2^k]$. Let $f \in X$ be defined by

$$f(t) := \exp(-(2^n + 2^k)^2) f(t - 2^n - 2^k),$$
$$t \in [2^n + 2^k, 2^n + 2^{k+1}]; \; k = 0, ..., n-1; n = 0, 1, 2, ...,$$

and $f(t) := 0$ for the remaining $t \geq 0$. It is easy to check that $f$ decays faster than any exponential. Therefore, the Laplace transform of $f$ extends to an entire function. By Theorem 5.3.4, the same is true for the local resolvent $\lambda \mapsto R(\lambda, A)f$. On the other hand, $f$ is constructed in such a way that each $f_k$ is contained in the closed linear span of the orbit of $f$. Hence the closed linear span of $U(\cdot)f$ is $C_{00}(\mathbb{R}_+)$ and $\sigma(B_f) = \sigma(B) = \{\operatorname{Re}\lambda \leq 0\}$.

## 5.4. Sets of spectral synthesis

As a corollary to (a special case of) the Arendt-Batty-Lyubich-Vũ theorem, in this section we prove the classical result that closed subsets of $\mathbb{R}$ whose boundary is countable are spectral. This fact plays an important role in Section 5.5 below.

We start with the introduction of the so-called Arveson spectrum of a uniformly bounded $C_0$-group **T**. As in Section 2.4, for $f \in L^1(\mathbb{R})$ we define the bounded operator $\hat{f}(\mathbf{T})$ by

$$\hat{f}(\mathbf{T})x := \int_\mathbb{R} f(t) T(t) x \, dt, \quad x \in X.$$

We define the *kernel* of **T** as

$$I_\mathbf{T} := \{f \in L^1(\mathbb{R}) : \hat{f}(\mathbf{T}) = 0\}.$$

The *Arveson spectrum*, notation $\operatorname{Sp}(\mathbf{T})$, is the hull of $I_\mathbf{T}$, i.e.,

$$\operatorname{Sp}(\mathbf{T}) := \{\omega \in \mathbb{R} : \hat{f}(\omega) = 0 \text{ for all } f \in I_\mathbf{T}\}.$$

The Arveson spectrum is related to $\sigma(A)$ by the following result of D.E. Evans:

**Theorem 5.4.1.** *Let* **T** *be a uniformly bounded* $C_0$-*group on a Banach space* $X$, *with generator* $A$. *Then* $\operatorname{Sp}(\mathbf{T}) = i\sigma(A)$.

*Proof:* First let $\omega \notin i\sigma(A)$. Noting that $\sigma(A) \subset i\mathbb{R}$, we choose a function $f \in L^1(\mathbb{R})$ whose Fourier transform is compactly supported and vanishes in a neighbourhood of $i\sigma(A)$ but not on $\omega$. By Lemma 2.4.3 (ii), $\hat{f}(\mathbf{T}) = 0$. But then $\hat{f}(\omega) \neq 0$ implies that $\omega \notin \operatorname{Sp}(\mathbf{T})$.

Conversely, for all $f \in L^1(\mathbb{R})$ and $\lambda \in \sigma(A) = \sigma_a(A)$,

$$\left| \int_\mathbb{R} e^{\lambda t} f(t)\, dt \right| \leq \|\hat{f}(\mathbf{T})\|. \tag{5.4.1}$$

The proof of this follows that of Lemma 5.2.4 verbatim and is therefore left to the reader. If $\omega \in i\sigma(A) = i\sigma_a(A)$, then (5.4.1) shows that

$$|\hat{f}(\omega)| = \left| \int_\mathbb{R} e^{-i\omega t} f(t)\, dt \right| \leq \|\hat{f}(\mathbf{T})\|,$$

so $\hat{f}(\omega) = 0$ for all $f \in I_\mathbf{T}$. Therefore, $\omega \in \operatorname{Sp}(\mathbf{T})$.  ////

Let **U** be the right translation group on $L^1(\mathbb{R})$, i.e.

$$U(t)f(s) := f(s - t), \quad t \in \mathbb{R}, \text{ a.a. } s \in \mathbb{R}.$$

Note that $\hat{g}(\mathbf{U})f = g * f$ for all $f, g \in L^1(\mathbb{R})$. Indeed, if both $f$ and $g$ are continuous and compactly supported, then for all $s$ we have

$$\begin{aligned}
\hat{g}(\mathbf{U})f(s) &= \left( \int_\mathbb{R} g(t) U(t) f\, dt \right)(s) \\
&= \left( \int_\mathbb{R} g(t) f(\cdot - t)\, dt \right)(s) \\
&= \int_\mathbb{R} g(t) f(s - t)\, dt \\
&= (g * f)(s)
\end{aligned}$$

and the general case follows from this by a density argument. By Theorem 5.4.1 and Lemma 2.4.3, the Arveson spectrum of a bounded $C_0$-group on a non-zero Banach space is always non-empty.

For a closed subset $E \subset \mathbb{R}$ we define

$$k(E) := \{f \in L^1(\mathbb{R}) : \hat{f} \text{ vanishes on } E\};$$
$$j_E := \{f \in L^1(\mathbb{R}) : \hat{f} \text{ vanishes in a neighbourhood of } E\};$$
$$J_E := \overline{j_E}.$$

Note that $k(E) \subset j_E \subset J_E$. Recall that a function $f \in L^1(\mathbb{R})$ is of *spectral synthesis* for $E$ if $f \in J_E$, and that the set $E$ is called *spectral* if $k(E) = J_E$.

The spaces $k(E)$ and $J_E$ are **U**-invariant. Hence we may define the induced $C_0$-groups $\mathbf{U}_{/J_E}$ on $L^1(\mathbb{R})/J_E$ and $\mathbf{U}_{k(E)/J_E}$ on $k(E)/J_E$ in the natural way by

$$U_{/J_E}(t)(f + J_E) := U(t)f + J_E, \quad f \in L^1(\mathbb{R}), t \in \mathbb{R};$$
$$U_{k(E)/J_E}(t)(f + J_E) := U(t)f + J_E, \quad f \in k(E), t \in \mathbb{R}.$$

Since **U** is isometric, so are $\mathbf{U}_{/J_E}$ and $\mathbf{U}_{k(E)/J_E}$. Clearly, $\hat{g}(\mathbf{U}_{/J_E})(f + J_E) = (g * f) + J_E$ and $\hat{g}(\mathbf{U}_{k(E)/J_E})(f + J_E) = (g * f) + J_E$ for all $g \in L^1(\mathbb{R})$ and all $f \in L^1(\mathbb{R})$ and $f \in k(E)$ respectively.

The following propositions gives some information about the spectra of the generators $B_{/J_E}$ and $B_{k(E)/J_E}$.

**Lemma 5.4.2.** *Let $E \subset \mathbb{R}$ be a closed set. Then $i\sigma(B_{/J_E}) \subset E$ and $i\sigma(B_{k(E)/J_E}) \subset \partial E$, the topological boundary of $E$.*

*Proof:* We start with the second inclusion. Let $f \in k(E)$ and $g \in j_{\partial E}$. By assumption, there exists an open neighbourhood $V \supset \partial E$ such that $\hat{g}$ vanishes on $V$. But then $(f * g)\hat{\,} = \hat{f} \cdot \hat{g}$ vanishes on the open neighbourhood $V \cup E$ of $E$, so $f * g \in J_E$. It follows that

$$\hat{g}(\mathbf{U}_{k(E)/J_E})(f + J_E) = (f * g) + J_E = 0.$$

Since this holds for all $f \in k(E)$, we have $\hat{g}(\mathbf{U}_{k(E)/J_E}) = 0$ for all $g \in j_{\partial E}$. Hence, $j_{\partial E} \subset I_{\mathbf{U}_{k(E)/J_E}}$, the kernel of $\mathbf{U}_{k(E)/J_E}$. Suppose $\omega \notin \partial E$. Choose a function $f \in L^1(\mathbb{R})$ whose Fourier transform is compactly supported, vanishes on a neighbourhood of $\partial E$, and satisfies $\hat{f}(\omega) = 1$. Then $f \in j_{\partial E}$, so $\hat{f}(\mathbf{U}_{k(E)/J_E}) = 0$, but $\hat{f}(\omega) = 1$. Therefore, $\omega \notin \mathrm{Sp}(\mathbf{U}_{k(E)/J_E})$, and hence $\omega \notin i\sigma(B_{k(E)/J_E})$ by Theorem 5.4.1.

The inclusion $i\sigma(B_{/J_E}) \subset E$ is proved similarly, this time observing that $f * g \in J_E$ for all $f \in L^1(\mathbb{R})$ and $g \in j_E$.  ////

**Theorem 5.4.3.** *If $E \subset \mathbb{R}$ is a closed set with countable boundary, then $E$ is a spectral set.*

*Proof:* We have to prove that $k(E) = J_E$. Consider the right translation group $\mathbf{U}$ on $L^1(\mathbb{R})$ and the induced isometric quotient group $\mathbf{U}_{k(E)/J_E}$ on $k(E)/J_E$ with generator $B_{k(E)/J_E}$. By Lemma 5.4.2, $i\sigma(B_{k(E)/J_E}) \subset \partial E$. In particular, $\sigma(B_{k(E)/J_E})$ is countable; it is also non-empty by Lemma 2.4.3. We are going to show that $\sigma_p(B^*_{k(E)/J_E}) = \emptyset$.

First, it is clear that
$$\sigma_p(B^*_{k(E)/J_E}) \subset \sigma(B^*_{k(E)/J_E}) = \sigma(B_{k(E)/J_E}) \subset -iE.$$
Let $\omega \in E$ and let $z^* \in D(B^*_{k(E)/J_E})$ be such that $B^*_{k(E)/J_E} z^* = -i\omega z^*$. We will show that $z^* = 0$.

Identifying $k(E)/J_E$ canonically with a closed subspace of $L^1(\mathbb{R})/J_E$, we choose an arbitrary Hahn-Banach extension $z_0^* \in (L^1(\mathbb{R})/J_E)^*$ of $z^*$. Let $\phi \in (BUC(\mathbb{R}_+))^*$ be a left invariant mean, and define $\psi \in (L^1(\mathbb{R}))^* = L^\infty(\mathbb{R})$ by
$$\langle \psi, f \rangle := \phi(e^{i\omega(\cdot)} \langle z_0^*, U_{L^1(\mathbb{R})/J_E}(\cdot)(f + J_E) \rangle), \qquad f \in L^1(\mathbb{R}).$$
Then it is easy to check that $U^*(t)\psi = e^{-i\omega t}\psi$ for all $t \geq 0$. But this means that the function $s \mapsto e^{i\omega s}\psi(s)$ is left translation invariant. The only left translation invariant functions in $L^\infty(\mathbb{R})$ are multiples of the constant one function. It follows that $\psi(s) = Ce^{-i\omega s}$ for almost all $s$ and some fixed constant $C$. But then for all $f \in k(E)$ we have, using that $\phi(\mathbf{1}) = 1$ and $U^*_{k(E)/J_E}(t)z^* = e^{-i\omega t}z^*$ for all $t$,
$$\begin{aligned}
\langle z^*, f + J_E \rangle &= \phi(\langle e^{i\omega(\cdot)} U^*_{k(E)/J_E}(\cdot) z^*, f + J_E \rangle) \\
&= \phi(e^{i\omega(\cdot)} \langle z^*, U_{k(E)/J_E}(\cdot)(f + J_E) \rangle) \\
&= \phi(e^{i\omega(\cdot)} \langle z_0^*, U_{L^1(\mathbb{R})/J_E}(\cdot)(f + J_E) \rangle) \\
&= \langle \psi, f \rangle = C \int_{-\infty}^\infty e^{-i\omega s} f(s)\, ds = C\hat{f}(\omega) = 0
\end{aligned}$$
since $\omega \in E$ and $f \in k(E)$. This proves that $z^* = 0$. Therefore, $\sigma_p(B^*_{k(E)/J_E}) = \emptyset$ as claimed.

We are in a position to apply the Arendt-Batty-Lyubich-Vũ theorem to $\mathbf{U}_{k(E)/J_E}$ and obtain that $\mathbf{U}_{k(E)/J_E}$ is uniformly stable. But this group is isometric at the same time, which is only possible if $k(E)/J_E = \{0\}$, i.e. $k(E) = J_E$.
////

Let us note that we do not need the full force of the Arendt-Batty-Lyubich-Vũ theorem. It suffices to know that for a uniformly bounded $C_0$-group on a non-zero Banach space, every isolated point in $\sigma(A)$ belongs to $\sigma_p(A)$ (and hence to $\sigma_p(A^*)$) by Proposition 5.1.15; this fact is easier to prove.

In Section 5.6 below, as another application of this technique we will prove that a function $f \in L^1(\mathbb{R}_+)$ satisfying
$$\int_0^\infty s|f(s)|\, ds < \infty$$
is of spectral synthesis with respect to the zero set of its Fourier transform.

## 5.5. A quantitative stability theorem

Let **T** be a uniformly bounded $C_0$-semigroup semigroup on $X$ with $\sigma(A) \cap i\mathbb{R}$ countable. In Section 5.1 we have seen that **T** is uniformly stable provided $\sigma_p(A^*) \cap i\mathbb{R} = \emptyset$. For contraction semigroups, in this section we prove a remarkable quantitative version of this result: the quantity $\lim_{t\to\infty} \|T(t)x\|$ can actually be computed in terms of the eigenvectors corresponding to the purely imaginary eigenvalues of $A^*$. In fact, we will show that

$$\lim_{t\to\infty} \|T(t)x\| = \inf\{\|x - y\| : y \in X_0\}$$
$$= \sup\{|\langle x^*, x\rangle| : x^* \in N\},$$

where $X_0$ is the closed subspace of all $y \in X$ such that $\lim_{t\to\infty} \|T(t)y\| = 0$ and $N$ is the weak*-closed linear span of all $x^* \in D(A^*)$ such that $A^*x^* \in i\omega x^*$ for some $\omega \in \mathbb{R}$. If $\sigma_p(A^*) \cap i\mathbb{R} = \emptyset$, then $N = \{0\}$ and we recover Theorem 5.1.5. The first identity even holds if only the local unitary spectrum of $x$ is assumed to be countable.

The proof relies heavily on the results of the previous sections and consists of two main steps. Firstly, we show that if **T** is a contraction semigroup on $X$ which has no non-zero orbits tending to zero strongly and $x_0 \in X$ has countable local spectrum, then the restriction of **T** to the closed linear span of the orbit of $x_0$ is isometric and extends to an isometric $C_0$-group.

Secondly, we show that the annihilator in $X$ of $N$ is $X_0$. Along the way, we obtain the following result about extendability of orbits: if **T** is uniformly bounded but not uniformly stable, then there exists an $x^\odot \in X^\odot$ whose orbit extends boundedly to negative time.

We start with some preliminaries from harmonic analysis; proofs and more details may be found in [Ka, Chapter VIII]. A function $\phi \in L^\infty(\mathbb{R})$ is *almost periodic* if for all $\epsilon > 0$ there exists a number $\tau_\epsilon > 0$ with the following property: every interval in $\mathbb{R}$ of length $\tau_\epsilon$ contains a $\tau$ such that

$$\sup_{s \in \mathbb{R}} |f(s) - f(s - \tau)| \leq \epsilon.$$

Equivalently, $\phi$ is almost periodic if it belongs to the closed linear span in $L^\infty(\mathbb{R})$ of the functions $t \mapsto e^{i\omega t}$, $\omega \in \mathbb{R}$. Clearly, if $f$ is almost periodic, then $f$ is uniformly continuous and $\|f\|_\infty = \|f|_{[t,\infty)}\|_\infty$ for all $t \in \mathbb{R}$.

In our first lemma we use the following result of L. Loomis: If $E \subset \mathbb{R}$ is closed and countable and $f \in BUC(\mathbb{R}) \cap J_E^\perp$, then $f$ is almost periodic. Recall from Section 5.4 that $J_E$ denotes the closed subspace in $L^1(\mathbb{R})$ of all functions that are of spectral synthesis with respect to $E$.

For a function $g \in L^1(\mathbb{R})$ we denote by $g_t$ the *right* translate over $t$: $g_t(s) := g(s-t)$, $s \in \mathbb{R}$. The reader should not confuse this notation with that of Section 5.3, where it was used to denote the left translates of functions in $BUC(\mathbb{R}_+, X)$.

**Lemma 5.5.1.** Let $E \subset \mathbb{R}$ be countable and closed. The map $\theta: L^1(\mathbb{R}_+)/J_E^+ \to L^1(\mathbb{R})/J_E$ defined by $\theta(f + J_E^+) := f + J_E$ is an isometric isomorphism.

*Proof:* Let $t > 0$. By taking an approximate identity supported in small compact neighbourhoods of 0 and translating it over $t$, we can find a sequence $(f_n) \subset L^1(\mathbb{R}_+)$ of compactly supported functions such that $\|f_n\|_1 = 1$ for all $n$ and $\lim_{n\to\infty} \|g * f_n - g_t\|_1 = 0$ for all $g \in L^1(\mathbb{R})$. There is some abuse of notation here: the subscript '$n$' indexes the sequence. Let $\tilde{f}_n(s) := f_n(-s)$, $s \in \mathbb{R}$, and let $\phi \in L^\infty(\mathbb{R})$. Then $(\phi * \tilde{f}_n)$ is weak*-convergent to $\phi_{-t}$. Hence,

$$\|\phi\|_\infty = \|\phi_{-t}\|_\infty \leq \liminf_{n\to\infty} \|\phi * \tilde{f}_n\|_\infty.$$

Moreover, $\phi * \tilde{f}_n$ is uniformly continuous for each $n$.

Suppose that $\phi \in J_E^\perp$. Then also $\phi * \tilde{f}_n \in J_E^\perp$. Since $E$ is countable, Loomis's theorem implies that $\phi * \tilde{f}_n$ is almost periodic.

Let $n \in \mathbb{N}$ be fixed an let $\tau_n > 0$ be so large that the support of $\tilde{f}_n$ is contained in $[-\tau_n, 0]$. Put $\phi_+ := \phi|_{\mathbb{R}_+}$. Then, for all $\tau \geq \tau_n$,

$$|(\phi * \tilde{f}_n)(\tau)| \leq \left|\int_{-\infty}^\infty \phi(s)\tilde{f}_n(\tau - s)\, ds\right| = \left|\int_\tau^{2\tau} \phi(s)\tilde{f}_n(\tau - s)\, ds\right|$$
$$= |(\phi_+ * \tilde{f}_n)(\tau)| \leq \|\phi_+ * \tilde{f}_n\|_\infty.$$

Hence,

$$\|(\phi * \tilde{f}_n)|_{[\tau_n,\infty)}\|_\infty \leq \|\phi_+ * \tilde{f}_n\|_\infty.$$

It follows that

$$\|\phi * \tilde{f}_n\|_\infty = \|(\phi * \tilde{f}_n)|_{[\tau_n,\infty)}\|_\infty \leq \|\phi_+ * \tilde{f}_n\|_\infty.$$

Combining everything, we obtain

$$\|\phi\|_\infty \leq \liminf_{n\to\infty} \|\phi * \tilde{f}_n\|_\infty \leq \liminf_{n\to\infty} \|\phi_+ * \tilde{f}_n\|_\infty \leq \|\phi_+\|_\infty \leq \|\phi\|_\infty.$$

This shows that the map $\theta^*: J_E^\perp \to (J_E^+)^\perp$, $\phi \mapsto \phi_+ = \phi|_{\mathbb{R}_+}$, is isometric. In particular, $\theta$ has dense range.

Since $\theta^*$ is isometric and weak*-continuous, it follows that its range has weak*-compact unit ball. Hence, by the Krein-Shmulyan theorem, the range of $\theta^*$ is weak*-closed in $(J_E^+)^\perp$. Therefore, the range of $\theta$ is closed by the closed range theorem.

It follows that $\theta$ is surjective, and since $\theta$ is also injective, it follows that $\theta$ is an isomorphism onto. Moreover, the adjoint $\theta^*$ is isometric, hence so is $\theta$. ////

We say that a $C_0$-semigroup **T** is *bounded away from* 0 if $\lim_{t\to\infty} \|T(t)x\| = 0$ implies that $x = 0$.

**Lemma 5.5.2.** Let **T** be a $C_0$-contraction semigroup on $X$ which is bounded away from 0. Let $E := i\sigma(A) \cap \mathbb{R}$ and let $f \in L^1(\mathbb{R}_+)$. Then $\|\hat{f}(\mathbf{T})\| \leq \|f + J_E^+\|$. If $E$ is countable, then $\|\hat{f}(\mathbf{T})\| \leq \|f + J_E\|$.

*Proof:* Let $g \in J_E^+$. By the Katznelson-Tzafriri theorem 5.2.3 we have $\lim_{t \to \infty} \|T(t)\hat{g}(\mathbf{T})\| = 0$. Since **T** is bounded away from 0, it follows that $\hat{g}(\mathbf{T})x = 0$ for all $x \in X$, so $\hat{g}(\mathbf{T}) = 0$. This implies that $\|\hat{f}(\mathbf{T})\| = \|\hat{f}(\mathbf{T}) - \hat{g}(\mathbf{T})\| \leq \|f - g\|_1$. By taking the infimum over all $g \in J_E^+$ the first part of the lemma follows. The second part follows from this and Lemma 5.5.1. ////

If **T** is a uniformly bounded $C_0$-semigroup on $X$, then we define $\mathcal{A}_\mathbf{T}$ as the closure with respect to the uniform operator topology in $\mathcal{L}(X)$ of the set $\{\hat{f}(\mathbf{T}) : f \in L^1(\mathbb{R}_+)\}$. With respect to the inherited uniform operator topology, $\mathcal{A}_\mathbf{T}$ is a commutative Banach algebra.

**Lemma 5.5.3.** Let **T** be a $C_0$-contraction semigroup on $X$ which is bounded away from 0. Assume that $E \subset \mathbb{R}$ is closed and countable and $x_0 \in X$ is such that

(i) The linear span of the orbit $T(\cdot)x_0$ is dense in $X$;
(ii) The map $\lambda \mapsto R(\lambda, A)x_0$ admits a holomorphic extension to some neighbourhood of $\{\operatorname{Re} \lambda \geq 0\} \setminus iE$.

Then there exists a contractive homomorphism $\xi : L^1(\mathbb{R}) \to \mathcal{A}_\mathbf{T}$ with the following properties:

(a) $\xi(f) = \hat{f}(\mathbf{T})$ for all $f \in L^1(\mathbb{R}_+)$;
(b) $\xi(f_t) = T(t)\xi(f)$ for all $f \in L^1(\mathbb{R})$ and $t \geq 0$;
(c) If $\hat{f} \equiv 0$ on $-E$, then $\xi(f) = 0$.

*Proof:* Let $(Y, \pi, \mathbf{U})$ be the isometric limit semigroup associated to **T**; note that $\pi$ is injective since **T** is bounded away from 0. Let $B$ be the generator of **U**. By Lemma 5.1.7, $F := -E = i\sigma(B) = i\sigma(B) \cap \mathbb{R}$ is countable.

Let $f \in L^1(\mathbb{R})$. By Lemma 5.5.1 there exists a $g \in L^1(\mathbb{R}_+)$ such that $f - g \in J_F$. By Lemma 5.5.2, $\hat{f}(\mathbf{U})\pi = \hat{g}(\mathbf{U})\pi = \pi\hat{g}(\mathbf{T})$. Hence $\hat{f}(\mathbf{U})\pi x \in \pi X$ for all $x \in X$, so we may define

$$\xi(f)x := \pi^{-1}\hat{f}(\mathbf{U})\pi x = \hat{g}(\mathbf{T})x, \quad x \in X.$$

Then $\|\xi(f)\| \leq \|g + J_F\| \leq \|f\|_1$, so $\xi$ is a contraction. It is easily verified that $\xi$ is a homomorphism and that (a) holds. If $\hat{f} \equiv 0$ on $F$, then $f \in J_F$ since $F$ is countable and hence spectral by Theorem 5.4.3. Hence, $\hat{f}(\mathbf{U}) = 0$ by Lemma 5.5.2 and therefore $\xi(f) = 0$. Finally, for $t \geq 0$, $g_t - f_t \in J_F$, so

$$\xi(f_t) = \hat{g}_t(\mathbf{T}) = T(t)\hat{g}(\mathbf{T}) = T(t)\xi(f).$$

////

Let **T** be a (not necessarily strongly continuous) semigroup on $X$ and let $x_0 \in X$. A map $\eta : \mathbb{R} \to X$ is called a *complete orbit through* $x_0$ if

(i) $\eta(0) = x_0$;
(ii) $T(t)\eta(s) = \eta(t+s)$ for all $t \geq 0$ and $s \in \mathbb{R}$.

It follows from (i) and (ii) that $\eta(t) = T(t)x_0$ for all $t \geq 0$, so $\eta$ is an extension of $T(\cdot)x_0$ to negative time.

Now we are in a position to prove the first main result of this section.

**Theorem 5.5.4.** Let **T** be a $C_0$–contraction semigroup on a Banach space $X$ which is bounded away from 0. Assume that $E \subset \mathbb{R}$ is closed and countable and $x_0 \in X$ is such that

(i) The linear span of the orbit $T(\cdot)x_0$ is dense in $X$;
(ii) The map $\lambda \mapsto R(\lambda, A)x_0$ admits a holomorphic extension to some neighbourhood of $\{\operatorname{Re}\lambda \geq 0\}\setminus iE$.

Then, **T** extends to an isometric $C_0$–group on $X$.

*Proof:* Fix an $f \in L^1(\mathbb{R})$ whose Fourier transform has compact support $K$. For $t \in \mathbb{R}$ and $y \in X$ we define
$$\eta_f(t) := \xi(f_t)y,$$
where $\xi$ is the map of the previous lemma. Note that $\eta_f(0) = \xi(f)y$.

Fix $\epsilon > 0$. By a well-known result of harmonic analysis we may choose $h \in L^1(\mathbb{R})$ such that $\hat{h}$ is compactly supported, $\hat{h} \equiv 1$ on $K$, and $\|h\|_1 \leq 1 + \epsilon$. Then $\hat{h}\hat{f} = \hat{f}$ and
$$\eta_f(t) = \xi(f_t)y = \xi((h*f)_t)y = \xi(h_t * f)y = \xi(h_t)\xi(f)y, \quad t \in \mathbb{R},$$
so
$$\sup_{t \in \mathbb{R}} \|\eta_f(t)\| \leq (1+\epsilon)\|\xi(f)y\|.$$

Since for $t \geq 0$ we have $\eta_f(t) = \xi(f_t)y = T(t)\xi(f)y$ it follows that $\eta_f$ is a complete bounded orbit through $\xi(f)y$ and
$$\sup_{t \in \mathbb{R}} \|\eta_f(t)\| = \|\xi(f)y\|.$$

Moreover, since **T** is bounded away from 0, this orbit is the *unique* extension of $T(\cdot)\xi(f)y$ to a complete bounded orbit.

We claim that there is a sequence $(f_n) \subset L^1(\mathbb{R})$ such that the Fourier transform of each $f_n$ is compactly supported and
$$\lim_{n \to \infty} \|y - \xi(f_n)y\| = 0.$$

To prove the claim, we may assume that $y \neq 0$. Fix $\epsilon_n \downarrow 0$. By taking suitable non-negative functions supported in a small interval $[0, \delta_n]$ we can find $h_n \in L^1(\mathbb{R}_+)$

with $\|h_n\|_1 = 1$ such that $\|y - \hat{h}_n(\mathbf{T})y\| \leq \epsilon_n/2$. Using the Fejér kernel, we find $g_n \in L^1(\mathbb{R})$ such that $\hat{g}_n$ has compact support and $\|h_n - h_n * g_n\|_1 \leq \epsilon_n/(2\|y\|)$. Then,

$$\|y - \xi(g_n * h_n)y\| \leq \|\xi(g_n * h_n - h_n)y\| + \|y - \xi(h_n)y\|$$
$$\leq \|g_n * h_n - h_n\|_1 \|y\| + \|y - \hat{h}_n(\mathbf{T})y\| \leq \epsilon_n.$$

This proves the claim, with $f_n := g_n * h_n$.

For each $n$, let $\eta_n(\cdot)$ denote the unique complete orbit through $\xi(f_n)y$ as constructed above. Then $\eta_n(\cdot) - \eta_m(\cdot)$ is the unique complete orbit through $\xi(f_n - f_m)y$ and hence

$$\sup_{t \in \mathbb{R}} \|\eta_n(t) - \eta_m(t)\| = \|\xi(f_n - f_m)y\|.$$

Noting that the right hand side tends to 0 as $n, m \to \infty$, we define $y(t) := \lim_{n \to \infty} \eta_n(t)$. Then $y(\cdot)$ is a complete bounded orbit through $\lim_{n \to \infty} \eta_n(0) = \lim_{n \to \infty} \xi(f_n)y = y$ and

$$\sup_{t \in \mathbb{R}} \|y(t)\| \leq \limsup_{n \to \infty} \|\xi(f_n)y\| = \|y\|.$$

Since $y(0) = y$ it follows that $\sup_{t \in \mathbb{R}} \|y(t)\| = \|y\|$. Define

$$U(t)y := y(t), \quad t \in \mathbb{R}.$$

Then $U(t)$ is a well-defined linear operator on $X$, $U(t)U(-t) = U(-t)U(t) = I$ and $\|U(t)\| = 1$ for all $t \in \mathbb{R}$, and $U(t) = T(t)$ for $t \geq 0$. It follows that $\mathbf{U}$ is an isometric group extending $\mathbf{T}$. ////

Our next goal is to exhibit a relationship between uniform stability and extendability of orbits of the adjoint semigroup to negative time. We start with two lemmas.

**Lemma 5.5.5.** *Let $\mathbf{T}$ be an isometric $C_0$-semigroup on a Banach space $X$. Then there exists a Banach space $\tilde{X}$ containing $X$ isometrically as a closed subspace, and an isometric $C_0$-group $\tilde{\mathbf{T}}$ on $X$ such that $\tilde{\mathbf{T}}|_X = \mathbf{T}$.*

*Proof:* Let $Z$ denote the closed linear span in $BUC(\mathbb{R}_+, X)$ of all functions $f$ for which there exists a $t_0 \geq 0$ such that $f(s+t) = T(t)f(s)$ for all $t \geq 0$ and $s \geq t_0$. Let $Z_0 = \{f \in Z : \lim_{t \to \infty} \|f(t)\| = 0\}$ and put $\tilde{X} := Z/Z_0$. For $f \in Z$ and $t \in \mathbb{R}$ we define $f_t \in Z$ by

$$f_t(s) := \begin{cases} f(0), & 0 \leq s \leq \max\{0, t\}; \\ f(s-t), & s \geq \max\{0, t\}. \end{cases}$$

On $\tilde{X}$, we define the $C_0$-group $\tilde{\mathbf{T}}$ by

$$\tilde{T}(t)(f + Z_0) := f_t + Z_0, \quad f \in Z, t \in \mathbb{R}.$$

This is well-defined and it follows easily from the definitions that $\tilde{\mathbf{T}}$ is isometric.

For all $t \geq 0$ and $x \in X$ the function

$$f_{t,x}(s) := \begin{cases} x, & 0 \leq s \leq t; \\ T(s-t)x, & s \geq t, \end{cases}$$

belongs to $Z$. Since $\mathbf{T}$ is isometric, the map $j : x \mapsto f_{0,x} + Z_0$ defines an isometric isomorphism of $X$ onto a closed subspace of $\tilde{X}$.

Finally, for all $x \in X$ and $t \geq 0$ we have

$$\tilde{T}(t)(jx) = \tilde{T}(t)(f_{0,x} + Z_0) = f_{t,x} + Z_0 = f_{0,T(t)x} + Z_0 = j(T(t)x),$$

so $\tilde{\mathbf{T}}$ extends $\mathbf{T}$. ////

**Lemma 5.5.6.** *Let $\mathbf{T}$ be a $C_0$-contraction semigroup on $X$, with generator $A$, and let $Y$ be a closed $\mathbf{T}$-invariant linear subspace of $X$. If $y^* \in D(A_Y^*)$ is such that $A_Y^* y^* = i\omega y^*$ for some $\omega \in \mathbb{R}$, then there exists an $x^* \in D(A^*)$ such that $\|x^*\| = \|y^*\|$, $x^*|_Y = y^*$, and $A^* x^* = i\omega x^*$.*

*Proof:* Let $x_0^* \in X^*$ be any Hahn-Banach extension of $y^*$ with $\|x_0^*\| = \|y^*\|$. Let $\phi$ be a left invariant mean on $BUC(\mathbb{R}_+)$ and define $x^* \in X^*$ by

$$\langle x^*, x \rangle := \phi(e^{-i\omega(\cdot)} \langle x_0^*, T(\cdot)x \rangle), \quad x \in X.$$

For all $y \in Y$ we have

$$\langle x^*, y \rangle = \phi(e^{-i\omega(\cdot)} \langle T_Y^*(\cdot)y^*, y \rangle) = \phi(\langle y^*, y \rangle \mathbf{1}) = \langle y^*, y \rangle.$$

Hence, $y^* = x^*|_Y$. Also,

$$\|x^*\| \leq \|\phi\| \, \|x_0^*\| = \|x_0^*\| = \|y^*\|$$

and hence $\|x^*\| = \|y^*\|$.

For all $x \in X$ we have

$$\begin{aligned}
\langle T^*(t)x^*, x \rangle &= \phi(e^{-i\omega(\cdot)} \langle x_0^*, T(t + \cdot)x \rangle) \\
&= e^{i\omega t} \phi(e^{-i\omega(t+\cdot)} \langle x_0^*, T(t + \cdot)x \rangle) \\
&= e^{i\omega t} \phi(e^{-i\omega(\cdot)} \langle x_0^*, T(\cdot)x \rangle) \\
&= e^{i\omega t} \langle x^*, x \rangle.
\end{aligned}$$

Hence $T^*(t)x^* = e^{i\omega t}x^*$, so $x^* \in D(A^*)$ and $A^* x^* = i\omega x^*$. ////

By $N$ we denote the weak*-closed linear span in $X^*$ of all $x^* \in D(A^*)$ such that $A^*x^* = i\omega x^*$ for some $\omega \in \mathbb{R}$, and by $M$ the weak*-closure in $X^*$ of the subspace of all $x^*$ admitting a bounded complete orbit. For any $x^* \in D(A^*)$ such that $A^*x^* = i\omega x^*$, $\eta(t) := e^{i\omega t}x^*$ is a bounded complete orbit through $x^*$, so $N \subset M$. If we want to stress that $N$ and $M$ derive from the operator $A$, we will also write $N(A)$ and $M(A)$.

If $Y \subset X^*$ is a linear subspace, then $Y_\perp$ denotes its annihilator in $X$:

$$Y_\perp := \{x \in X : \langle x^*, x\rangle = 0 \text{ for all } x^* \in Y\}.$$

**Theorem 5.5.7.** *Let $\mathbf{T}$ be a uniformly bounded $C_0$–semigroup on a Banach space $X$ and let $X_0 = \{x \in X : \lim_{t \to \infty} \|T(t)x\| = 0\}$.*
(i) $X_0 = M_\perp \subset N_\perp$;
(ii) *If $\sigma(A) \cap i\mathbb{R}$ is countable, then $M_\perp = N_\perp = X_0$.*

*Proof:* (i): Let $x^* \in M$ and let $\eta$ be a bounded complete orbit through $x^*$. For all $x \in X$ and $t \geq 0$,

$$|\langle x^*, x\rangle| = |\langle \eta(0), x\rangle| = |\langle T^*(t)\eta(-t), x\rangle| \leq \left(\sup_{s \in \mathbb{R}} \|\eta(s)\|\right) \|T(t)x\|.$$

If $x \in X_0$, the right hand side tends to 0 as $t \to \infty$, so $\langle x^*, x\rangle = 0$. This proves that $X_0 \subset M_\perp$.

For the converse, we may renorm $X$ and assume that $\mathbf{T}$ is contractive. Let $(Y, \pi, \mathbf{U})$ be the isometric limit semigroup associated to $\mathbf{T}$. Let $(\tilde{Y}, \tilde{\mathbf{U}})$ be the extension of $\mathbf{U}$ to an isometric group of Lemma 5.5.5; let $j : Y \to \tilde{Y}$ be the inclusion map.

If $x \notin X_0$, then $\pi x \neq 0$ and we may choose $\tilde{y}^* \in \tilde{Y}^*$ such that $\langle \tilde{y}^*, j\pi x\rangle \neq 0$. Put $\eta(t) := \pi^* j^* \tilde{U}^*(t)\tilde{y}^*$, $t \in \mathbb{R}$. Then $\eta$ is a bounded $X^*$-valued map and

$$T^*(t)\eta(s) = T^*(t)\pi^* j^* \tilde{U}^*(s)\tilde{y}^* = \pi^* j^* \tilde{U}^*(t+s)\tilde{y}^* = \eta(t+s).$$

Hence $\eta$ is a bounded complete orbit through $x^* := \eta(0)$. Moreover, $\langle x^*, x\rangle = \langle \pi^* j^* \tilde{y}^*, x\rangle \neq 0$, so $x \notin M_\perp$. This proves the inclusion $M_\perp \subset X_0$.

If $A^*x^* = i\omega x^*$, then

$$\|x^*\| \|T(t)x\| \geq |\langle x^*, T(t)x\rangle| = |e^{i\omega t}\langle x^*, x\rangle| = |\langle x^*, x\rangle|.$$

Hence $X_0 \subset N_\perp$.

(ii): The subspace $Y := N_\perp$ is closed and $\mathbf{T}$-invariant. Let $A_Y$ be the part of $A$ in $Y$ and let $y^* \in D(A_Y^*)$ be such that $A_Y^* y^* = i\omega y^*$. By Lemma 5.5.6 there exists a Hahn-Banach extension $x^* \in D(A^*)$ of $y^*$ such that $A^*x^* = i\omega x^*$. Since $x^* \in N$, it follows that $\langle x^*, y\rangle = 0$ for all $y \in N_\perp = Y$. Hence, $y^* = x^*|_Y = 0$. It follows that $\sigma_p(A_Y^*) \cap i\mathbb{R} = \emptyset$. Also, $\sigma(A_Y) \cap i\mathbb{R}$ is countable by Proposition 1.1.7. Thus, the Arendt-Batty-Lyubich-Vũ theorem implies that $\lim_{t \to \infty} \|T(t)y\| = 0$ for all $y \in Y$, which shows that $N_\perp = Y \subset X_0$. ////

**Corollary 5.5.8.** *If* **T** *is a uniformly bounded $C_0$–semigroup on $X$ which is not uniformly stable, then there exists a non-zero $x^\odot \in X^\odot$ admitting a bounded complete orbit.*

*Proof:* By Theorem 5.5.7 (i), there exists a bounded complete orbit $\eta$ through some non-zero $x^* \in X^*$. Then for all $\lambda \in \varrho(A)$, $t \mapsto R(\lambda, A^*)\eta(t)$ defines a bounded complete orbit through $R(\lambda, A^*)x^* \in X^\odot$.  ////

If **T** is $C_0$–contraction semigroup on $X$ which is bounded away from zero with $\sigma(A) \cap i\mathbb{R}$ countable, then Theorem 5.5.7 implies that the linear span of the unimodular eigenvectors is weak*-dense in $X^*$. We are going to prove next the stronger result that the weak*-closure of this span actually norms $X$.

**Lemma 5.5.9.** *Let* **T** *be a $C_0$–contraction semigroup on $X$ and let* **T**$_1$ *be the quotient semigroup on $X_1 := X/X_0$. Then for all $x \in X$,*

$$\lim_{t \to \infty} \|T_1(t)(x + X_0)\| = \lim_{t \to \infty} \|T(t)x\|.$$

*In particular,* **T**$_1$ *is bounded away from 0.*

*Proof:* Let $x \in X$ and put

$$l_x := \lim_{t \to \infty} \|T_1(t)(x + X_0)\| = \lim_{t \to \infty} \|T(t)x + X_0\|.$$

For all $\epsilon > 0$ there exist a $y \in X_0$ and $s \geq 0$ such that $\|T(s)x - y\| \leq l_x + \epsilon$. For all $t \geq 0$,

$$\|T(t+s)x\| \leq \|T(t+s)x - T(t)y\| + \|T(t)y\| \leq l_x + \epsilon + \|T(t)y\|.$$

Hence $\lim_{t \to \infty} \|T(t)x\| \leq l_x + \epsilon$ for all $\epsilon > 0$, so $\lim_{t \to \infty} \|T(t)x\| \leq l_x$. On the other hand, $\|T(t)(x + X_0)\| \leq \|T(t)x\|$ for all $t \geq 0$, hence also $l_x \leq \lim_{t \to \infty} \|T(t)x\|$. ////

After these preparations we come to the second main result of this section.

**Theorem 5.5.10.** *Let* **T** *be a $C_0$–contraction semigroup on a Banach space $X$. Assume that $E \subset \mathbb{R}$ is closed and countable and $x_0 \in X$ is such that $\lambda \mapsto R(\lambda, A)x_0$ admits a holomorphic extension to some neighbourhood of $\{\operatorname{Re}\lambda \geq 0\}\backslash iE$. Then,*

$$\lim_{t \to \infty} \|T(t)x_0\| = \inf\{\|x_0 - x\| : x \in X_0\}.$$

*If $\sigma(A) \cap i\mathbb{R}$ is countable, then*

$$\lim_{t \to \infty} \|T(t)x_0\| = \sup\{|\langle x^*, x_0\rangle| : x^* \in N, \|x^*\| \leq 1\}.$$

*Proof:* Let $Y = X_{x_0}$ denote the closed linear span of the orbit of $x_0$ and let **S** denote the restriction of **T** to $Y$. Let **S**$_1$ denote the quotient semigroup on $Y_1 = Y/Y_0$, where $Y_0 = \{y \in Y : \lim_{t \to \infty} \|S(t)y\| = 0\}$. Since $\mathbf{S}_1$ is bounded away from 0 by Lemma 5.5.9, $\mathbf{S}_1$ extends to an isometric $C_0$–group by Theorem 5.5.4. Hence by Lemma 5.5.9,

$$\lim_{t\to\infty} \|T(t)x_0\| = \lim_{t\to\infty} \|S(t)x_0\|$$
$$= \lim_{t\to\infty} \|S_1(t)(x_0 + Y_0)\|$$
$$= \|x_0 + Y_0\|.$$

On the other hand, since $\mathbf{T}$ is contractive, for all $x \in X_0$ we have

$$\lim_{t\to\infty} \|T(t)x_0\| = \lim_{t\to\infty} \|T(t)(x_0 + x)\| \leq \|x_0 + x\|.$$

Taking the infimum over all $x \in X_0$, it follows that

$$\lim_{t\to\infty} \|T(t)x_0\| \leq \|x_0 + X_0\|.$$

Hence

$$\|x_0 + X_0\| \leq \|x_0 + Y_0\| = \lim_{t\to\infty} \|T(t)x_0\| \leq \|x_0 + X_0\|.$$

This proves the first part of the theorem. The second part follows from the first by invoking Theorem 5.5.7 (ii): in view of $X_0 = N_\perp$ and the weak*-closedness of $N$, we have $(X/X_0)^* = (X_0)^\perp = (N_\perp)^\perp = N$. /////

**Example 5.5.11.** Consider the Schrödinger operator

$$A = \frac{1}{2}\Delta - V$$

on $\mathbb{R}^n$, where $V \geq 0$ is measurable. Then $A$ generates a $C_0$–contraction semigroup on $L^1(\mathbb{R}^n)$ [Si] and $\sigma(A) \subset (-\infty, 0]$ [HV]. It was shown in [Ba2] that the only solutions in $L^\infty(\mathbb{R}^n)$ of $A^*g = 0$ are scalar multiples of

$$g(x) := \lim_{t\to\infty} (T^*(t)\mathbf{1})(x) = \mathbb{E}^x \left( \exp\left( -\int_0^\infty V(B(s))\, ds \right) \right),$$

where $B(s)$ is Brownian motion and $\mathbb{E}$ is expectation with respect to Wiener measure. One has either $g = 0$ or $\|g\|_\infty = 1$, so Theorem 5.5.10 shows that

$$\lim_{t\to\infty} \|T(t)f\|_1 = \left| \int_{\mathbb{R}^n} f(x)g(x)\, dx \right|, \quad \forall f \in L^1(\mathbb{R}^n).$$

**Example 5.5.12.** Consider a symmetric, purely second order uniformly elliptic operator on $\mathbb{R}^n$ with bounded measurable coefficients:

$$A = \sum_{i,j=1}^{n} \frac{\partial}{\partial x_i}\left(a_{ij}\frac{\partial}{\partial x_j}\right).$$

Then $A$ generates a $C_0$-contraction semigroup $\mathbf{T}$ in $L^1(\mathbb{R}^n)$ [Da2] and $\sigma(A) \subset (-\infty, 0]$ [Ar1]. Moreover, the only solutions in $L^\infty(\mathbb{R}^n)$ of $A^*g = 0$ are constants [Fr, Appendix, Thm. 3]. Thus, Theorem 5.5.10 shows that

$$\lim_{t\to\infty} \|T(t)f\|_1 = \left|\int_{\mathbb{R}^n} f(x)\,dx\right|, \quad \forall f \in L^1(\mathbb{R}^n).$$

The following simple example shows that Theorem 5.5.10 fails if the countability assumption is dropped.

**Example 5.5.13.** Let $X = C_0(\mathbb{R})$ with the norm

$$\|f\| := \max\{\|f|_{\mathbb{R}_+}\|, \frac{1}{2}\|f|_{\mathbb{R}_-}\|\}.$$

Let $\mathbf{U}$ be the left translation group on $X$. Then $\mathbf{U}$ is contractive, $X_0 = \{0\}$, and if $f \in X$ is any function with support in $\mathbb{R}_+$, then

$$\lim_{t\to\infty} \|U(t)f\| = \frac{1}{2}\|f\| = \frac{1}{2}\|f\| = \frac{1}{2}\|f + X_0\|.$$

## 5.6. A Tauberian theorem for the Laplace transform

In this section we derive stability results for $C_0$-semigroups as a special case of Tauberian theorems for the Laplace transform of bounded functions.

The basic idea is easily explained. Suppose $f : \mathbb{R}_+ \to X$ is a bounded, strongly measurable function whose Laplace transform $g = \mathcal{L}f$ extends holomorphically to a neighbourhood of the origin. We will prove that, under certain conditions on the singularities of $g$ on the rest of the imaginary axis,

$$\lim_{t\to\infty} \int_0^t f(s)\,ds = g(0).$$

Now if $f$ is uniformly continuous, then the convergence of its indefinite integral implies $\lim_{t\to\infty} \|f(t)\| = 0$. Applied to orbits of semigroups, we obtain the following type of stability results: if the orbit $T(\cdot)x_0$ is bounded and uniformly continuous, then $\lim_{t\to\infty}\|T(t)x_0\| = 0$ provided some assumptions on the behaviour of $R(\lambda, A)x_0$ near the imaginary axis are satisfied. These assumptions are different from those of Section 5.3.

Suppose $f : \mathbb{R}_+ \to X$ is a bounded, strongly measurable function whose Laplace transform $g$ extends holomorphically across the entire imaginary axis and suppose we want to estimate the difference

$$\int_0^t f(s)\,ds - g(0).$$

Defining the entire function $g_t(z) := \int_0^t e^{-sz} f(s)\,ds$, by Cauchy's theorem we have

$$\int_0^t f(s)\,ds - g(0) = g_t(0) - g(0) = \frac{1}{2\pi i}\int_\Gamma (g_t(z) - g(z))\frac{1}{z}dz,$$

where $\Gamma$ is a suitable contour around 0. If one tries to estimate this integral directly, one is left with terms that grow as $t \to \infty$. Indeed, the term $e^{-sz}$ in the definition of $g_t$ and the term $z^{-1}$ in the integral cause problems on the part of $\Gamma$ in the left half-plane. This can be fixed with a simple but clever trick: one adds an extra term $h(z)e^{tz}$ in the integral, where $h$ is holomorphic in a neighbourhood of $z = 0$ and satisfies $h(0) = 1$. Then, again by Cauchy's theorem,

$$\int_0^t f(s)\,ds - g(0) = \frac{1}{2\pi i}\int_\Gamma h(z)(g_t(z) - g(z))\frac{e^{tz}}{z}dz.$$

The choice

$$h(z) := 1 + \frac{z^2}{R^2}$$

brings the problematic term $z^{-1}$ under control and the extra $e^{tz}$ takes care of the term $e^{-zs}$.

This idea also works in the less elementary situation that $g$ is singular in certain points of $i\mathbb{R}$, to which we turn next.

**Lemma 5.6.1.** *Let $X$ be a Banach space and let $f : \mathbb{R}_+ \to X$ be a bounded, strongly measurable function. Let $g$ denote the Laplace transform of $f$ and let $iE$ be the set of singular points of $g$ on $i\mathbb{R}$. Suppose that $0 \notin E$. Let $R > 0$ and let $x_j \in \mathbb{R}$ and $\epsilon_j > 0$, $j = 1, ..., n$, be such that the intervals $(-\infty, R)$, $(R, \infty)$, and $(\xi_j - \epsilon_j, \xi_j + \epsilon_j)$, $j = 1, ..., n$, are disjoint, cover $E$, and do not contain 0. Suppose further that for all $j = 1, ..., n$ there exists some $\omega_j \in (\xi_j - \epsilon_j, \xi_j + \epsilon_j)$ such that*

$$M_j := \sup_{t \geq 0} \left\|\int_0^t e^{-i\omega_j s} f(s)\,ds\right\| < \infty.$$

Then,

$$\limsup_{t\to\infty} \left\| \int_0^t f(s)\,ds - g(0) \right\| \leq \frac{2M}{R} \prod_{j=1}^n a_j$$
$$+ 12 \sum_{j=1}^n \frac{M_j \epsilon_j \xi_j^2}{(|\xi_j| - \epsilon_j)(\xi_j^2 - \epsilon_j^2)} \prod_{k=1,\,k\neq j}^n b_{jk},$$

where $M := \sup_{t\geq 0} \|f(t)\|$,

$$a_j = \frac{\xi_j^2}{\xi_j^2 - \epsilon_j^2} \left(1 + \frac{\epsilon_j^2}{(R - |\xi_j|)^2}\right),$$

$$b_{jk} = \frac{\xi_k^2}{\xi_k^2 - \epsilon_k^2} \left(1 + \frac{\epsilon_k^2}{(|\xi_j - \xi_k| - \epsilon_j)^2}\right), \quad k \neq j.$$

*Proof:* By renumbering we may arrange that

$$-R \leq \xi_1 - \epsilon_1 < \xi_1 + \epsilon_1 < \xi_2 - \epsilon_2 < \ldots < \xi_n + \epsilon_n \leq R.$$

The function $g$ extends holomorphically to a simply connected open set $U$ containing $\{\operatorname{Re} z \geq 0, z \notin iE\}$. Take a contour $\Gamma$ in $U$ consisting of the right hand half of the circle $|z| = R$, the right hand halves of the circles $|z - i\xi_j| = \epsilon_j$, and smooth paths $\Gamma_j$ joining $-iR$ to $i(\xi_1 - \epsilon_1)$ $(j = 0)$, $i(\xi_j + \epsilon_j)$ to $i(\xi_{j+1} - \epsilon_{j+1})$ $(j = 1, \ldots, n-1)$, and $i(\xi_n + \epsilon_n)$ to $iR$ $(j = n)$ lying entirely (except at the endpoints) within $U \cap \{\operatorname{Re} z < 0\}$. Then $\Gamma$ is a closed contour, which may be taken to be simple, with 0 in its interior.

Define the functions

$$h_j(z) := \frac{\xi_j^2}{\xi_j^2 - \epsilon_j^2}\left(1 + \frac{\epsilon_j^2}{(z - i\xi_j)^2}\right);$$

$$h(z) := \left(1 + \frac{z^2}{R^2}\right) \prod_{j=1}^n h_j(z);$$

$$g_t(z) := \int_0^t e^{-sz} f(s)\,ds, \quad t \in \mathbb{C},\, t > 0.$$

Then $h_j(0) = h(0) = 1$. Hence, by Cauchy's theorem,

$$g(0) - g_t(0) = \frac{1}{2\pi i} \int_\Gamma h(z)(g(z) - g_t(z)) \frac{e^{tz}}{z}\,dz.$$

We estimate the integral on the different parts of $\Gamma$.

(i) On $|z| = R$, $\operatorname{Re} z > 0$. If $z = Re^{i\theta}$, $-\frac{\pi}{2} < \theta < \frac{\pi}{2}$, then

$$\|(g(z) - g_t(z))e^{tz}\| = \left\|\int_t^\infty e^{-(s-t)z} f(s)\, ds\right\| = \left\|\int_0^\infty e^{-rz} f(r+t)\, dr\right\|$$

$$\leq M \int_0^\infty e^{-r \operatorname{Re} z}\, dr = \frac{M}{R \cos\theta}.$$

Also, $|1 + z^2/R^2| = 2\cos\theta$ and $|h_j(z)| \leq a_j$. Hence,

$$\left\|\int_{|z|=R,\,\operatorname{Re} z>0} h(z)(g(z) - g_t(z)) \frac{e^{tz}}{z}\, dz\right\| \leq \frac{2\pi M}{R} \prod_{j=1}^n a_j. \tag{5.6.1}$$

(ii) On $|z - i\xi_j| = \epsilon_j$, $\operatorname{Re} z > 0$. If $z = i\xi_j + \epsilon_j e^{i\theta}$, $-\frac{\pi}{2} < \theta < \frac{\pi}{2}$, then letting $F_j(t) := \int_0^t \exp(-i\omega_j s) f(s)\, ds$ and performing a partial integration we obtain

$$\|(g(z) - g_t(z))e^{tz}\| = \left\|e^{tz} \int_t^\infty \exp(-s(i(\xi_j - \omega_j) + \epsilon_j e^{i\theta})) \exp(-i\omega_j s) f(s)\, ds\right\|$$

$$= \left\|e^{tz}\Bigl(-\exp(-t(i(\xi_j - \omega_j) + \epsilon_j e^{i\theta})) F_j(t)\right.$$

$$\left. + (i(\xi_j - \omega_j) + \epsilon_j e^{i\theta}) \int_t^\infty \exp(-s(i(\xi_j - \omega_j) + \epsilon_j e^{i\theta})) F_j(s)\, ds\Bigr)\right\|$$

$$\leq M_j + 2\epsilon_j M_j \int_t^\infty e^{-(s-t)\epsilon_j \cos\theta}\, ds$$

$$= M_j\left(1 + \frac{2}{\cos\theta}\right) \leq \frac{3M_j}{\cos\theta}.$$

Also, $|1 + z^2/R^2| \leq 2$, $|h_j(z)| = (2\cos\theta)\xi_j^2(\xi_j^2 - \epsilon_j^2)^{-1}$, $|h_k(z)| \leq b_{jk}$ ($k \neq j$), and $|z^{-1}| \leq (|\xi_j| - \epsilon_j)^{-1}$. Hence,

$$\left\|\int_{|z-i\xi_j|=\epsilon_j,\,\operatorname{Re} z>0} h(z)(g(z) - g_t(z)) \frac{e^{tz}}{z}\, dz\right\|$$

$$\leq \epsilon_j \pi \cdot 12 M_j \frac{\xi_j^2}{(|\xi_j| - \epsilon_j)(\xi_j^2 - \epsilon_j^2)} \prod_{k=1,\, k \neq j}^n h_{jk}. \tag{5.6.2}$$

(iii) On $\Gamma_j$. By the dominated convergence theorem,

$$\lim_{t\to\infty} \int_{\Gamma_j} h(z) g(z) \frac{e^{tz}}{z}\, dz = 0. \tag{5.6.3}$$

(iv) Since $g_t$ is an entire function,

$$\int_{\Gamma_0 \cup \ldots \cup \Gamma_n} h(z) g_t(z) \frac{e^{tz}}{z}\, dz = \int_{|z|=R,\,\operatorname{Re} z<0} h(z) g_t(z) \frac{e^{tz}}{z}\, dz$$

$$+ \sum_{j=1}^n \int_{|z-i\xi_j|=\epsilon_j,\,\operatorname{Re} z<0} h(z) g_t(z) \frac{e^{tz}}{z}\, dz.$$

If $z = Re^{i\theta}$, $\frac{\pi}{2} < \theta < \frac{3\pi}{2}$, then

$$\|g_t(z)e^{tz}\| = \left\|\int_0^t e^{-(s-t)z} f(s)\, ds\right\|$$
$$\leq M \int_0^t e^{-(s-t)R\cos\theta}\, ds$$
$$\leq \frac{M}{R|\cos\theta|}.$$

So estimating as in (i) we obtain

$$\left\|\int_{|z|=R,\, \operatorname{Re} z<0} h(z) g_t(z) \frac{e^{tz}}{z}\, dz\right\| \leq \frac{2\pi M}{R} \prod_{j=1}^n a_j. \tag{5.6.4}$$

For $z = i\xi_j + \epsilon_j e^{i\theta}$, $\frac{\pi}{2} < \theta < \frac{3\pi}{2}$, we have

$$\|g_t(z)e^{tz}\| = \left\|e^{tz}\int_0^t \exp(-s(i(\xi_j - \omega_j) + \epsilon_j e^{i\theta})) \exp(-i\omega_j s) f(s)\, ds\right\|$$
$$= \left\|e^{it\omega_j} F_j(t) + (i(\xi_j - \omega_j) + \epsilon_j e^{i\theta})\right.$$
$$\left. \cdot e^{tz} \int_0^t \exp(-s(i(\xi_j - \omega_j) + \epsilon_j e^{i\theta})) F_j(s)\, ds\right\|$$
$$\leq M_j + 2\epsilon_j M_j \int_0^t e^{-(s-t)\epsilon_j \cos\theta}\, ds \leq \frac{3M_j}{|\cos\theta|}.$$

Estimating as in (ii) we obtain

$$\left\|\int_{|z-i\xi_j|=\epsilon_j,\, \operatorname{Re} z<0} h(z) g_t(z) \frac{e^{tz}}{z}\, dz\right\|$$
$$\leq \epsilon_j \pi \cdot 12 M_j \frac{\xi_j^2}{(|\xi_j| - \epsilon_j)(\xi_j^2 - \epsilon_j^2)} \prod_{k=1,\, k\neq j}^n b_{jk}. \tag{5.6.5}$$

Now the lemma follows from (5.6.1) - (5.6.5).  ////

As a special case, if $E \cap [-R, R] = \emptyset$, then

$$\limsup_{t\to\infty} \left\|\int_0^t f(s)\, ds - g(0)\right\| \leq \frac{2M}{R}. \tag{5.6.6}$$

Of course, in this case the proof of the lemma simplifies considerably. This special case leads to the following simplified proof of Corollary 5.3.7.

*Second proof of Corollary 5.3.7:* By (5.6.6) applied to $f(t) = T(t)x$,

$$\limsup_{t\to\infty} \left\|\int_0^t T(s)x_0\, ds - g(0)\right\| = 0.$$

Since $T(\cdot)x_0$ is bounded and uniformly continuous, this implies that $\lim_{t\to\infty} \|T(t)x_0\| = 0$ (cf. the end of the proof of Corollary 4.3.3).  ////

Our next goal is to prove a Tauberian theorem for Laplace transforms. We start with a covering lemma.

**Lemma 5.6.2.** *Let $E \subset \mathbb{R}$ be a compact set of measure zero. Then for all $\epsilon > 0$ there exists an $n \in \mathbb{N}$ and a $\theta \in (0, \epsilon/n)$ such that $E$ can be covered by $n$ disjoint open intervals of length $\theta$.*

*Proof:* The regularity of the Lebesgue measure implies that there is an open set $O$ containing $E$ with $\operatorname{meas} O \leq \frac{1}{2}\epsilon$. Let $(O_\alpha)_\alpha$ be an arbitrary open cover of $E$ and define $V_\alpha := O_\alpha \cap O$. Then $V := \cup_\alpha V_\alpha$ is a countable union of disjoint open intervals, say $V = \cup_j W_j$. Then $(W_j)_j$ is an open cover of $E$ and $\cup_j W_j \subset O$. Since $E$ is compact there are $I_1 := W_{j_1}, ..., I_m := W_{j_m}$ such that $E \subset \cup_{i=1}^m I_i$. Also, since $I_1, ..., I_m$ are disjoint, we have $\sum_{i=1}^m \operatorname{meas} I_i \leq \operatorname{meas} O \leq \frac{1}{2}\epsilon$.

Fix $0 < \theta < \epsilon/4m$. Choose $x \in \mathbb{R}$ such that $x + k\theta \notin E$ for all $k \in \mathbb{Z}$ and let $K := \{k \in \mathbb{Z} : (x + k\theta, x + (k+1)\theta) \cap E \neq \emptyset\}$. Note that $\#K < \infty$ since $E$ is compact. The pairwise disjoint intervals $J_k := (x + k\theta, x + (k+1)\theta)$, $k \in K$, cover $E$.

Let $K_0 := \{k \in K : J_k \subset \cup_{i=1}^m I_i\}$ and $K_1 := K \setminus K_0$. For all $k \in K$ we have $J_k \cap \cup_{i=1}^m I_i \neq \emptyset$ since both share at least one common point of $E$. Therefore, if some $J_k$ is not contained in $I := \cup_{i=1,...,m} I_i$, it must contain at least one boundary point of $I$. But $I$ has at most $2m$ boundary points, and hence $\#K_1 \leq 2m$. By considering the total length of the intervals $J_k$ defining $K$, $K_0$, and $K_1$, we see that $\theta \cdot \#K \leq \frac{1}{2}\epsilon + 2m\theta \leq \epsilon$; so $\theta < \epsilon/\#K$. This proves the lemma, with $n := \#K$ and intervals $J_k$, $k = 1, ..., n$. ////

Now we can state and prove the main result of this section. All subsequent stability results are derived from it.

**Theorem 5.6.3.** *Let $f : \mathbb{R}_+ \to X$ be a bounded, strongly measurable function. Let $g$ be the Laplace transform of $f$ and denote by $iE$ the set of all singularities of $g$ on $i\mathbb{R}$. If $E$ is null, $0 \notin E$, and*

$$M_0 := \sup_{t \geq 0} \sup_{\omega \in E} \left\| \int_0^t e^{-i\omega s} f(s) \, ds \right\| < \infty,$$

*then*

$$\lim_{t \to \infty} \int_0^t f(s) \, ds = g(0).$$

*Proof:* Let $R > 0$ such that $\pm R \notin E$. Let $\delta > 0$ such that $|\xi| \geq 2\delta$ and $R - |\xi| \geq 2\delta$ for all $\xi \in E \cap [-R, R]$. Let $\epsilon \in (0, \delta/2)$ be arbitrary and fixed. By Lemma 5.6.2, there exist $\xi_1 < \xi_2 < .... < \xi_n \in \mathbb{R}$ and $\theta \in (0, \epsilon/n)$ such that the intervals $(\xi_j - \theta, \xi_j + \theta)$ are pairwise disjoint and cover $E \cap [-R, R]$. We may assume that $(\xi_j - \theta, \xi_j + \theta) \cap E \neq \emptyset$ for all $j = 1, ..., n$. Then $|\xi_j| \geq \delta$ and $R - |\xi_j| \geq \delta$, $j = 1, ..., n$. We apply Lemma 5.6.1 using the notation there. We have

$$\frac{\xi_j^2}{\xi_j^2 - \theta^2} = \frac{1}{1 - \theta^2/\xi_j^2} \leq \frac{1}{1 - \theta^2/\delta^2}$$

$$\leq 1 + 2\theta^2/\delta^2 \leq 1 + \theta/\delta \leq e^{\theta/\delta},$$

using that $\theta/\delta \leq \frac{1}{2}$. Also,

$$1 + \frac{\theta^2}{(R - |\xi_j|)^2} \leq 1 + \theta^2/\delta^2 \leq e^{\theta/\delta}.$$

Hence, $a_j \leq e^{2\theta/\delta}$ and

$$\prod_{j=1}^n a_j \leq e^{2n\theta/\delta} \leq e^{2\epsilon/\delta} \leq e$$

using that $n\theta \leq \epsilon$ and $\epsilon \leq \delta/2$. Moreover, since

$$\xi_{j+m} - \xi_j - \theta \geq (2m-1)\theta, \quad m = 1, ..., n-j,$$
$$\xi_j - \xi_{j-m} - \theta \geq (2m-1)\theta, \quad m = 1, ..., j-1,$$

we have

$$\prod_{k=1, k \neq j}^n b_{j,k} = \prod_{k=1, k \neq j}^n \frac{\xi_k^2}{\xi_k^2 - \theta^2} \left(1 + \frac{\theta^2}{(|\xi_j - \xi_k| - \theta)^2}\right) \leq Ce^{(n-1)\theta/\delta},$$

where $C := \prod_{m=1}^\infty (1 + (2m-1)^{-2})^2$. So from Lemma 5.6.1 we obtain

$$\limsup_{t \to \infty} \left\| \int_0^t f(s)\,ds - g(0) \right\|$$
$$\leq \frac{2Me}{R} + 12n \cdot M_0 \cdot \theta \cdot \frac{2}{\delta} \cdot e^{\theta/\delta} \cdot Ce^{(n-1)\theta/\delta}$$
$$\leq \frac{2Me}{R} + \epsilon \cdot \frac{24M_0 Ce^{\epsilon/\delta}}{\delta}$$
$$\leq \frac{2Me}{R} + \epsilon \cdot \frac{24M_0 Ce}{\delta}.$$

Since $\epsilon \in (0, \delta/2)$ was arbitrary, it follows that

$$\limsup_{t \to \infty} \left\| \int_0^t f(s)\,ds - g(0) \right\| \leq \frac{2Me}{R}$$

whenever $\pm R \notin E$, $R > 0$. Since there exist arbitrarily large such $R$, the theorem follows. ////

If $f : \mathbb{R}_+ \to X$ is bounded and uniformly continuous, the convergence of the integrals $\int_0^t f(s)\,ds$ implies $\lim_{t \to \infty} \|f(t)\| = 0$. Applying this to the orbit of a $C_0$-semigroup, Theorem 5.6.3 yields the following result.

**Theorem 5.6.4.** *Let* **T** *be a* $C_0$*-semigroup on a Banach space* $X$. *Let* $x_0 \in X$ *and let* $E \subset \mathbb{R}$ *be a closed null set such that:*

(i) *The map* $t \mapsto T(t)x_0$ *is bounded and uniformly continuous;*
(ii) *The map* $\lambda \mapsto R(\lambda, A)x_0$ *admits a holomorphic extension to a neighbourhood of* $\{\operatorname{Re}\lambda \geq 0\} \setminus iE$;
(iii) $\displaystyle\sup_{t\geq 0}\sup_{\omega\in E} \left\| \int_0^t e^{-i\omega s} T(s)x_0\, ds \right\| < \infty.$

*Then* $\lim_{t\to\infty} \|T(t)x_0\| = 0$.

This theorem is an analogue of Theorem 5.3.6: the countability assumption is replaced by nullity but a stronger assumption is made on the behaviour of the integrals $\int_0^t e^{-i\omega s} T(s)x_0\, ds$.

Our first application is a Tauberian form of the Katznelson-Tzafriri theorem. If **T** is a $C_0$-semigroup on $X$ and the orbit of an $x_0 \in X$ is bounded, for $f \in L^1(\mathbb{R}_+)$ we denote by $\hat{f}(\mathbf{T})x_0$ the element

$$\hat{f}(\mathbf{T})x := \int_0^\infty f(t)T(t)x_0\, dt.$$

The map $f \mapsto \hat{f}(\mathbf{T})x_0$ is continuous as a map from $L^1(\mathbb{R}_+)$ to $X$.

**Theorem 5.6.5.** *Let* **T** *be a* $C_0$*-semigroup on a Banach space* $X$. *Let* $x_0 \in X$ *and let* $E \subset \mathbb{R}$ *be a closed set. Assume*

(i) *The orbit* $t \mapsto T(t)x_0$ *is bounded;*
(ii) *The map* $\lambda \mapsto R(\lambda, A)x_0$ *admits a holomorphic extension to a neighbourhood of* $\{\operatorname{Re}\lambda \geq 0\} \setminus iE$.

*If* $f \in L^1(\mathbb{R}_+)$ *satisfies*

$$\int_0^\infty s|f(s)|\, ds < \infty,$$

*and the Fourier transform* $\hat{f}$ *vanishes on* $-E$, *then*

$$\lim_{t\to\infty} \|T(t)\hat{f}(\mathbf{T})x_0\| = 0.$$

*Proof:* If $\hat{f}$ is identically zero, then $f = 0$ and there is nothing to prove. Otherwise, the function

$$g(z) := \hat{f}\left(\frac{z+1}{z-1}\right), \quad |z| < 1,$$

defines a non-zero holomorphic function on the unit disc with continuous extension to the boundary, i.e. an element of the disc algebra $A(D)$. The zero set on $\Gamma$ of such functions has measure zero (cf. [Ho]), and therefore the zero set of $\hat{f}$ on $\mathbb{R}$ has measure zero. Since $\hat{f}$ vanishes on $-E$, it follows that $E$ is a null set.

Define the bounded function $g_0$ by

$$g_0(t) := T(t)\hat{f}(\mathbf{T})x_0 = \int_0^\infty f(s)T(t+s)x_0\,ds.$$

By Lemma 5.3.8, the singular set of the Laplace transform $\mathcal{L}g_0$ on $i\mathbb{R}$ is contained in $iE$; in particular it is null.

For $\omega \in E$ we have

$$\lim_{t\to\infty} \int_0^t e^{i\omega s} f(s)\,ds = 0$$

since $\hat{f}(-\omega) = 0$. Using this and a partial integration, we have

$$\int_0^t e^{-i\omega s} g_0(s)\,ds = \int_0^t \int_0^\infty e^{-i\omega s} f(u)T(s+u)x_0\,du\,ds$$

$$= \int_0^\infty \left( \int_0^t e^{-i\omega(s+u)} T(s+u)x_0\,ds \right) e^{i\omega u} f(u)\,du$$

$$= -\lim_{r\to\infty} \int_0^r e^{-i\omega(u+t)} T(u+t)x_0 \left( \int_0^u e^{i\omega v} f(v)\,dv \right) du.$$

With $M := \sup_{t\geq 0} \|T(t)x_0\|$, it follows that

$$\sup_{t\geq 0} \left\| \int_0^t e^{-i\omega s} g_0(s)\,ds \right\| \leq M \int_0^\infty \left| \int_0^u e^{i\omega v} f(v)\,dv \right| du$$

$$= M \int_0^\infty \left| \int_u^\infty e^{i\omega v} f(v)\,dv \right| du$$

$$\leq M \int_0^\infty \int_u^\infty |f(v)|\,dv\,du$$

$$= M \int_0^\infty \int_0^v |f(v)|\,du\,dv$$

$$= M \int_0^\infty v|f(v)|\,dv.$$

Therefore we can apply Corollary 5.6.4 to $\hat{f}(\mathbf{T})x_0$, noting that its orbit is uniformly continuous (by the proof of Theorem 5.3.9). ////

The following result gives a global version of Theorem 5.6.5 parallel to Theorem 5.2.3.

**Corollary 5.6.6.** *Let $\mathbf{T}$ be a uniformly bounded $C_0$–semigroup on a Banach space $X$ with and let $f \in L^1(\mathbb{R}_+)$ satisfy*

$$\int_0^\infty s|f(s)|\,ds < \infty.$$

Furthermore, assume that $\hat{f}$ vanishes on $i\sigma(A) \cap \mathbb{R}$. Then,
$$\lim_{t\to\infty} \|T(t)\hat{f}(\mathbf{T})\| = 0.$$

*Proof:* Theorem 5.6.5 implies that
$$\lim_{t\to\infty} \|T(t)\hat{f}(\mathbf{T})x\| = 0, \quad \forall x \in X.$$
By Lemma 5.2.2, we may apply this to the semigroup $\mathcal{T}$ on $\mathcal{L}_0(X)$ and obtain
$$\lim_{t\to\infty} \|T(t)\hat{f}(\mathbf{T})\hat{g}(\mathbf{T})\| = 0, \quad \forall g \in L^1(\mathbb{R}_+).$$
Using an approximate identity argument as in the proof of Theorem 5.2.3, the corollary follows from this. ////

We are going to apply this corollary to show that the Tauberian hypothesis on $f$ actually implies that $f$ is of spectral synthesis with respect to the zero set of its Fourier transform. Thus, *a posteriori* Theorem 5.6.5 and its corollary appear as a special case of Theorem 5.3.9.

**Theorem 5.6.7.** *If $f \in L^1(\mathbb{R}_+)$ satisfies*
$$\int_0^\infty s|f(s)|\, ds < \infty,$$
*then $f$ is of spectral synthesis with respect to the zero set of its Fourier transform.*

*Proof:* We use the notation of the previous lemma, with $E = \{\hat{f} = 0\}$. We have to prove that $f \in J_E$.

From Corollary 5.6.6 and the first part of Lemma 5.4.2 we see that for all $g \in L^1(\mathbb{R})$ we have
$$\lim_{t\to\infty} \|U_{/J_E}(t)\hat{f}(\mathbf{U}_{/J_E})(g + J_E)\| = 0,$$
where $\mathbf{U}$ is the right translation group on $L^1(\mathbb{R})$. Since $\mathbf{U}$ is an isometric group, the same is true for the quotient group $\mathbf{U}_{/J_E}$. Therefore,
$$\hat{f}(\mathbf{U}_{/J_E})(g + J_E) = 0.$$
But
$$\hat{f}(\mathbf{U}_{/J_E})(g + J_E) = \int_0^\infty f(s) U_{/J_E}(s)(g + J_E)\, ds$$
$$= \left(\int_0^\infty f(s) U(s) g\, ds\right) + J_E$$
$$= \left(\int_0^\infty f(s) g(\cdot - s)\, ds\right) + J_E$$
$$= (f * g) + J_E.$$
We have proved that $f * g \in J_E$ for all $g \in L^1(\mathbb{R})$. Taking for $g$ an approximate identity, the closedness of $J_E$ implies that also $f \in J_E$. ////

We close this section with some improvements of Corollary 5.2.7. Let us first note that the assumption on the spectrum made there, viz. $\sigma(A) \cap i\mathbb{R} \subset \{0\}$, is rather restrictive. Indeed, in view of the spectral mapping theorems, a natural way to relax this condition would be to require that $\sigma(A) \cap i\mathbb{R} \subset i\omega\mathbb{Z}$ for some $\omega > 0$; under this assumption one expects

$$\lim_{t\to\infty} \|T(t + \frac{2\pi}{\omega})R(\lambda, A) - T(t)R(\lambda, A)\| = 0.$$

In fact, we have the following more general result.

**Theorem 5.6.8.** *Let* **T** *be a $C_0$-semigroup on a Banach space $X$, with generator $A$. Let $x_0 \in X$ be such that*
  (i) *The orbit $t \mapsto T(t)x_0$ is bounded and uniformly continuous;*
  (ii) *For some $\omega > 0$, the map $\lambda \mapsto R(\lambda, A)x_0$ admits a holomorphic extension to a neighbourhood of $\{\operatorname{Re}\lambda \geq 0\}\setminus i\omega\mathbb{Z}$.*

*Then,*

$$\lim_{t\to\infty} \|T(t + \frac{2\pi}{\omega})x_0 - T(t)x_0\| = 0.$$

*Proof:* We are going to verify the condition of Theorem 5.6.4 for $y := T(\frac{2\pi}{\omega})x_0 - x_0$. For $\operatorname{Re}\lambda > 0$ we have $R(\lambda, A)y = (T(\frac{2\pi}{\omega}) - I)R(\lambda, A)x_0$ and therefore the singular set of $R(\lambda, A)y$ on the imaginary axis is contained in $i\omega\mathbb{Z}$. For all $k \in \mathbb{Z}$ we have

$$\left\| \int_0^t e^{-i\omega ks} \left( T(s + \frac{2\pi}{\omega})x_0 - T(s)x_0 \right) ds \right\|$$

$$= \left\| \int_{\frac{2\pi}{\omega}}^{t+\frac{2\pi}{\omega}} e^{-i\omega ks} T(s)x_0\, ds - \int_0^t e^{-i\omega ks} T(s)x_0\, ds \right\|$$

$$= \left\| \int_t^{t+\frac{2\pi}{\omega}} e^{-i\omega ks} T(s)x_0\, ds - \int_0^{\frac{2\pi}{\omega}} e^{-i\omega ks} T(s)x_0\, ds \right\|$$

$$\leq \frac{4\pi M}{\omega},$$

where $M := \sup_{t \geq 0} \|T(t)x_0\|$. /////

By the above estimates, this theorem could have been derived alternatively as a consequence of Theorem 5.3.6.

If an orbit $T(\cdot)x_0$ is bounded, then $T(\cdot)\hat{f}(\mathbf{T})x_0$ is bounded and uniformly continuous. By Lemma 5.3.8, the map $\lambda \mapsto R(\lambda, A)\hat{f}(\mathbf{T})x_0$ extends holomorphically along with $R(\lambda, A)x_0$. Hence we may replace (i) by mere boundedness of $T(t)x_0$ and obtain that

$$\lim_{t\to\infty} \|T(t + \frac{2\pi}{\omega})\hat{f}(\mathbf{T})x_0 - T(t)\hat{f}(\mathbf{T})x_0\| = 0$$

for all $f \in L^1(\mathbb{R}_+)$. Finally, we consider the global case:

**Corollary 5.6.9.** Let $\mathbf{T}$ be a uniformly bounded $C_0$-semigroup on a Banach space $X$, with generator $A$. If $\sigma(A) \cap i\mathbb{R} \subset i\omega\mathbb{Z}$ for some $\omega > 0$, then for all $f \in L^1(\mathbb{R}_+)$ we have

$$\lim_{t\to\infty} \|T(t + \frac{2\pi}{\omega})\hat{f}(\mathbf{T}) - T(t)\hat{f}(\mathbf{T})\| = 0.$$

*Proof:* Fix $\mu \in \varrho(A)$. By the above remark we have

$$\lim_{t\to\infty} \|T(t + \frac{2\pi}{\omega})\hat{f}(\mathbf{T})x - T(t)\hat{f}(\mathbf{T})x\| = 0, \quad \forall x \in X.$$

We apply this to the semigroup $\mathcal{T}$ on $\mathcal{L}_0(X)$ as in the proof of Theorem 5.2.3.
////

**Corollary 5.6.10.** Let $\mathbf{T}$ be a uniformly bounded $C_0$-semigroup on a Banach space $X$, with generator $A$. If $\sigma(A) \cap i\mathbb{R} \subset \{0\}$, then for all $s > 0$ and $f \in L^1(\mathbb{R}_+)$ we have

$$\lim_{t\to\infty} \|T(t+s)\hat{f}(\mathbf{T}) - T(t)\hat{f}(\mathbf{T})\| = 0.$$

## 5.7. The splitting theorem of Glicksberg and DeLeeuw

In Section 5.1 we saw that a necessary condition for a $C_0$-semigroup $\mathbf{T}$ to be uniformly stable is that the unitary point spectrum of the generator $A$ should be empty. For weakly almost periodic semigroups, in this section we prove a more detailed result known as the *Glicksberg-DeLeeuw theorem*: if $\mathbf{T}$ is a weakly almost periodic $C_0$-semigroup on $X$, then $X$ can be decomposed into a direct sum of two $\mathbf{T}$-invariant closed subspaces $X_0 \oplus X_1$, where $X_0$ is the closed linear span of all eigenvectors of $A$ with purely imaginary eigenvalues, and $X_1$ is the subspace of all $x \in X$ such that 0 belongs to the weak closure of the orbit $T(\cdot)x$. If $\mathbf{T}$ is almost periodic, the restriction of $\mathbf{T}$ to $X_1$ is uniformly stable. The decomposition of $X$ will be induced by a projection whose existence is a consequence of a general algebraic existence theorem for idempotents in topological semigroups.

A pair $(G, \phi)$, where $G$ is a set and $\phi : G \times G \to G$ is a mapping, is called an *semigroup* if $\phi(\phi(g_0, g_1), g_2) = \phi(g_0(\phi(g_1, g_2)))$ for all $g_0, g_1, g_2 \in G$. The map $\phi$ is referred to as the *multiplication* of $G$. Whenever $\phi$ is understood, we denote the semigroup $(G, \phi)$ simply by $G$ and write $g_0 g_1$ for $\phi(g_0, g_1)$.

An element $e \in G$ is called a *unit* if $eg = ge = g$ for all $g \in G$. If a unit exists, it is unique: for if $e_0, e_1$ are units, then $e_0 = e_0 e_1 = e_1$. A semigroup with unit $e$ is called a *group* if for each $g \in G$ there exists an element $g^{-1}$, the *inverse of $g$*, such that $gg^{-1} = g^{-1}g = e$. The inverse of an element $g$ is necessarily unique: for if $h_0, h_1$ are inverse to $g$, then $h_0 = h_0 e = h_0(gh_1) = (h_0 g)h_1 = eh_1 = h_1$.

A semigroup $G$ is *abelian* if $g_0 g_1 = g_1 g_0$ for all $g_0, g_1 \in G$.

A semigroup $G$ with unit $e$, endowed with a Hausdorff topology, is called a *semitopological semigroup* if the multiplication is separately continuous, i.e. for each $g \in G$ the maps $h \mapsto hg$ and $h \mapsto gh$ are continuous. A semitopological semigroup is a *topological semigroup* if the multiplication is jointly continuous, i.e. if the map $(g, h) \mapsto gh$ is continuous. A topological semigroup is a *topological group* if it is a group and the mapping $g \mapsto g^{-1}$ is continuous.

A non-empty subset $I$ of a semigroup $G$ is an *ideal* if $IG \subset I$ and $GI \subset I$. An ideal is *minimal* if it contains no ideal other than itself.

**Lemma 5.7.1.** *Let $G$ be a compact abelian semigroup and let $K$ be the intersection of all closed ideals in $G$. Then $K$ is the unique minimal ideal in $G$, and*

$$K = \cap_{g \in G} gG.$$

*Furthermore $K$ is a group, and $K = uG$ where $u$ is the unit of $K$.*

*Proof:* The intersection of finitely many ideals $I_1, ..., I_n$ contains the product $I_1 I_2 \cdot ... \cdot I_n$, which is non-empty. Since $G$ is compact, by the finite intersection property the intersection $K$ of all closed ideals in $G$ is non-empty. If $I \subset K$ is an ideal and $g \in I$, then $gG$ is a closed ideal contained in $I$; we use that $G$ is compact and abelian. Hence, $gG \subset I \subset K \subset gG$, $I = K$, and $K$ is minimal. If $L$ is another minimal ideal, then $K \subset gG \subset L$ for all $g \in L$, and since $K$ is an ideal it follows that $K = L$ by the minimality of $L$. This shows that $G$ has only one minimal ideal.

As each $gG$ is a closed ideal, we have $K \subset \cap_{g \in G} gG$, and if $I$ is any closed ideal and $g \in I$, then $gG \subset I$ so that $\cap_{g \in G} gG \subset \cap \{I : I \text{ closed ideal in } G\} = K$.

For each $g \in K$, $gK$ is an ideal contained in $K$ and therefore $gK = K$ by minimality. It follows that there is an element $u_g \in K$ such that $gu_g = g$. Similarly, for each $h \in K$ there exists an element $v_{gh} \in K$ such that $gv_{gh} = h$. Hence

$$hu_g = gv_{gh} u_g = v_{gh} gu_g = v_{gh} g = gv_{gh} = h$$

for each two $g, h \in K$. Taking $u_h$ instead of $h$ in this identity, we see that $u_h u_g = u_h$ for any two $g, h \in K$. By reversing the roles of $g$ and $h$ we also have $u_g u_h = u_g$. It follows that $u_g = u_g u_h = u_h u_g = u_h$. Let $u$ denote this common element. Then $ug = gu = g$ for all $g \in K$, so $u$ is a unit for $K$. The identity $gK = K$ and the fact that $u \in K$ imply the existence of an element $g^{-1}$ such that $gg^{-1} = u$. Applying this to $g^{-1}$, there is an element $h$ such that $g^{-1}h = u$, and then $h = uh = gg^{-1}h = gu = g$, so $g^{-1}g = u$ and $g^{-1}$ is inverse to $g$. This proves that $K$ is a group. The last assertion follows from the inclusions $uG \subset KG \subset K$ and $K = uK \subset uG$.
////

The group $K$ is called the *Sushkevich kernel* of $G$.

We are going to apply the lemma to subsemigroups of $\mathcal{L}(X)$ endowed with the weak operator topology or the strong operator topology. The *weak operator topology* is the topology on $\mathcal{L}(X)$ generated by the sets

$$V_{x,x^*,T,\epsilon} := \{S \in \mathcal{L}(X) : |\langle x^*, (T-S)x \rangle| < \epsilon\},$$

with $\epsilon > 0$ and $x$, $x^*$, and $T$ ranging over $X$, $X^*$, and $\mathcal{L}(X)$, respectively. It is trivial to verify that a net $(T_\alpha)$ converges to $T$ in this topology if and only if $\lim_\alpha \langle x^*, T_\alpha x \rangle = \langle x^*, Tx \rangle$ for all $x \in X$ and $x^* \in X^*$. From this it easily follows that multiplication is separately continuous in the weak operator topology. Also, if $G \subset (\mathcal{L}(X), \text{wo})$ is an semitopological semigroup, then so is its weak operator closure $\overline{G}^{wo}$; this is an easy consequence of the separate continuity of the multiplication in $(\mathcal{L}(X), \text{wo})$. If $G$ is abelian, then so is $\overline{G}^{wo}$.

The *strong operator topology* is the topology on $\mathcal{L}(X)$ generated by the sets

$$V_{x,T,\epsilon} := \{S \in \mathcal{L}(X) : \|(T-S)x\| < \epsilon\},$$

with $\epsilon > 0$ and $x$ and $T$ ranging over $X$ and $\mathcal{L}(X)$, respectively. It is trivial to verify that a net $(T_\alpha)$ converges to $T$ in this topology if and only if $\lim_\alpha T_\alpha x = Tx$ strongly for all $x \in X$. From this it easily follows that multiplication is separately continuous in this topology. Also, if $G \subset (\mathcal{L}(X), \text{so})$ is an semitopological semigroup, then so is its strong operator closure $\overline{G}$; this is an easy consequence of the separate continuity of the multiplication in $(\mathcal{L}(X), \text{so})$. If $G$ is abelian, then so is $\overline{G}$.

A semigroup $G \subset \mathcal{L}(X)$ is called *(weakly) almost periodic* if for each $x \in X$ the set $\{Tx : T \in G\}$ is relatively (weakly) compact in $X$.

**Lemma 5.7.2.** *Let $G$ be a semigroup in $\mathcal{L}(X)$.*

*(i) $G$ is weakly almost periodic if and only if $G$ is a relatively compact subset of $\mathcal{L}(X)$ in the weak operator topology. Moreover, for all $x \in X$ the weak closure of $Gx$ is precisely $\overline{G}^{wo} x$.*

*(ii) $G$ is almost periodic if and only if $G$ is a relatively compact subset of $\mathcal{L}(X)$ in the strong operator topology. Moreover, for all $x \in X$ the closure of $Gx$ is precisely $\overline{G}x$.*

*Proof:* We only prove (i); mutatis mutandis the same argument proves (ii).

If $G$ is relatively weak operator compact, then $\overline{G}^{wo}$ is weak operator compact in $\mathcal{L}(X)$. Since for each $x \in X$ the map $x : (\mathcal{L}(X), \text{wo}) \to (X, \text{weak})$, $T \mapsto Tx$, is continuous, $\overline{G}^{wo} x$ is weakly compact and $Gx$ is relatively weakly compact in $X$ for all $x \in X$.

Conversely, assume that $Gx$ is relatively weakly compact for all $x \in X$. Then $Gx$ is bounded in $X$ for all $x \in X$ and by the uniform boundedness theorem $G$ is bounded, i.e. $\sup\{\|T\| : T \in G\} =: M < \infty$. For each $x \in X$ let $E_x := \overline{Gx}^{weak}$, and let $E$ be the Cartesian product of all $E_x$. By Tychonoff's theorem, $E$ is compact with respect to the product topology. Let $(T_\alpha)$ be a net in $G$. Then $((T_\alpha x)_{x \in X})$ can be regarded as a net in $E$ and by compactness there exists a subnet $((T_\beta x)_{x \in X})$ converging in $E$. This means that weak-$\lim_\beta T_\beta x$ exists for all $x \in X$. Define the operator $T$ by $Tx := $ weak-$\lim_\beta T_\beta x$. Then $T$ is linear and bounded of norm $\leq M$. By its definition, $T$ belongs to the weak operator closure of $G$. This proves that $G$ is relatively weak operator compact.

To prove the final assertion, observe that the weak compactness of $\overline{G}^{wo} x$ implies that $\overline{Gx}^{weak} \subset \overline{G}^{wo} x$, and the reverse inclusion follows from the definition of the weak operator topology. ////

Let $G$ be a weakly almost periodic semigroup in $\mathcal{L}(X)$. Then the weak operator closure $\overline{G}^{wo}$ is a compact semitopological semigroup by Lemma 5.7.2. If in addition $G$ is abelian we can apply Lemma 5.7.1 to $\overline{G}^{wo}$. Let $K$ be the Sushkevich kernel of $\overline{G}^{wo}$ and let $U$ be its unit. Then $U^2 = U$, so $U$ is a projection. Accordingly, $X$ admits a direct sum decomposition

$$X = X_0 \oplus X_1$$

where $X_0 := UX$ and $X_1 := (I - U)X$. The following theorem gives a characterization of the spaces $X_0$ and $X_1$ in terms of $G$.

**Theorem 5.7.3.** *Let $G \subset \mathcal{L}(X)$ be an abelian weakly almost periodic semigroup. Then there exists a direct sum decomposition $X = X_0 \oplus X_1$ of $G$-invariant subspaces, where*

$$X_0 = \{x \in X : 0 \in \overline{Gx}^{weak}\}$$

*and*

$$X_1 = \{x \in X : x \in \overline{Gy}^{weak} \text{ whenever } y \in \overline{Gx}^{weak}\}.$$

*Moreover, the restriction of $\overline{G}^{wo}$ to $X_1$ is a weakly almost periodic group.*

*Proof:* Let $X = UX \oplus (I - U)X$ as in the preceding discussion and let the spaces $X_0$ and $X_1$ be defined as in the statement of the theorem; we shall prove that $X_0 = (I - U)X$ and $X_1 = UX$.

Let $x \in X$ be such that $x \in \overline{Gy}^{weak}$ whenever $y \in \overline{Gx}^{weak}$. Since the weak closures of $Gx$ and $Gy$ are $\overline{G}^{wo} x$ and $\overline{G}^{wo} y$ respectively, the condition on $x$ means that for any $T \in \overline{G}^{wo}$ there exists an $S \in \overline{G}^{wo}$ such that $STx = x$. In particular, there is an $S \in \overline{G}^{wo}$ such that $SUx = x$. Since $\overline{G}^{wo}$ is abelian it follows that $x = SUx = USx$, proving that $x \in UX$. Conversely, let $x \in UX$, say $x = Ux_1$ for some $x_1 \in X$, and assume that $y \in \overline{Gx}^{weak} = \overline{G}^{wo} x$. Then $y = Tx$ for some $T \in \overline{G}^{wo}$, so $y = TUx_1$. Since the Sushkevich kernel $K$ of $\overline{G}^{wo}$ is an ideal in $\overline{G}^{wo}$ and $U \in K$, we have $TU \in K$, and since $K$ is a group there exists an $S \in K$ inverse to $TU$. But then $Sy = STUx_1 = Ux_1 = x$. This means that $x \in \overline{G}^{wo} y = \overline{Gy}^{weak}$.

As for $(I - U)X$, we shall prove that this space consists precisely of those $x \in X$ such that $0 \in \overline{Gx}^{weak}$. If $0 \in \overline{Gx}^{weak} = \overline{G}^{wo} x$, then $Tx = 0$ for some $T \in \overline{G}^{wo}$. Since $TU \in K$ we can choose $S \in K$ with $STU = U$. Since $x = Ux + (I - U)x$ and

$\overline{G}^{wo}$ is abelian, it follows that $x = STUx + (I-U)x = SUTx + (I-U)x = (I-U)x$. This proves that $x \in (I-U)X$. Conversely, let $x \in (I-U)X$. Since $I-U$ is a projection, we have $x = (I-U)x$ and $Ux = U(I-U)x = 0$. This proves that $0 = Ux \in \overline{G}^{wo}x = \overline{Gx}^{weak}$.

Next we prove that the restriction to $UX$ of $\overline{G}^{wo}$ is a group. Let $T \in \overline{G}^{wo}$ be arbitrary. Then $TU \in K$. Letting $S$ be the inverse of $TU$ in $K$, for all $x \in X_0 = UX$ we have $STx = STUx = x$. Thus, the restriction of $S$ to $UX$ is the inverse of the restriction of $T$ to $UX$. It follows that the restriction of $\overline{G}^{wo}$ to $UX$ is a group. Since $\overline{G}^{wo}$ is weakly almost periodic, so is its restriction to $UX$. ////

The characterization of $X_1$ is not a very practical one. We are going to prove another characterization of this space in terms of eigenvalues of $G$. The proof is based on a basic result due to R. Ellis, which states that every compact semitopological semigroup is a topological semigroup.

Let $G$ be a subsemigroup of $\mathcal{L}(X)$. A non-zero vector $x \in X$ is called an *eigenvector* for $G$ if there is a map $\lambda : G \to \mathbb{C}$ such that $Tx = \lambda(T)x$ for all $T \in G$. If $T_\alpha \to T$ with respect to the weak operator topology and $x \in X$ is an eigenvector for $G$, then for all $x^* \in X^*$ we have

$$\langle x^*, Tx \rangle = \lim_\alpha \langle x^*, T_\alpha x \rangle = \lim_\alpha \lambda(T_\alpha) \langle x^*, x \rangle.$$

Hence $\lambda(T) := \lim_\alpha \lambda(T_\alpha)$ exists and $Tx = (\lim_\alpha \lambda(T_\alpha))x$. It follows from this that $x$ is also an eigenvector of $\overline{G}^{wo}$ and that the map $\lambda$ is continuous and extends continuously to $\overline{G}^{wo}$. An eigenvector $x$ of $G$ is called *unimodular* if for the associated map $\lambda$ we have $|\lambda(T)| = 1$ for all $T \in G$; in that case $x$ is also a unimodular eigenvector for $\overline{G}^{wo}$. In order to avoid trivialities we also regard $0$ as a unimodular eigenvector.

**Theorem 5.7.4.** *Let $G \subset \mathcal{L}(X)$ be an abelian weakly almost periodic semigroup. Then there exists a direct sum decomposition $X = X_0 \oplus X_1$ of $G$-invariant subspaces, where*

$$X_0 = \{x \in X : 0 \in \overline{Gx}^{weak}\}$$

*and $X_1$ is the closed linear span of all unimodular eigenvectors of $G$. Moreover, the restrictions of $\overline{G}^{wo}$ and $\overline{G}$ to $X_1$ are weakly almost periodic and almost periodic groups, respectively.*

*Proof:* Let $x \in X$ be a unimodular eigenvector for $G$. Then $x$ is also a unimodular eigenvector for $\overline{G}^{wo}$.

Fix $T \in \overline{G}^{wo}$ arbitrary. The identity $T^n x = (\lambda(T))^n x$ shows that there is a sequence $n_k \to \infty$ such that $\lim_{k \to \infty} T^{n_k} x = x$. It follows that for any finite set $x_1^*, ..., x_n^* \in X^*$ the set $\cap_{i=1}^n \{S \in \overline{G}^{wo} : |\langle x_i^*, STx - x \rangle| \le \epsilon\}$ is non-empty. By the compactness of $\overline{G}^{wo}$, the intersection of all these sets is non-empty. Therefore, there exists an $S_0 \in \overline{G}^{wo}$ such that $|\langle x^*, S_0 Tx - x \rangle| = 0$ for all $x^* \in X^*$, i.e., $S_0 Tx = x$. As we have seen in the proof of Theorem 5.7.3, this implies that $x \in X_1$.

The converse is more difficult to prove. By Ellis's theorem, the Sushkevich kernel $K$ of $\overline{G}^{wo}$, being a compact semitopological group, is a topological group. Let $\mu$ be its normalized Haar measure and let $K'$ be the character group of $K$.

For given $\gamma \in K'$ and $x \in X$ we define the element $T_\gamma x \in X^{**}$ by

$$\langle T_\gamma x, x^* \rangle := \int_K \gamma(T) \langle x^*, Tx \rangle \, d\mu(T), \quad x^* \in X^*.$$

In this way we obtain a bounded operator $T_\gamma : X \to X^{**}$, of norm $\leq \sup\{\|T\| : T \in K\}$. We shall show that $T_\gamma$ actually maps $X$ into itself; we identify $X$ with its canonical image in $X^{**}$.

As the map $T \mapsto \gamma(T)Tx$ is continuous from $K$ in its weak operator topology into $X$ in its weak topology, the set $\{\gamma(T)Tx : T \in K\}$ is weakly compact in $X$. By the Krein-Shmulyan theorem, so is its closed convex hull $H := \overline{\text{co}}\,\{\gamma(T)Tx : T \in K\}$. As a set in $X^{**}$, $H$ is weak*-compact. Now, if for some $x_0^* \in X^*$ and $\alpha \in \mathbb{R}$ we have

$$\operatorname{Re} \gamma(T) \langle x_0^*, Tx \rangle \geq \alpha, \quad \forall T \in K,$$

then

$$\operatorname{Re} \langle T_\gamma x, x_0^* \rangle = \operatorname{Re} \int_K \gamma(T) \langle x_0^*, Tx \rangle \, d\mu(T) \geq \int_K \alpha \, d\mu(T) = \alpha$$

since $\mu(K) = 1$. By the Hahn-Banach separation theorem, this implies that $T_\gamma x$ belongs to the weak*-closed convex hull in $X^{**}$ of the set $\{\gamma(T)Tx : T \in K\}$, which, as we have seen, is $H$. Therefore, $T_\gamma x \in H \subset X$.

It follows that $T_\gamma$ is a bounded operator on $X$. For all $S \in K$, $x \in X$, and $x^* \in X^*$ we have

$$\langle x^*, ST_\gamma x \rangle = \langle S^* x^*, T_\gamma x \rangle = \int_K \gamma(T) \langle S^* x^*, Tx \rangle \, d\mu(T)$$
$$= \overline{\gamma}(S) \int_K \gamma(ST) \langle x^*, STx \rangle \, d\mu(T)$$
$$= \overline{\gamma}(S) \int_K \gamma(ST) \langle x^*, STx \rangle \, d\mu(ST)$$
$$= \overline{\gamma}(S) \langle x^*, T_\gamma x \rangle.$$

Here $\overline{\gamma}(T)$ denotes the complex conjugate of $\gamma(T)$; we used the translation invariance of the Haar measure. Since this holds for all $x \in X$ and $x^* \in X^*$, it follows that $ST_\gamma = \overline{\gamma}(S) T_\gamma$. For the unit $U$ of $K$ we have $\overline{\gamma}(U) = \gamma(U) = 1$. Hence in particular we obtain $UT_\gamma = \overline{\gamma}(U) T_\gamma = T_\gamma$.

Now let $S \in G$ be arbitrary. Applying the above to the operator $SU \in K$ it follows that $ST_\gamma = SUT_\gamma = \overline{\gamma}(SU) T_\gamma = \overline{\gamma}(S) T_\gamma$ for all $S \in G$. This means that $T_\gamma X$ consists of unimodular eigenvectors of $G$.

To finish the proof, it suffices to show that $X_1 = UX$ is contained in the closed linear span $Y$ of $\{T_\gamma x : \gamma \in K', x \in X\}$. Suppose some $x_0^* \in X^*$ annihilates $Y$. Then

$$\int_K \gamma(T)\langle x_0^*, Tx\rangle\, d\mu(T) = 0, \quad \forall \gamma \in K', x \in X. \tag{5.7.1}$$

Assume, for a contradiction, that $\langle x_0^*, Ux_0\rangle \neq 0$ for some $x_0 \in X$. Since the map $T \mapsto \langle x_0^*, Tx_0\rangle$ is continuous on $K$, there is a neighbourhood $F$ of $U$ such that

$$\int_K \chi_F(T)\langle x_0^*, Tx_0\rangle\, d\mu(T) \neq 0. \tag{5.7.2}$$

If $g \in L^2(K, \mu)$ is orthogonal to all characters $\chi \in K'$, then $g = 0$ by the injectivity of the Fourier-Plancherel transform for locally compact abelian groups. In other words, the characters of $K$, viewed as elements of $L^2(K, \mu)$, span a dense linear subspace of $L^2(K, \mu)$.

It follows that there are finite linear combinations $\phi_n$ of characters approximating $\chi_F$ in $L^2(K, \mu)$. Since $\mu(K) = 1$, we also have $\phi_n \to \chi_F$ in $L^1(K, \mu)$ and hence, using (5.7.1) and (5.7.2),

$$0 = \lim_n \int_K \phi_n(T)\langle x_0^*, Tx_0\rangle\, d\mu(T) = \int_K \chi_F(T)\langle x_0^*, Tx_0\rangle\, d\mu(T) \neq 0.$$

This contradiction shows that $\langle x_0^*, Ux\rangle = 0$ for all $x \in X$. By the Hahn-Banach theorem, this implies that $Ux \in Y$ for all $x \in X$.

Finally, consider the restriction $G_1$ of $G$ to $UX = X_1$ in the strong operator topology of $X_1$. It remains to prove that $G_1$ is an almost periodic group. For each unimodular eigenvector $x_1$ of $G$, the orbit $Gx_1$ is obviously relatively compact in $X_1$. We claim that all orbits in $X_1$ are relatively compact.

To see this, fix $x \in X_1$ and $(T_n) \subset G_1$. Let $x_n \to x$ with each $x_n$ in the linear span of the unimodular eigenvectors. There is a subsequence $(n_k)$ and a $y_1 \in X_1$ such that $T_{n_k} x_1 \to y_1$. This subsequence has a further subsequence $(n_{k_j})$ such that $T_{n_{k_j}} x_2 \to y_2$ for some $y_2 \in X_1$. Continuing is this way, a diagonal argument yields a sequence $(n_i)$ and vectors $y_m \in X_1$ such that $T_{n_i} x_m \to y_m$ for all $m = 1, 2, \ldots$ Now the estimate

$$\|y_m - y_{m'}\| \leq \|y_m - T_{n_i} x_m\| + \|T_{n_i} x_m - T_{n_i} x_{m'}\| + \|T_{n_i} x_{m'} - y_{m'}\|$$

and the uniform boundedness of $G_1$ show that $(y_m)$ is a Cauchy sequence in $X_1$, say with limit $y$. But then from

$$\|T_{n_i} x - y\| \leq \|T_{n_i} x - T_{n_i} x_m\| + \|T_{n_i} x_m - y_m\| + \|y_m - y\|$$

it follows that $T_{n_i} x \to y$. Thus, the orbit $G_1 x$ is relatively compact and the claim is proved.

Hence, $G_1$ is almost periodic as a subsemigroup of $\mathcal{L}(X_1)$. Letting $K_1$ be the Sushkevich kernel of $\overline{G_1}$, the proof that $\overline{G_1}$ is a group proceeds as in the weak operator case. ////

This theorem is usually referred to as the *Glicksberg-DeLeeuw theorem*.

The characterization of the space $X_1$ can also be improved if $G$ is almost periodic.

**Theorem 5.7.5.** *Let $G \subset \mathcal{L}(X)$ be an abelian almost periodic semigroup. Then there exists a direct sum decomposition $X = X_0 \oplus X_1$ of $G$-invariant subspaces, where*

$$X_0 = \{x \in X : 0 \in \overline{Gx}\}$$

*and $X_1$ is the closed linear span of all unimodular eigenvectors of $G$. Moreover, the restriction of $\overline{G}$ to $X_1$ is an almost periodic group.*

*Proof:* We only have to prove that $0 \in \overline{Gx}$ whenever $0 \in \overline{Gx}^{weak}$. So let $x \in X$ be such that $0 \in \overline{Gx}^{weak}$. Choose a net $(T_\alpha)$ in $G$ such that weak-$\lim_\alpha T_\alpha x = 0$ Since $G$ is almost periodic, the set $\{T_\alpha x : \alpha \in I\}$ is relatively compact; $I$ is the index set for the net. Hence there is a subnet $(T_\beta x)$ converging strongly to some $y \in X$. Then also $T_\beta x \to y$ weakly, and therefore $y = 0$. It follows that 0 belongs to the strong closure of $\{Tx : T \in G\}$. ////

Next we are going to apply these results to $C_0$-semigroups.

**Theorem 5.7.6.** *Let $\mathbf{T}$ be a weakly almost periodic $C_0$-semigroup on a Banach space $X$. Then there exists a direct sum decomposition $X = X_0 \oplus X_1$ of $\mathbf{T}$-invariant subspaces, where $X_0$ is the closed subspace of all $x \in X$ such that 0 belongs to the weak closure of the orbit of $x$ and $X_1$ is the closed linear span of all eigenvectors of the generator $A$ with purely imaginary eigenvalues Moreover, the restriction of $\mathbf{T}$ to $X_1$ extends to an almost periodic $C_0$-group. If $\mathbf{T}$ is contractive, this group is isometric.*

*Proof:* The representation of $X_1$ follows from the observation that an $x \in X$ is an eigenvector for $A$ if and only if $x$ is an eigenvector for $\mathbf{T}$ in the sense defined above.

The restriction to $X_1$ of the strong operator closure $\overline{\mathbf{T}}$ is a group, which is almost periodic by Theorem 5.7.4. It contains the inverses in $X_1$ of the restrictions $T_1(t)$ of the operators $T(t)$. Therefore, the restriction $\mathbf{T}_1$ of $\mathbf{T}$ to $X_1$ extends to an almost periodic $C_0$-group $\mathbf{T}_1$ on $X_1$.

If $\mathbf{T}$ is contractive, i.e. $\|T(t)\| \leq 1$ for all $t \geq 0$, then also $\|T\| \leq 1$ for all $T \in \overline{\mathbf{T}}$. Therefore, the inverses of the operators $T_1(t)$ are contractive. Now if there were an $x_1 \in X_1$ such that $\|T_1(t)x_1\| < \|x_1\|$ for some $t \in \mathbb{R}$, then $\|T_1(-t)\| > 1$, a contradiction. Therefore, $\mathbf{T}_1$ is isometric. ////

Similarly, in the almost periodic case we have:

**Theorem 5.7.7.** *Let $\mathbf{T}$ be an almost periodic $C_0$-semigroup on a Banach space $X$. Then there exists a direct sum decomposition $X = X_0 \oplus X_1$ of $\mathbf{T}$-invariant subspaces, where*

$$X_0 = \{x \in X : \lim_{t \to \infty} \|T(t)x\| = 0\}$$

and $X_1$ is the closed linear span of all eigenvectors of the generator $A$ with purely imaginary eigenvalues Moreover, the restriction of **T** to $X_1$ extends to an almost periodic $C_0$−group on $X_1$. If **T** is contractive, this group is isometric.

*Proof:* We only have to prove that $0 \in \overline{\{T(t)x : t \geq 0\}}$ implies $\lim_{t\to\infty} \|T(t)x\| = 0$. Let $(t_n) \subset \mathbb{R}_+$ be a sequence such that $\lim_{n\to\infty} \|T(t_n)x\| = 0$. Since **T** is uniformly bounded, say $\|T(t)\| \leq M$ for all $t \geq 0$, for all $n$ and all $t \geq t_n$ we have $\|T(t)x\| \leq M\|T(t_n)x\|$. Since the latter becomes arbitrarily small as $n \to \infty$, the desired result follows. ////

Thus, an almost periodic semigroup can be decomposed into a bounded $C_0$−group and a uniformly stable semigroup.

**Corollary 5.7.8.** *Let* **T** *be a uniformly bounded $C_0$−semigroup on a Banach space $X$. Assume there exists a compact operator $K$ on $X$ with dense range, commuting with each $T(t)$. Then the following assertions are equivalent:*

 (i) **T** *is uniformly stable;*
 (ii) $\sigma_p(A) \cap i\mathbb{R} = \emptyset$
 (iii) $\sigma_p(A^*) \cap i\mathbb{R} = \emptyset$.

*Proof:* For each $x = Ky \in \text{range } K$ the set $\{T(t)x : t \geq 0\} = \{KT(t)y : t \geq 0\}$ is relatively compact. Since the range of $K$ is dense, it follows that $\{T(t)x : t \geq 0\}$ is relatively compact for all $x \in X$; cf. the proof of Theorem 5.7.4. Thus **T** is almost periodic and we can apply Theorem 5.7.7 to obtain the equivalence (i) $\Leftrightarrow$(ii). The equivalence (ii) $\Leftrightarrow$(iii) follows from Proposition 5.1.15.  ////

We have the following generalization of Theorem 5.1.11:

**Theorem 5.7.9.** *Let* **T** *be a uniformly bounded $C_0$−semigroup on a Banach space $X$, with generator $A$. Let $x_0 \in X$ be such that $\lambda \mapsto R(\lambda, A)x_0$ admits a holomorphic extension to a neighbourhood of $\{\text{Re}\,\lambda \geq 0\}\setminus iE$, where $E \subset \mathbb{R}$ is closed and countable. Then the following assertions are equivalent:*

 (i) *The set $\{T(t)x_0 : t \geq 0\}$ is relatively weakly compact;*
 (ii) *The set $\{T(t)x_0 : t \geq 0\}$ is relatively compact.*
 (iii) *For all $\omega \in E$ the limit $\lim_{t\to\infty} \dfrac{1}{t}\int_0^t e^{-i\omega s}T(s)x_0\,ds$ exists.*

*Proof:* (ii)$\Rightarrow$(i) is trivial.
 (i)$\Rightarrow$(iii): Fix $\omega \in E$. Since the closed absolute convex hull of the orbit of $x_0$ is weakly compact by the Krein-Shmulyan theorem, it follows that the net $(t^{-1}\int_0^t e^{-i\omega s}T(s)x_0\,ds)_{t>0}$ has a weak cluster point $y$. Then

$$\lim_{t\to\infty} \frac{1}{t}\int_0^t e^{-i\omega s}T(s)x_0\,ds = y$$

by the mean ergodic theorem for $C_0$−semigroups.

(iii)⇒(ii): By renorming we may assume that **T** is contractive. Let $X_c$ be the closed subspace in $X$ of all $y \in X$ whose orbit $T(\cdot)y$ is relatively compact. Let **S** denote the quotient semigroup on $X/X_c$ and let $B$ be its generator. If $F(\lambda)$ is a holomorphic extension of $R(\lambda, A)x_0$ to a connected neighbourhood $V$ of $\{\operatorname{Re}\lambda \geq 0\}\backslash iE$, then $qF(\lambda)$ is a holomorphic extension to $V$ of $R(\lambda, B)qx_0$.

On the other hand, if $x_\omega := \lim_{t \to \infty} \frac{1}{t}\int_0^t e^{-i\omega s}T(s)x_0\,ds$, then $T(t)x_\omega = e^{i\omega t}x_\omega$ for all $t \geq 0$, so $x_\omega \in X_c$. Hence $qx_\omega = 0$ and

$$\lim_{t \to \infty} \frac{1}{t}\int_0^t e^{-i\omega s}S(s)qx_0\,ds = q\left(\lim_{t \to \infty} \frac{1}{t}\int_0^t e^{-i\omega s}T(s)(x_0 - x_\omega)\,ds\right) = 0.$$

We are in a position to apply Theorem 5.1.11 and obtain $\lim_{t \to \infty} \|qT(t)x_0\| = \|S(t)qx_0\| = 0$. Thus, for a given $\epsilon > 0$ there exist $t_0 > 0$ and $y \in X_c$ such that $\|T(t_0)x_0 - y\| < \epsilon$. Then $\|T(t+t_0)x_0 - T(t)y\| < \epsilon$ for all $t \geq 0$, and consequently

$$\{T(t)x_0 : t \geq 0\} \subset K_\epsilon + \epsilon B_X,$$

where $K_\epsilon$ is the compact set $\{T(t)x_0 : 0 \leq t \leq t_0\} \cup \overline{\{T(t)y : t \geq 0\}}$. Since $\epsilon > 0$ is arbitrary, it follows that $\{T(t)x_0 : t \geq 0\}$ is relatively compact. ////

As is shown by Example 5.1.12, the uniform boundedness assumption on the orbit cannot be weakened to boundedness; the orbit $T(\cdot)y$ constructed in this example tends to 0 weakly by Corollary 5.3.11 but not strongly. Thus $T(\cdot)y$ is not relatively compact by the argument in the proof of Theorem 5.7.7.

The proof of (i)⇒(iii) is valid for any $i\omega \in i\mathbb{R}$. It follows that in (iii) we may replace 'for all $\omega \in E$' by 'for all $\omega \in \mathbb{R}$'.

**Theorem 5.7.10.** *Let **T** be a uniformly bounded $C_0$–semigroup on a Banach space $X$, with generator $A$, and assume that $\sigma(A) \cap i\mathbb{R}$ is countable. Then the following assertions are equivalent:*

(i) **T** *is weakly almost periodic;*
(ii) **T** *is almost periodic;*
(iii) *For all $x \in X$ and $i\omega \in \sigma(A) \cap i\mathbb{R}$ the limit $\lim_{t \to \infty} \frac{1}{t}\int_0^t e^{-i\omega s}T(s)x\,ds$ exists;*
(iv) *For all $i\omega \in \sigma(A) \cap i\mathbb{R}$, $\operatorname{range}(A - i\omega) + \ker(A - i\omega)$ is dense in $X$.*

*Proof:* The equivalence of (i), (ii), and (iii) is an immediate consequence of the preceding theorem.

(ii)⇒(iv): Let $X = X_0 \oplus X_1$ be the Glicksberg-DeLeeuw decomposition. Then $\mathbf{T}_0$ is uniformly stable.

Fix $i\omega \in \sigma(A) \cap i\mathbb{R}$ and let $x_0^* \in X^*$ vanish on $\operatorname{range}(A - i\omega) + \ker(A - i\omega)$. Then $\langle x_0^*, (A - i\omega)x\rangle = 0$ for all $x \in D(A)$, so $x_0^* \in D(A^*)$ and $A^*x_0^* = i\omega x_0^*$. This implies $T^*(t)x_0^* = e^{i\omega t}x_0^*$ for all $t \geq 0$. We are going to prove that $x_0^*$ annihilates both $X_0$ and $X_1$.

If $x \in X_0$, then
$$0 = \lim_{t \to \infty} \langle x_0^*, T(t)x \rangle = \lim_{t \to \infty} e^{i\omega t} \langle x_0^*, x \rangle,$$
which is only possible if $\langle x_0^*, x \rangle = 0$. Thus, $x_0^*$ annihilates $X_0$.

Let $x \in D(A)$ be such that $Ax = i\lambda x$ for some $\lambda \in \mathbb{R}$. If $\lambda = \omega$, then $x \in \ker(A - i\omega)$ and $\langle x_0^*, x \rangle = 0$ by assumption. If $\lambda \neq \omega$, then $T(t)x = e^{i\lambda t}x$ for all $t \geq 0$ and
$$\langle x_0^*, x \rangle = e^{-i\lambda t} \langle x_0^*, T(t)x \rangle = e^{-i\lambda t} \langle T^*(t)x_0^*, x \rangle = e^{-i(\lambda - \omega)t} \langle x_0^*, x \rangle,$$
which can only be true for all $t \geq 0$ if $\langle x_0^*, x \rangle = 0$. Thus, $x_0^*$ annihilates all eigenvectors of $A$ with purely imaginary eigenvalues. Since these span $X_1$, it follows that $x_0^*$ also annihilates $X_1$.

It follows that $x_0^*$ annihilates all of $X$, so $x_0^* = 0$ and (iv) follows.

(iv)$\Rightarrow$(iii): Let $i\omega \in \sigma(A) \cap i\mathbb{R}$ and let $x \in \text{range}\,(A - i\omega)$, say $x = (A - i\omega)y$. Then
$$\lim_{t \to \infty} \frac{1}{t} \int_0^t e^{-i\omega s} T(s)x\, ds = \lim_{t \to \infty} \frac{1}{t}(e^{-i\omega t}T(t)y - y) = 0$$
since $\mathbf{T}$ is uniformly bounded. If $x \in \ker(A - i\omega)$, then $T(t)x = e^{i\omega t}x$ for all $t \geq 0$, so
$$\lim_{t \to \infty} \frac{1}{t} \int_0^t e^{-i\omega s} T(s)x\, ds = x.$$
It follows that the limit in (5.7.3) exists for all $x \in \text{range}\,(A - i\omega) + \ker(A - i\omega)$, which is a dense subspace by assumption. Hence by a density argument the limit exists for all $x \in X$. ////

As in Theorem 5.7.9, in (iii) and (iv) the condition 'for all $i\omega \in \sigma(A) \cap i\mathbb{R}$' may be replaced by 'for all $\omega \in \mathbb{R}$'.

**Notes.** In this chapter we have presented two approaches to the theory of individual stability: the limit isometric group approach and the Tauberian approach. The main results are Theorems 5.3.6 and 5.3.9, and Theorems 5.6.4 and 5.6.5, respectively. It is possible to prove a Tauberian theorem similar to 5.6.3 from which Theorem 5.3.6 immediately follows; cf. [BNR2]. On the other hand, Theorem 5.6.5 can be deduced from Theorem 5.3.9; indeed, it can be proved directly that the Tauberian assumption on $f$ in 5.6.5 implies that $f$ is of spectral synthesis with respect to the zero set of its Fourier transform [AM], [Po]. We do not know whether Theorem 5.3.9 can be derived from Tauberian results, or whether Theorem 5.6.4 can be derived via limit isometric groups.

Theorem 5.1.2 is taken from [Vu1].

Lemma 5.1.4 is a classical result of I.M. Gelfand; the proof presented here is due to G.R. Allan and T.J. Ransford [AR].

Theorem 5.1.5 was proved independently and simultaneously by W. Arendt and C.J.K. Batty [AB] and Yu.I. Lyubich and Vũ Quôc Phóng [LV1]. Another proof, based on Theorem 5.2.3 and the fact that countable sets are sets of spectral synthesis for $L^1(\mathbb{R})$ was given by J. Esterle, E. Strouse, and F. Zouakia [ESZ1]. The proof presented here follows Lyubich and Vũ. For uniformly continuous semigroups, the result had been obtained earlier by G.M. Skylar and V.Ya. Shirman [SS]. The special case that a uniformly bounded semigroup whose generator has no unitary spectrum on the imaginary axis is uniformly stable was proved by Falun Huang [HF1]; an English translation of the proof appeared in [HF5]. His proof, however, seems to contain a gap concerning the use of approximate eigenvectors in two different norms.

Example 5.1.6 is taken from [AB].

Lemma 5.1.7, Theorem 5.1.11 and its corollaries, and Example 5.1.12 (we were suggested to look at this semigroup by R. deLaubenfels) are taken from [BNR1]. Special cases of Corollary 5.1.13 had been obtained earlier by F.L. Huang [HF4] and C.J.K. Batty [Ba4]. Lemma 5.1.9 is a special case of ergodic theorems for $C_0$-semigroups; cf. [Kr, Chapter 2].

Example 5.1.14 is taken from [BV1].

The first part of Proposition 5.1.15 is taken from [AB].

A short proof of de Pagter's [Pa] theorem can be found [Ne1]. Grothendieck's lemma is proved, e.g., in [Di].

Theorem 5.2.1 is taken from [BNR1]. Theorem 5.2.3 is due to J. Esterle, E. Strouse and F. Zouakia [ESZ2] and Vũ Quôc Phóng [Vu1]. We follow the (easier) proof of [Vu1]. Theorem 5.2.3 is the semigroup analogue of the following theorem due to Y. Katznelson and L. Tzafriri for power bounded operators $T$: If $f(z) = \sum_{n=0}^{\infty} c_n z^n$ is holomorphic in the unit disc with $\sum_{n=0}^{\infty} |c_n| < \infty$ and if $f$ is of spectral synthesis with respect to the unitary spectrum of $T$, then $\lim_{n\to\infty} \|T^n \hat{f}(T)\| = 0$; here $\hat{f}(T) := \sum_{n=0}^{\infty} c_n T^n$.

For contractions in Hilbert space, the Katznelson-Tzafriri theorem can be improved as follows [ESZ1]: If $T$ is a contraction on a Hilbert space $H$ and $f \in A(D)$ vanishes on $\sigma(T) \cap \Gamma$, then $\lim_{n\to\infty} \|T^n \hat{f}(T)\| = 0$; $A(D)$ is the disc algebra. Under these assumptions, the unitary spectrum $\sigma(T) \cap \Gamma$ is necessarily null. The proof is based on the famous inequality of J. von Neumann that $\|p(T)\| \leq \|p\|_{C(\Gamma)}$ for all polynomials $p(z) = \sum_{k=0}^{n-1} a_k z^k$. In this connection we mention the following theorem of B. Sz.-Nagy and C. Foias [NF]: if $T$ is a completely non-unitary contraction on a Hilbert space $H$ and $\sigma(T) \cap \Gamma$ is null, then $\lim_{n\to\infty} \|T^n x\| = 0$ for all $x \in H$. The proof is based on unitary dilation arguments. An analogous result holds for $C_0$-semigroups.

A result much more general than Theorem 5.2.5 has been proved by Sen-Zhong Huang [HS2].

The implication (iii)⇒(ii) in Corollary 5.2.6 is proved in [Vu1], where also the special case $\hat{f}(\mathbf{T}) = R(\lambda, A)$, $\operatorname{Re} \lambda > 0$, of Corollary 5.6.10 is obtained as a consequence of Theorem 5.2.3 and the fact that singletons are sets of spectral synthesis.

The main results of Section 5.3 are taken from [BNR2]. Corollary 5.3.7 is implicit in [BV1] with a different proof. The idea of studying bounded uniformly continuous orbits via the translation semigroup in $BUC(\mathbb{R}_+, X)$ is due to S. Kantorovitz [Kn] and was used in [DV] to obtain results similar to the ones presented here under global spectral assumptions. For real line, results similar to those presented in Section 5.3 have been obtained by W.M. Ruess and Vũ Quôc Phóng [RV].

The Arveson spectrum was introduced in [Av]. Theorem 5.4.1 is due to D.E. Evans [Ev]. In [Ne8] it is applied to obtain an elementary proof of Wiener's Tauberian theorem. The inclusion $i\sigma(B_{/J_E}) \subset E$ in Lemma 5.4.2 is taken from [BG], where it is proved that actually equality $i\sigma(B_{/J_E}) = E$ holds. The proof of of Theorem 5.4.3 is inspired by [HNR], where a similar technique was used to prove that a certain class of Banach subalgebras of $L^1(G)$ satisfies the so-called Ditkin condition. As is well-known, cf. [Ka, Theorem 8.5.6], the conclusion of Theorem 5.4.3 is valid in every regular semisimple Banach algebra with the Ditkin property.

The main results and examples in Section 5.5 are taken from [BBG], where the global case is considered. The extension to individual orbits is from [BNR1], whose presentation of Theorem 5.5.4 we followed here. Lemma 5.5.1 is due to [ESZ2]; see also [BBG]. The discrete case was obtained earlier by J.-P. Kahane and Y. Katznelson [KK]. The invertibility of isometric $C_0$-semigroups whose generator has countable spectrum on $i\mathbb{R}$ is due to D.A. Greenfield [Gf]. Lemma 5.5.5 was proved in a more general setting by R.G. Douglas [Do]. Theorem 5.5.7 extends a result of Vũ Quôc Phóng [Vu2], who proved the same under additional assumptions.

The asymptotic behaviour of the Schrödinger semigroup of Example 5.5.11 is investigated in [ABB] and [Ba2]. In these papers the following results are proved:

(i) If $0 \leq V \in L^1_{loc}(\mathbb{R}^n)$ and $1 < p < \infty$, then $\frac{1}{2}\Delta - V$ generates a uniformly stable $C_0$-semigroup on $L^p(\mathbb{R}^n)$;

(ii) If $n \in \{1, 2\}$ and $0 \leq V \in L^1_{loc}(\mathbb{R}^n)$, $V \not\equiv 0$, then $\frac{1}{2}\Delta - V$ generates a uniformly stable $C_0$-semigroup on $L^1(\mathbb{R}^n)$;

(iii) If $n \geq 3$ and $0 \leq V \in L^1_{loc}(\mathbb{R}^n)$ satisfies
$$\int_{|y| \geq 1} \frac{V(y)}{|y|^{n-2}} \, dy < \infty,$$
then $\frac{1}{2}\Delta - V$ generates a $C_0$-semigroup on $L^1(\mathbb{R}^n)$ which is not uniformly stable;

(iv) If $n \geq 3$ and $0 \leq V \in L^1_{loc}(\mathbb{R}^n)$ is radial and satisfies
$$\int_{|y| \geq 1} \frac{V(y)}{|y|^{n-2}} \, dy = \infty,$$

then $\frac{1}{2}\Delta - V$ generates a uniformly stable $C_0$-semigroup on $L^1(\mathbb{R}^n)$.

The proofs in [ABB] are analytical, whereas in [Ba2] more detailed results are obtained by probabilistic methods.

Most of the results of Sections 5.1, 5.2, 5.3, 5.4, and 5.5 admit generalizations to uniformly bounded strongly continuous representations **T** in $\mathcal{L}(X)$ of certain subsemigroups $S$ of locally compact abelian groups; we refer to [BV2], [BG], [BBG], [HNR], the review paper [Ba3], and the references given there. The *spectrum* Sp(**T**) is then defined as the set of all non-zero, continuous, bounded, complex-valued homomorphisms $\phi$ of $S$ for which the obvious generalization of the inequality of Lemma 5.2.4 holds:

$$\left| \int_S \phi(s) f(s)\, d\mu(s) \right| \leq \|T(f)\|, \quad \forall f \in L^1(G).$$

Here we use the notation of the notes at the end of Chapter 2. This notion of spectrum was introduced in the paper [DL]. Furthermore, the unitary spectrum $\text{Sp}_u(\mathbf{T})$ is defined as the set of all univalent homomorphisms in $\phi \in \text{Sp}(\mathbf{T})$, i.e. $|\phi(s)| = 1$ for all $s \in S$. In the case $S = G$ this definition of spectrum can be shown to agree with the Arveson spectrum. In this abstract setting the role of the spectral projection corresponding to an isolated point in $\sigma(B)$ in Theorem 5.1.5 is taken over by Shilov's idempotent theorem. If $A$ generates a bounded $C_0$-semigroup **T**, then $\sigma(A) \subset \text{Sp}(\mathbf{T})$ and $\text{Sp}_u(\mathbf{T}) = i\sigma(A) \cap \mathbb{R}$ under identification of the homomorphism $\lambda : t \mapsto e^{i\lambda t} \in \text{Sp}(\mathbf{T})$ with $i\lambda \in \sigma(A)$. This result is due to C.J.K. Batty and Vũ Quôc Phóng [BV2] and may be regarded as a semigroup extension of Evans's theorem.

Strongly continuous representations of locally compact abelian groups satisfying a non-quasianalytic growth condition were studied in detail by Sen-Zhong Huang [HS3].

In [Vu3], Theorems 5.1.2, 5.1.5, and 5.2.3 are generalized to $C_0$-semigroups whose growth is dominated by a one-sided non-quasianalytic weight function $\omega$; rather than strong convergence to 0 one now typically obtains $o(\omega(t))$-estimates.

Lemma 5.6.1 is taken from [AB]. In that paper, it is shown that Theorem 5.1.5 can be derived from it via an argument involving transfinite induction. Actually, close inspection of this argument shows that it leads to the special case of Corollary 5.1.13 where the set $E$ is independent of $y \in Y$.

The special case (5.6.6) of Lemma 5.6.1 can be found in a paper by J. Korevaar [Kv]; the technique of the proof is due to D.J. Newman [Nm]. The case $\sigma(A) \cap i\mathbb{R} = \emptyset$ is contained in a Tauberian theorem of A.E. Ingham [In].

Theorem 5.6.3 is due to W. Arendt and C.J.K. Batty [AB]; it is the semigroup version of an analogous result for power bounded operators due to G.R. Allan, A.G. O'Farrell and T.J. Ransford [AOR]. A different version of this theorem, with boundedness replaced by slow oscillation, was applied to the study of asymptotic behaviour of Volterra equations in [AP]. Further results along this line are proved in [Ba1].

Theorem 5.6.5, taken from [BNR2], is an individual version of Theorem 7.1 in [Ba1]. This, in turn, is the semigroup version of a result in [AOR].

The proof of Theorem 5.6.7 is due to Sen-Zhong Huang and initiated the work in [HNR]; the result itself is due to S. Agmon and S. Mandelbrojt [AM]. A stronger version was proved by A. Beurling and H. Pollard [Po].

Our presentation of the Glicksberg-DeLeeuw theorem follows [Kr]. The proof of Ellis's theorem is given in [El]. Corollary 5.7.8 is due to Vũ Quôc Phóng [Vu4]. Theorem 5.7.9 is taken from [BNR1]. Theorem 5.7.10 is proved in [VL]; a different proof of the equivalence of (i), (ii), and (iv) is given in [BV1].

Many authors have made significant recent contributions to the theory of almost periodic operators and $C_0$-semigroups; among others we mention Yu.I. Lyubich, W.M. Ruess, W.H. Summers, and Vũ Quôc Phóng. A review of some of these developments is given in [Vu5].

Theorem 5.7.10 has been developed further by F. Räbiger, R. Nagel, Sen-Zhong Huang, and M. Wolff [NR], [HR], [RW]. Their results are most easily described for the powers of a single bounded operator $T$ on a Banach space $X$. For each free ultrafilter $\mathcal{U}$ on $\mathbb{N}$, the ultrapower $X_\mathcal{U}$ is defined as the quotient space $l^\infty(X)/c_\mathcal{U}(X)$, where $c_\mathcal{U}(X)$ is the space of all sequences $(x_n) \subset X$ such that $\lim_\mathcal{U} x_n = 0$. As in Section 2.3, $T$ can be extended in the natural way to a bounded operator $T_\mathcal{U}$ on $X_\mathcal{U}$, for which we have $\|T\| = \|T_\mathcal{U}\|$ and $\sigma(T) = \sigma(T_\mathcal{U})$. In particular, if $T$ is power bounded, then so is $T_\mathcal{U}$ and if $\sigma(T) \cap \Gamma$ is countable, the same is true for $T_\mathcal{U}$.

Now assume that $T$ is a power bounded operator on a superreflexive space $X$ such that $\sigma(T) \cap \Gamma$ is countable. Then for each free ultrafilter $\mathcal{U}$, the ultrapower $X_\mathcal{U}$ is reflexive and $\sigma(T_\mathcal{U}) \cap \Gamma$ is countable. Therefore, by the operator analogue of Theorem 5.7.10 the semigroup $\{T^n\}_{n \in \mathbb{N}}$ is *super almost periodic*, i.e. for each free ultrafilter $\mathcal{U}$ the ultrapower semigroup $\{T_\mathcal{U}^n\}_{n \in \mathbb{N}}$ is almost periodic on $X_\mathcal{U}$. The main result of [NR] (for the case $X$ superreflexive) and [RW] is the following converse to this observation: if $T$ is a super almost periodic operator on an arbitrary Banach space $X$, then $T$ is power bounded and $\sigma(T) \cap \Gamma$ is countable.

For $C_0$-semigroups, there is the following analogue [HR] (for the case $X$ superreflexive) and [RW]: if **T** is a super almost periodic $C_0$-semigroup on an arbitrary Banach space $X$, then **T** is uniformly bounded and $\sigma(A) \cap i\mathbb{R}$ is countable.

The $C_0$-semigroup case can be reduced to the operator case, and the operator case follows from the fact that if $T$ has uncountable unitary spectrum, then each ultrapower of $T$ can be restricted to an appropriate subspace on which its action resembles the multiplication operator $M$ on $L^2(E, \mu)$, $M : f(\lambda) \mapsto \lambda f(\lambda), \lambda \in E$. Here $E$ is an appropriately chosen uncountable subset of $\sigma(T) \cap \Gamma$ and $\mu$ is a diffuse probability measure whose support is $E$ (cf. Example 5.1.6). The operator $M$ is easily seen not to be almost periodic, and since almost periodicity is inherited by passing to invariant subspaces, it follows that none of the ultrapowers of $\{T^n\}_{n \in \mathbb{N}}$ is almost periodic. It is also shown in [RW] that if one ultrapower of $T$ is almost periodic, then all ultrapowers of $T$ are almost periodic.

# Appendix

## A.1. Fractional powers

In this section we collect some properties of fractional powers of sectorial operators. Detailed proofs and further information can be found in the books of Pazy [Pz] (for the case $A$ is the generator of a holomorphic semigroup) and Triebel [Tr].

A densely defined linear operator $A$ on a Banach space $X$ is called *sectorial* if $(0, \infty) \subset \varrho(A)$ and there exists a constant $M > 0$ such that

$$\|R(\lambda, A)\| \leq M(1+\lambda)^{-1}, \quad \forall \lambda > 0.$$

This terminology is justified by the following simple proposition, which is an easy consequence of Lemma 2.3.4.

**Proposition A.1.1.** *Let $A$ be a sectorial operator on a Banach space $X$. Then there exist $\theta \in (0, \frac{\pi}{2})$, $\epsilon > 0$, and a constant $M' \geq M$ such that $\Sigma_{\theta,\epsilon} := \{\lambda \in \mathbb{C} : |\arg \lambda| \leq \theta\} \cup \{\lambda \in \mathbb{C} : |\lambda| \leq \epsilon\} \subset \varrho(A)$ and*

$$\|R(\lambda, A)\| \leq M'(1+|\lambda|)^{-1}, \quad \forall \lambda \in \Sigma_{\theta,\epsilon}. \tag{A.1.1}$$

The supremum of all $\theta$ such that the proposition holds for some $\epsilon$ and $M'$ is called the *opening angle* of $A$. For a sectorial operator $A$ and a positive real number $\alpha > 0$ we define the operator $(-A)^{-\alpha}$ by

$$(-A)^{-\alpha}x := \frac{1}{2\pi i} \int_{\Gamma_{\theta,\epsilon}} (-\lambda)^{-\alpha} R(\lambda, A) x \, d\lambda, \quad x \in X, \tag{A.1.2}$$

where $\Gamma_{\theta,\epsilon} = \Gamma_1 \cup \Gamma_2 \cup \Gamma_3$ is the upwards oriented oriented path consisting of

$$\Gamma_1 = \{\arg \lambda = -\theta, |\lambda| \geq \epsilon\};$$
$$\Gamma_2 = \{|\arg \lambda| > \theta, |\lambda| = \epsilon\};$$
$$\Gamma_3 = \{\arg \lambda = \theta, |\lambda| \geq \epsilon\}.$$

Here, $\theta \in (0, \frac{\pi}{2})$ and $\epsilon > 0$ are chosen such that $\Sigma_{\theta,\epsilon} \subset \varrho(A)$ and the estimate (A.1.1) holds. We use the branch of $(-\lambda)^{-\alpha}$ that yields positive values for negative real $\lambda$.

By (A.1.1), the integral in (A.1.2) is absolutely convergent for all $x \in X$ and defines a bounded operator $(-A)^{-\alpha}$ on $X$. The operators $(-A)^{-\alpha}$ are injective for all $\alpha > 0$. Thus, we can define $(-A)^\alpha$ as the inverse of $(-A)^{-\alpha}$; further we put $A^0 := I$. The operators $(-A)^\alpha$, $\alpha \in \mathbb{R}$, have the following properties.

(i) For $k \in \mathbb{Z}$, the definition of $(-A)^k$ agrees with the usual one;
(ii) For all $\alpha > 0$, $(-A)^\alpha$ is a closed, densely defined operator and $D((-A)^\alpha) = \{(-A)^{-\alpha}x : x \in X\}$;
(iii) For all $\alpha_1 \geq \alpha_0 \geq 0$ we have $D((-A)^{\alpha_1}) \subset D((-A)^{\alpha_0})$;
(iv) For all $\alpha_0, \alpha_1 \in \mathbb{R}$ and $x \in D((-A)^{\max\{\alpha_0, \alpha_1, \alpha_0+\alpha_1\}})$ we have $(-A)^{\alpha_0+\alpha_1}x = (-A)^{\alpha_0}((-A)^{\alpha_1}x)$.

For $0 < \alpha < 1$ we can collapse the integration path to the positive real axis and obtain the following representation for $(-A)^{-\alpha}$:

$$(-A)^\alpha x = \frac{\sin \pi \alpha}{\pi} \int_0^\infty t^{-\alpha} R(t, A) x \, dt, \quad x \in X. \tag{A.1.3}$$

Let $A$ be a sectorial operator. Then for all $\omega \geq 0$, $A_\omega := A - \omega$ is sectorial with the same opening angle. Thus, the fractional powers of $-A_\omega$ are defined. For $\alpha \leq 0$ these are bounded operators and for $\alpha > 0$ these are closed, densely defined operators. The following result due to H. Komatsu [Ko] compares the domains of $(-A_\omega)^\alpha$ and $(-A)^\alpha$:

**Proposition A.1.2.** *Let $A$ be a sectorial operator on a Banach space $X$. Then for all $\omega \geq 0$ and $\alpha > 0$ we have $D((-A)^\alpha) = D((-A_\omega)^\alpha)$.*

## A.2. Interpolation theory

In this section we collect some facts from interpolation theory. We refer to the book [Tr] for the proofs and more detailed information; see also [KPS] and [BS].

Two Banach spaces $X_0, X_1$ are said to be a *interpolation couple* if there exists a Hausdorff topological vector space $\mathcal{X}$ in which both $X_0$ and $X_1$ are continuously embedded. If $(X_0, X_1)$ is an interpolation couple, the *intersection* $X_0 \cap X_1$ is the set of all $x \in \mathcal{X}$ that belong both to $X_0$ and $X_1$; with the norm

$$\|x\|_{X_0 \cap X_1} = \max\{\|x\|_{X_0}, \|x\|_{X_1}\}$$

this space is a Banach space. Similarly, the *sum* $X_0 + X_1$ is defined as the set of all $x \in \mathcal{X}$ that admit a decomposition $x = x_0 + x_1$ with $x_j \in X_j$, $j = 0, 1$; with the norm

$$\|x\|_{X_0+X_1} := \sup\{\|x_0\|_{X_0} + \|x_1\|_{X_1} : x = x_0 + x_1, \, x_j \in X_j, \, j = 0, 1\}$$

# Appendix

this space is also a Banach space. A Banach space $Y$ is called an *intermediate space* of the couple $(X_0, X_1)$ if $X_0 \cap X_1 \subseteq Y \subseteq X_0 + X_1$ with continuous inclusions. An intermediate space $Y$ is an *interpolation space* if the following holds: whenever $T$ is a linear operator from $X_0 + X_1$ to $X_0 + X_1$ which restricts to a bounded operator on each of the $X_j$, $j = 0, 1$, then $Y$ is $T$-invariant and $T$ restricts to a bounded operator on $Y$. An interpolation space $Y$ is called *exact* if for all such $T$ we have

$$\|T|_Y\|_Y \leq \max\{\|T|_{X_0}\|_{X_0}, \|T|_{X_0}\|_{X_0}\}.$$

Two classes of interpolation spaces are of importance to us. The first class arises from the so-called *real interpolation method*. Let $(X_0, X_1)$ be an interpolation couple. For $x \in X_0 + X_1$ and $t > 0$, we define the quantity $K(t, x) = K(t, x; X_0, X_1)$ by

$$K(t, x) := \inf\{\|x_0\|_{X_0} + t\|x_1\|_{X_1} : x = x_0 + x_1, x_j \in X_j, j = 0, 1\}$$

and for $0 < \theta \leq 1$ we define

$$(X_0, X_1)_{\theta,\infty} := \{x \in X_0 + X_1 : \sup_{0<t<\infty} t^{-\theta} K(t, x) < \infty\};$$

with the norm

$$\|x\|_{(X_0, X_1)_{\theta,\infty}} := \sup_{0<t<\infty} t^{-\theta} K(t, x),$$

this space is an exact interpolation space for the couple $(X_0, X_1)$.

For all $0 < \theta < 1$ there exists a constant $C_\theta$ such that for all $x \in X_0 \cap X_1$ we have

$$\|x\|_{(X_0, X_1)_{\theta,\infty}} \leq C_\theta \|x\|_{X_0}^{1-\theta} \|x\|_{X_1}^{\theta}. \tag{A.2.1}$$

The second class of interpolation spaces arises from the so-called *complex interpolation method*. Let $S$ denote the strip $\{z \in \mathbb{C} : 0 < \operatorname{Re} z < 1\}$. We define $F(X_0, X_1)$ as the set of all uniformly bounded $(X_0 + X_1)$-valued holomorphic functions on $S$ with continuous extension to the closure $\overline{S}$ such that

$$f(j + it) \in X_j, \quad j = 0, 1; \, t \in \mathbb{R};$$

with the norm

$$\|f\|_{F(X_0, X_1)} := \max\{\sup_{t \in \mathbb{R}} \|f(it)\|_{X_0}, \sup_{t \in \mathbb{R}} \|f(1 + it)\|_{X_1}\}$$

this is a Banach space. For $0 < \theta < 1$ we define

$$[X_0, X_1]_\theta := \{x \in X_0 + X_1 : \exists f \in F(X_0, X_1) \text{ with } f(\theta) = x\}.$$

With the norm

$$\|x\|_{[X_0, X_1]_\theta} := \inf_{f(\theta)=x} \|f\|_{F(X_0, X_1)},$$

where the infimum is taken over all $f \in F(X_0, X_1)$ such that $f(\theta) = x$, this is an exact interpolation space for the couple $(X_0, X_1)$.

If $X_1$ is continuously embedded in $X_0$, then [Tr, p. 64] we have continuous inclusions

$$[X_0, X_1]_{\theta_1} \subseteq (X_0, X_1)_{\theta_1, \infty} \subseteq [X_0, X_1]_{\theta_0}, \quad 0 < \theta_0 < \theta_1 < 1. \tag{A.2.2}$$

If $A$ is a sectorial operator on a Banach space $X$, then for all $\alpha \geq 0$ the couple $(X, D((-A)^\alpha))$ is an interpolation couple. The following proposition, taken from [WW], gives some information about interpolation spaces of this couple.

**Proposition A.2.1.** *Let $A$ be a sectorial operator on a Banach space $X$, let $0 \leq \alpha_0 < \alpha_1$ and $0 < \theta < 1$, and put $\alpha_\theta := (1-\theta)\alpha_0 + \theta \alpha_1$. Then we have a continuous inclusion*

$$D((-A)^{\alpha_\theta}) \subseteq ((D(-A)^{\alpha_0}), D((-A)^{\alpha_1}))_{\theta, \infty}, \tag{A.2.3}$$

*and for all $0 < \theta' < \theta$ a continuous inclusion*

$$D((-A)^{\alpha_\theta}) \subseteq [(D(-A)^{\alpha_0}), D((-A)^{\alpha_1})]_{\theta'}. \tag{A.2.4}$$

*Proof:* In view of (A.2.2), (A.2.4) follows from (A.2.3). Choosing an integer $m > \alpha_1 - \alpha_0$, by [Tr, Thm. 1.15 (d) and (f)] we have

$$D((-A)^{\alpha_\theta - \alpha_0}) \subseteq (X, D(A^m))_{\frac{\alpha_\theta - \alpha_0}{m}, \infty} = (X, D((-A)^{\alpha_1 - \alpha_0}))_{\theta, \infty}.$$

Since $(-A)^{\alpha_0}$ is an isomorphism from $D((-A)^{\beta + \alpha_0})$ onto $D((-A)^\beta)$ for all $\beta \geq 0$, and since the real interpolation method is functorial, (A.2.3) follows from this. ////

By combining (A.2.1) and (A.2.3), it follows that for all $0 \leq \alpha \leq 1$ there exists a constant $C_\alpha$ such that for all $x \in D(A)$ we have

$$\|(-A)^\alpha x\| \leq C_\alpha \|x\|^{1-\alpha} \|Ax\|^\alpha. \tag{A.2.5}$$

Let $(\Omega, \Sigma, \mu)$ be a positive $\sigma$-finite measure and let $(X_0, X_1)$ be an interpolation couple of Banach spaces. The following proposition [Tr, 1.18.4] describes the complex interpolation spaces between the Lebesgue-Bochner spaces $L^p(\mu, X_0)$.

**Proposition A.2.2.** *For all $1 \leq p_0, p_1 < \infty$ and $0 < \theta < 1$ we have*

$$[L^{p_0}(\mu, X_0), L^{p_1}(\mu, X_1)]_\theta = L^{p_\theta}(\mu, [X_0, X_1]_\theta),$$

*where $p_\theta = (1-\theta)p_0 + \theta p_1$.*

In particular, if $(\tilde\Omega, \tilde\Sigma, \tilde\mu)$ is another positive $\sigma$-finite measure, then

$$[L^{p_0}(\mu, L^{p_0}(\tilde\mu)), L^{p_1}(\mu, L^{p_1}(\tilde\mu))]_\theta = L^{p_\theta}(\mu, (L^{p_\theta}(\tilde\mu)).$$

Let $w : (\Omega, \Sigma, \mu) \to (0, \infty)$ be a measurable function. For $1 \leq p < \infty$ we define the Banach space $L^p(w\,d\mu, X)$ as the set of all strongly measurable $X$-valued functions $f$ of $\Omega$ such that

$$\|f\|_{L^p(w\,d\mu, X)} := \left(\int_\Omega \|f(s)\|^p w(s)\, d\mu(s)\right)^{\frac{1}{p}} < \infty.$$

The following proposition [Tr, Thm. 1.18.5] describes certain complex interpolation spaces between these spaces.

**Proposition A.2.3.** *For all $1 \leq p_0, p_1 < \infty$ and $0 < \theta < 1$ we have*

$$[L^{p_0}(w^{p_0}\,d\mu, X), L^{p_1}(w^{p_1}\,d\mu, X)]_\theta = L^{p_\theta}(w^{p_\theta}\,d\mu, X),$$

*where $p_\theta = (1-\theta)p_0 + \theta p_1$.*

If $\mu$ is the Lebesgue measure on $\mathbb{R}$ or $\mathbb{R}_+$, we also write $L^p_w(\mathbb{R}, X)$ and $L^p_w(\mathbb{R}_+, X)$ instead of $L^p(w\,d\mu, X)$ and $L^p(w\,d\mu, X)$.

## A.3. Banach lattices

In this section we recall some elementary facts about real Banach lattices and their complexifications; for more details we refer to the books [Sf] and [MN].

A *real Banach lattice* is a partially ordered real Banach space $(X, \leq)$ with the following properties:
(B1) $x \leq y$ implies $x + z \leq y + z$ for all $x, y, z \in X$;
(B2) $ax \geq 0$ for all $0 \leq a \in \mathbb{R}$ and $0 \leq x \in X$;
(B3) For all $x, y \in X$ the least upper bound $x \vee y$ and the greatest lower bound $x \wedge y$ exist;
(B4) For all $x, y \in X$ satisfying $|x| \leq |y|$ we have $\|x\| \leq \|y\|$.

Here, $|x| := x \vee -x$ is the *modulus* of $x$; the vectors $x_+ := x \vee 0$ and $x_- := (-x) \vee 0$ are the *positive part* and the *negative part* of $x$, respectively. We have $x = x_+ - x_-$ and $|x| = x_+ + x_-$. Two elements $x, y \in X$ are *disjoint*, notation $x \perp y$, if $|x| \wedge |y| = 0$. For all $x \in X$ we have $x_+ \perp x_-$.

A linear subspace $Y$ of $X$ is an *ideal* if for any two $x, y \in X$, $|x| \leq |y|$ and $y \in Y$ implies $x \in Y$. An ideal $Y$ is a *band* if for any subset of $Y$ its least upper bound, if it exists, belongs to $Y$. Every band in a real Banach lattice is closed. A band $Y$ is a *projection band* if there exists a band $Y_\perp$ such that there is a direct sum decomposition $X = Y \oplus Y_\perp$. In this case, $x \perp y$ for all $x \in Y$ and $y \in Y_\perp$. A projection associated with a projection band is a *band projection*.

A real Banach lattice has *order continuous norm* if $0 \leq x_\alpha \uparrow x$ implies that $\lim_\alpha \|x - x_\alpha\| = 0$. Every reflexive real Banach lattice and every real $L^1$-space has order continuous norm. Every closed ideal in a real Banach lattice with order continuous norm is a projection band.

The Banach space dual $X^*$ of a real Banach lattice $X$ is a real Banach lattice with respect to the partial ordering $\leq$ defined by $x^* \geq 0$ if and only if $\langle x^*, x \rangle \geq 0$ for all $0 \leq x \in X$.

A bounded operator $T$ on a real Banach lattice $X$ is *positive*, notation $T \geq 0$, if $Tx \geq 0$ for all $0 \leq x \in X$. If $T$ is positive, then so is its adjoint $T^*$. For a positive operator $T$ on $X$ we have $|Tx| \leq T|x|$ for all $x \in X$.

The *complexification* of a real Banach lattice $X$ is the complex Banach space $X_{\mathbb{C}}$ whose elements are pairs $(x, y) \in X \times X$, with addition and scalar multiplication defined by $(x_0, y_0) + (x_1, y_1) := (x_0 + x_1, y_0 + y_1)$ and $(a + bi)(x, y) := (ax - by, ay + bx)$, and norm

$$\|(x, y)\| := \left\| \sup_{0 \leq \theta \leq 2\pi} (x \sin \theta + y \cos \theta) \right\|;$$

one can show that the supremum in the above definition indeed exists in $X$. By identifying $(x, 0) \in X_{\mathbb{C}}$ with $x \in X$, $X$ is isometrically isomorphic to a real-linear subspace of $X_{\mathbb{C}}$. The partial ordering of $X$ is extended to $X_{\mathbb{C}}$ by defining $(x_0, y_0) \leq (x_1, y_0)$ if $x_0 \leq x_1$ and $y_0 = y_1$. Note that $x \geq 0$ in $X_{\mathbb{C}}$ if and only if $x \in X$ and $x \geq 0$ in $X$.

A *complex Banach lattice* is a partially ordered complex Banach space $(X_{\mathbb{C}}, \leq)$ that arises as the complexification of a real Banach lattice $X$. The underlying real Banach lattice $X$ is called the *real part* of $X_{\mathbb{C}}$ and is uniquely determined as the closed linear span of all $0 \leq x \in X_{\mathbb{C}}$.

Instead of the cumbersome notation $(x, y)$ for elements of $X_{\mathbb{C}}$, we usually write $x + iy$. The *complex conjugate* of an element $z = x + iy \in X_{\mathbb{C}}$ is the element $\bar{z} := x - iy$. The modulus $|\cdot|$ of $X$ is extended to $X_{\mathbb{C}}$ by defining

$$|x + iy| := \sup_{0 \leq \theta \leq 2\pi} (x \sin \theta + y \cos \theta).$$

All concepts introduced for real Banach lattice above have a natural extension to complex Banach lattice. For example, an *ideal* of a complex Banach lattice $X_{\mathbb{C}}$ is a linear subspace $Y$ with the property that $|x| \leq |y|$ and $y \in Y$ implies $x \in Y$. An ideal $Y$ in $X_{\mathbb{C}}$ is a *band* if its real part is a band in $X$ and a *projection band* if there exists another band $Y_\perp$ such that $X = Y \oplus Y_\perp$. In that case, $|x| \perp |y|$ for all $x \in Y$ and $y \in Y_\perp$. A complex Banach lattice has *order continuous norm* if its real part has; every closed ideal in such a space is a projection band. Every reflexive complex Banach lattice and every complex $L^1$-space has order continuous norm. The dual of a complex Banach lattice is a complex Banach lattice in the natural way and one has a natural identification $(X_\mathbb{C})^* \simeq (X^*)_\mathbb{C}$, where $X$ is the real part of $X_\mathbb{C}$. A bounded operator $T$ on a complex Banach lattice $X_\mathbb{C}$ is *positive*, notation $T \geq 0$, if $Tx \geq 0$ for all $x \geq 0$, in which case we have $|Tx| \leq T|x|$ for all $x \in X$; moreover, $T^* \geq 0$.

Conform the convention at the beginning of Chapter 1 that all Banach spaces are complex, in these notes all Banach lattices are complex unless otherwise stated.

The complexification of a Banach lattice can also be defined in a functorial way in terms of the so-called *l*-tensor product $\otimes_l$; the result is that $X_\mathbb{C}$, as a real Banach lattice, is isometrically lattice isomorphic to $X \otimes_l \mathbb{R}^2$ in a natural way; for details we refer to [Ne7].

## A.4. Banach function spaces

In this section we recall some facts about Banach function spaces, rearrangement invariant Banach function spaces, and Orlicz spaces. For the proofs we refer the reader to [Za1], [KPS], and [Za2], respectively.

Throughout, let $(\Omega, \mu)$ be a positive $\sigma$-finite measure space. By $M(\mu)$ we denote the linear space of $\mu$-measurable functions $\Omega \to \mathbb{C}$, identifying functions which are equal $\mu$-a.e. A *Banach function norm* is a function $\rho : M(\mu) \to [0, \infty]$ with the following properties:

(N1) $\rho(f) = 0$ if and only $f = 0$ a.e.;
(N2) if $|f| \leq |g|$ $\mu$-a.e., then $\rho(f) \leq \rho(g)$;
(N3) $\rho(af) = |a|\rho(f)$ for all scalars $a \in \mathbb{C}$ and all $\rho(f) < \infty$;
(N4) $\rho(f + g) \leq \rho(f) + \rho(g)$ for all $f, g \in M(\mu)$.

Let $E = E_\rho$ be the set $\{f \in M(\mu) : \|f\|_E := \rho(f) < \infty\}$. Then $E$ is easily seen to be a normed linear space. If $E$ is complete, then $E$ is called a *Banach function space* over $(\Omega, \Sigma, \mu)$. Note that $E$ is an ideal in $M(\mu)$: if $|f| \leq |g|$ $\mu$-a.e. with $g \in E$, then also $f \in E$ (and $\|f\|_E \leq \|g\|_E$).

Let $E$ be a Banach function space over $(\Omega, \Sigma, \mu)$. We say that $E$ is *carried* by a subset $\Omega'$ of $\Omega$ if the following is true: whenever $H \subset \Omega'$ is a measurable set of positive measure, then there exists a function $f \in E$ that is not zero $\mu$-a.e. on $H$. In order to exclude the pathological situation that $\Omega$ is larger than the 'joint support' of the functions in $E$, we will always assume that $E$ is carried by $\Omega$. This is no loss of generality, because there always exists a maximal subset $\Omega'$ of $\Omega$ such that $E$ is carried by $\Omega'$.

If $f_n \to f$ in norm in $E$, then there is a subsequence $(f_{n_k})$ converging to $f$ pointwise $\mu$-a.e.

The *associate function norm* $\rho' : M(\mu) \to [0, \infty]$ of the function norm $\rho$ is defined by

$$\rho'(g) := \sup\left\{\left|\int_\Omega fg\,d\mu\right| : f \in M(\mu), \rho(f) \leq 1\right\}.$$

The map $\rho$ is a function norm and the *associate space* $E' := \{g \in M(\mu) : \rho'(g) < \infty\}$ is a Banach function space. We have Hölder's inequality

$$\left| \int_\Omega fg \, d\mu \right| \leq \|f\|_E \|g\|_{E'}.$$

The associate of the space $L^p(\mu)$ is $L^q(\mu)$, $1 \leq p \leq \infty$, $\frac{1}{p} + \frac{1}{q} = 1$.

Next we discuss rearrangement invariant Banach function spaces. Since we shall only be interested in spaces over $\mathbb{R}_+$ (with the Lebesgue measure), we restrict ourselves to this measure space.

Two non-negative measurable functions $f, g$ on $\mathbb{R}_+$ are called *equimeasurable* if meas $\{f > t\}$ = meas $\{g > t\}$ for all $t \geq 0$. A function norm $\rho$ is called *rearrangement invariant* if $\rho(f) = \rho(g)$ whenever $|f|$ and $|g|$ are equimeasurable. A Banach function space $E$ over $\mathbb{R}_+$ is called *rearrangement invariant* if it its norm arises from a rearrangement invariant function norm. The associate space of a rearrangement invariant function space is rearrangement invariant.

The *fundamental function* of a rearrangement invariant Banach function space $E$ is defined by $\varphi_E(t) := \|\chi_{H_t}\|_E$, where $H_t \subset \mathbb{R}_+$ is any measurable subset of measure $t$ and $\chi_{H_t}$ denotes its characteristic function. By the rearrangement invariance, this function is well-defined. We have the following relation between the fundamental functions of $E$ and $E'$:

$$\varphi_E(t)\varphi_{E'}(t) = t, \quad t \geq 0. \tag{A.4.1}$$

If $E$ and $F$ are two rearrangement invariant Banach function spaces over $\mathbb{R}_+$, then $(E, F)$ is an interpolation couple (both are continuously embedded in the space of all measurable functions with the topology of convergence in measure), and therefore the intersection $E \cap F$ and the sum $E + F$ are well-defined. It is a consequence of the definitions of these spaces that both are rearrangement invariant again.

For every rearrangement Banach function space $E$ over $\mathbb{R}_+$, we have inclusions

$$L^1(\mathbb{R}_+) \cap L^\infty(\mathbb{R}_+) \subset E \subset L^1(\mathbb{R}_+) + L^\infty(\mathbb{R}_+).$$

In particular, bounded functions of compact support belong to $E$.

**Proposition A.4.1.** *Every rearrangement invariant Banach function space over $\mathbb{R}_+$ with order continuous norm is an interpolation space between $L^1(\mathbb{R}_+)$ and $L^\infty(\mathbb{R}_+)$).*

Furthermore, the following assertions are equivalent:

(i) $E$ has order continuous norm;
(ii) $E$ is separable;
(iii) $\varphi_E(0+) = 0$ and the simple functions are dense in $E$.

A *simple function* is a finite linear combination of characteristic functions of measurable sets of finite measure.

A certain class of rearrangement invariant Banach function spaces, the Orlicz spaces, is of special interest.

Let $\phi : \mathbb{R}_+ \to [0, \infty]$ be a function which is non-decreasing, left-continuous, and not identically 0 or $\infty$ on $(0, \infty)$. Define

$$\Phi(t) := \int_0^t \phi(s)\, ds.$$

A function $\Phi$ of this form is called a *Young function*.

Let $f : \mathbb{R}_+ \to \mathbb{C}$ be a measurable function. Let $\Phi$ be a Young function. We define

$$M^\Phi(f) := \int_0^\infty \Phi(|f(s)|)\, ds.$$

The set $L^\Phi$ of all $f$ for which there exists a $k > 0$ such that $M^\Phi(kf) < \infty$ is easily checked to be a linear space. With the norm

$$\rho^\Phi(f) := \inf\{k : M^\Phi(\tfrac{1}{k}f) \leq 1\} \qquad (A.4.2)$$

the space $(L^\Phi, \rho^\Phi)$ becomes a rearrangement invariant Banach function space over $\mathbb{R}_+$. Spaces of this type are called *Orlicz spaces* over $\mathbb{R}_+$.

Trivial examples of Orlicz spaces are the spaces $L^p(\mathbb{R}_+)$, $1 \leq p \leq \infty$. They are obtained from $\phi(t) = pt^{p-1}$ $(1 \leq p < \infty)$ and

$$\phi(t) = \begin{cases} 0, & 0 \leq t \leq 1, \\ \infty, & t > 1, \end{cases} \qquad (p = \infty).$$

The following proposition follows easily from (A.4.2).

**Proposition A.4.2.** *Let $\Phi$ be a Young function, $\Phi(t) = \int_0^t \phi(s)\, ds$, and let $L^\Phi$ be the associated Orlicz space over $\mathbb{R}_+$. If $0 < \phi(t) < \infty$ for all $t > 0$, then $\Phi$ is one-to-one, and the fundamental function $\varphi_{L^\Phi}(t)$ can be expressed in terms of the inverse function $\Phi^{-1}$ by*

$$\varphi_{L^\Phi}(t) = \left(\Phi^{-1}\left(\frac{1}{t}\right)\right).$$

*In particular, $\lim_{t \to \infty} \varphi_{L^\Phi}(t) = \infty$.*

# References

[AM]  S. Agmon and S. Mandelbrojt, Une généralisation de théorème Tauberien de Wiener, Acta Sci. Math. Szeged **12** (1950), 167-176.

[AOR]  G.R. Allan, A.G. O'Farrell, and T.J. Ransford, A Tauberian theorem arising in operator theory, Bull. London Math. Soc. **19** (1987), 537-545.

[AR]  G.R. Allan and T.J. Ransford, Power-dominated elements in a Banach algebra, Studia Math. **94** (1989), 63-79.

[Ar1]  W. Arendt, Gaussian estimates and interpolation of the spectrum in $L^p$, Diff. Integral Eq. **7** (1994), 1153-1168.

[Ar2]  W. Arendt, Spectrum and growth of positive semigroups, in: G. Ferreyra, G. Ruiz Goldstein and F. Neubrander (eds): *Evolution Equations*, Lect. Notes in Pure and Appl. Math. 168, Marcel Dekker Inc. (1995), 21-28.

[AB]  W. Arendt and C.J.K. Batty, Tauberian theorems for one-parameter semigroups, Trans. Am. Math. Soc. **306** (1988), 837-852.

[ABB]  W. Arendt, C.J.K. Batty, and Ph. Bénilan, Asymptotic stability of Schrödinger semigroups on $L^1(\mathbb{R}^N)$, Math. Z. **209** (1992), 511-518.

[AG]  W. Arendt and G. Greiner, The spectral mapping theorem for one-parameter groups of positive operators on $C_0(X)$, Semigroup Forum **30** (1984), 297-330.

[AP]  W. Arendt and J. Prüss, Vector-valued Tauberian theorems and asymptotic behavior of linear Volterra equations, SIAM J. Appl. Math. **23** (1992), 412-448.

[Av]  W. Arveson, On groups of automorphisms of operator algebras, J. Func. Anal. **15** (1974), 217-243.

[Ba1]  C.J.K. Batty, Tauberian theorems for the Laplace-Stieltjes transform, Tr. Am. Math. Soc. **322** (1990), 783-804.

[Ba2]  C.J.K. Batty, Asymptotic stability of Schrödinger semigroups: path integral methods, Math. Ann. **292** (1992), 457-492.

[Ba3]  C.J.K. Batty, Asymptotic behaviour of semigroups of linear operators, in: J. Zemánek, (ed.), *Functional Analysis and Operator Theory*, Banach Centre Publications **30** (1994), 35-52.

[Ba4]  C.J.K. Batty, Spectral conditions for stability of one - parameter semigroups, to appear in: J. Diff. Eq.

[BBG] C.J.K. BATTY, Z. BRZEŹNIAK, AND D.A. GREENFIELD, A quantitative asymptotic theorem for contraction semigroups with countable unitary spectrum, preprint.

[BD] C.J.K. BATTY AND E.B. DAVIES, Positive semigroups and resolvents, J. Operator Theory **10** (1982), 357-363.

[BG] C.J.K. BATTY AND D.A. GREENFIELD, On the invertibility of isometric semigroup representations, Studia Math. **110** (1994), 235-250.

[BNR1] C.J.K. BATTY, J.M.A.M. VAN NEERVEN, AND F. RÄBIGER, Local spectra and individual stability of uniformly bounded $C_0$-semigroups, submitted.

[BNR2] C.J.K. BATTY, J.M.A.M. VAN NEERVEN, AND F. RÄBIGER, Tauberian theorems and stability of solutions of the Cauchy problem, submitted.

[BV1] C.J.K. BATTY AND VŨ QUÔC PHÓNG, Stability of individual elements under one-parameter semigroups, Trans. Am. Math. Soc. **322** (1990), 805-818.

[BV2] C.J.K. BATTY AND VŨ QUÔC PHÓNG, Stability of strongly continuous representations of abelian semigroups, Math. Z. **209** (1992), 75-88.

[Be] A. BEAUZAMY, *Introduction to Operator Theory and Invariant Subspaces*, North Holland, Amsterdam (1988).

[BS] C. BENNETT, R. SHARPLEY, *Interpolation of Operators*, Pure and Applied Mathematics 129, Academic Press (1988).

[BP] O. BLASCO AND A. PELCZYNSKI, Theorems of Hardy and Paley for vector-valued analytic functions and related classes of Banach spaces, Trans. Am. Math. Soc. **323** (1991), 335-367.

[Bo] J. BOURGAIN, Vector-valued Hausdorff-Young inequalities and applications, Springer Lect. Notes in Math. 1317, Springer-Verlag (1988), 239-249.

[Bu] A.V. BUKHVALOV, Hardy spaces of vector-valued functions, Zap. Nauchn. Sem. Leningr. Otd. Mat. Inst. Akad. Nauk. SSSR **65** (1976), 5-16.

[BD] A.V. BUKHVALOV AND A.A. DANILEVICH, Boundary properties of analytic and harmonic functions with values in Banach space, Math. Notes **31** (1976), 104-110.

[CPY] S.R. CARRADUS, W.E. PFAFFENBERGER, AND B. YOOD, *Calkin Algebras and Algebras of Operators on Banach Spaces*, Lect. Notes in Pure and Appl. Math. 9, Marcel Dekker, New York (1974).

[vC] J. VAN CASTEREN, *Generators of Strongly Continuous Semigroups*, Pitman, Boston-London-Melbourne (1985).

[Ch] R. CHILL, *Tauberche Sätze und Asymptotik des Abstrakten Cauchy Problems*, Diplomarbeit, University of Tübingen (1995).

[Dk] R. DATKO, Extending a theorem of A.M. Liapunov to Hilbert space, J. Math. Anal. Appl. **32** (1970), 610-616.

[Da1] E.B. DAVIES, *One-Parameter Semigroups*, Academic Press, London-New York-San Fransisco (1980).

[Da2] E.B. DAVIES, *Heat kernels and spectral theory*, Cambridge University Press, Cambridge (1989).

[DV] R. DELAUBENFELS AND VŨ QUÔC PHÓNG, Stability and almost periodicity of solutions of ill-posed abstract Cauchy problems, to appear in: Proc. Am. Math. Soc.

[De] R. DERNDINGER, Über das Spektrum positiver Generatoren, Math. Z. **172** (1980), 281-293.

[Di] J. DIESTEL, *Sequences and Series in Banach Spaces*, Grad. Texts in Math. 92, Springer-Verlag (1984).

[DU] J. DIESTEL AND J.J. UHL, *Vector Measures*, Math. Surveys nr. 15, Amer. Math. Soc., Providence, R.I. (1977).

[DL] Y. DOMAR AND L.-A. LINDAHL, Three spectral notions for representations of commutative Banach algebras, Ann. Inst. Fourier Grenoble **25** (1975), 1-32.

[Do] R.G. DOUGLAS, On extending a commutative semigroup of isometries, Bull. London Math. Soc. **1** (1969), 157-159.

[El] R. ELLIS, Locally compact transformation groups, Duke Math. J. **24** (1957), 119-126.

[EE] O. EL-MENNAOUI AND K.-J. ENGEL, On the characterization of eventually norm continuous semigroups in Hilbert spaces, Arch. Math. (Basel) **63** (1994), 437-440.

[ESZ1] J. ESTERLE, E. STROUSE, AND F. ZOUAKIA, Theorems of Katznelson-Tzafriri type for contractions, J. Func. Anal. **94** (1990), 273-287.

[ESZ2] J. ESTERLE, E. STROUSE, AND F. ZOUAKIA, Stabilité asymptotique de certains semigroupes d'opérateurs et ideaux primaires, J. Operator Th. **28** (1992), 203-228.

[Ev] D.E. EVANS, On the spectrum of a one-parameter strongly continuous representation, Math. Scand. **39** (1976), 80-82.

[Fa] H.O. FATTORINI, *The Abstract Cauchy Problem*, Addison-Wesley, Reading (Mass.) (1983).

[Fr] A. FRIEDMAN, *Partial Differential Equations of Parabolic Type*, Prentice-Hall, New York (1964).

[Ge] L. GEARHART, Spectral theory for contraction semigroups on Hilbert spaces, Trans. Am. Math. Soc. **236** (1978), 385-394.

[Go] J. A. GOLDSTEIN, *Semigroups of Operators and Applications*, Oxford University Press (1985).

[Gf] D.A. GREENFIELD, *Semigroup Representations: An Abstract Approach*, Ph. D. Thesis, Oxford (1994).

[Gr] G. GREINER, Some applications of Fejér's theorem to one-parameter semigroups, Semesterbericht Funktionalanalysis, Tübingen (1984/85), 33-50.

[GNa] G. GREINER AND R. NAGEL, On the stability of strongly continuous semigroups of positive operators on $L^2(\mu)$, Ann. Scuola Norm. Sup. Pisa **10** (1983), 257-262.

[GVW] G. GREINER, J. VOIGT, AND M. WOLFF, On the spectral bound of the generator of semigroups of positive operators, J. Operator Th. **5** (1981), 245-256.

[GNe] U. GROH AND F. NEUBRANDER, Stabilität starkstetiger positiver Operatorhalbgruppen auf $C^*$-Algebren, Math. Ann. **256** (1981), 129-173.

[Hk] J.A.P. HEESTERBEEK, $R_0$, Ph.D. Thesis, Leiden (1992).

[HV] R. HEMPEL AND J. VOIGT, The spectrum of a Schrödinger operator in $L^p(\mathbb{R}^\nu)$ is $p$-independent, Comm. Math. Phys. **104** (1986), 243-250.

[He] I.W. HERBST, The spectrum of Hilbert space semigroups, J. Operator Th. **10** (1983), 87-94.

[Hb] M. HIEBER, Spectral theory for positive semigroups generated by differential operators, Archiv Math. (Basel) **63** (1994), 333-340.

[Hi] E. HILLE, Une généralisation du problème de Cauchy, Ann. Inst. Fourier Grenoble **4** (1952), 31-48.

[HP] E. HILLE AND R.S. PHILLIPS, *Functional Analysis and Semi-Groups*, Amer. Math. Soc. Colloq. Publ., vol. 31, rev. ed., Providence, R.I. (1957).

[Ho] K. HOFFMAN, *Banach Spaces of Analytic Functions*, Englewood Cliffs, New York (1962).

[Hw] J.S. HOWLAND, On a theorem of Gearhart, Integral Eq. Operator Th. **7** (1984), 138-142.

[HF1] F.L. HUANG, Strong asymptotic stability of linear dynamical systems in Banach spaces, Kexue Tongbao **10** (1983), 584-586.

[HF2] F.L. HUANG, Characteristic conditions for exponential stability of linear dynamical systems in Hilbert spaces, Ann. Diff. Eq. **1** (1985), 43-56.

[HF3] F.L. HUANG, Exponential stability of linear systems in Banach spaces, Chin. Ann. Math. **10B** (1989), 332-340.

[HF4] F.L. HUANG, Spectral properties and stability of one-parameter semigroups, J. Diff. Eq. **104** (1993), 182-195.

[HF5] F.L. HUANG, Strong asymptotic stability of linear dynamical systems in Banach spaces, J. Diff. Eq. **104** (1993), 307-324.

[HL] F.L. HUANG AND K.S. LIU, A problem of exponential stability for linear dynamical systems in Hilbert spaces, Kexue Tongbao **33** (1988), 460-462.

[HS1] S.Z. HUANG, An equivalent description of non-quasianalyticity through spectral theory of $C_0$−groups, J. Operator Theory **32** (1994), 299-309.

[HS2] S.Z. HUANG, Stability properties characterizing the spectra of operators on Banach spaces, J. Func. Anal. **132** (1995), 361-382.

[HS3] S.Z. HUANG, *Spectral Theory for Non-Quasianalytic Representations of Locally Compact Abelian Groups*, monograph, in preparation.

[HS4] S.Z. HUANG, Spectra and asymptotic behaviour of propagators of linear differential equations in Hilbert spaces, preprint.

[HN] S.Z. HUANG AND J.M.A.M. VAN NEERVEN, $B$-convexity, the analytic Radon-Nikodym property, and individual stability of $C_0$−semigroups, submitted.

[HNR] S.Z. HUANG AND J.M.A.M. VAN NEERVEN, AND F. RÄBIGER, Ditkin's condition for certain Beurling algebras, submitted.

[HR] S.Z. HUANG AND F. RÄBIGER, Superstable $C_0$−semigroups on Banach spaces, in: Lect. Notes in Pure and Appl. Math. 155, Marcel Dekker (1994), 291-300.

[In] A.E. INGHAM, On Wiener's method in Tauberian theorems, Proc. London Math. Soc. **38** (1935), 458-480.

[JJ] W.B. JOHNSON AND L. JONES, Every $L_p$-operator is an $L_2$-operator, Proc. Am. Math. Soc. **72** (1978), 309-312.

[KV] M.A. KAASHOEK AND S.M. VERDUYN LUNEL, An integrability condition for hyperbolicity of the semigroup, J. Diff. Eq. **112** (1994), 374-406.

[KK] J.-P. KAHANE AND Y. KATZNELSON, Sur les algèbres de restriction des séries de Taylor absolument convergentes à un fermé du cercle, J. Analyse Math.**23** (1970), 185-197.

[Kn] S. KANTOROVITZ, The Hille-Yosida space of an arbitrary operator, J. Math. Anal. Appl.**136** (1988), 107-111.

[Ka] Y. KATZNELSON, *An Introduction to Harmonic Analysis*, Dover Publications, 2nd ed., Inc., New York (1976).

[KT] Y. KATZNELSON AND L. TZAFRIRI, On power bounded operators, J. Func. Anal. **68** (1986), 313-328.

[Ko] H. KOMATSU, Fractional powers of operators, Pacific J. Math. **19** (1966), 285-346.

[Kv] J. KOREVAAR, On Newman's quick way to the prime number theorem, Math. Intelligencer **4** (1982), 108-115.

[KPS] S.G. KREIN, YU.I. PETUNIN, E.M. SEMENOV, *Interpolation of Linear Operators*, Transl. Math. Monogr. 54, Am. Math. Soc., Providence (1982).

[Kr] U. KRENGEL, *Ergodic Theorems*, De Gruyter, Berlin (1985).

[Kw] S. KWAPIEN, Isomorphic characterizations of inner product spaces by orthogonal series with vector valued coefficients, Studia Math. **44** (1972), 583-595.

[LM] Y. LATUSHKIN AND S. MONTGOMERY-SMITH, Evolutionary semigroups and Lyapunov theorems in Banach spaces, J. Func. Anal. **127** (1995), 173-197.

[Li] W. LITTMAN, A generalization of a theorem of Datko and Pazy, in: Lect. Notes in Control and Inform. Sci. 130, Springer-Verlag (1989), 318-323.

[LN] W.A.J. LUXEMBURG AND J.M.A.M. VAN NEERVEN, Estimates for the Laplace transform in rearrangement invariant Banach function spaces, in preparation.

[Ly] YU. I. LYUBICH, *Introduction to the Theory of Banach Representations of Groups*, Operator Theory: Advances and Applications, Vol. 30, Birkhäuser (1988).

[LV1] YU. I. LYUBICH AND VŨ QUÔC PHÓNG, Asymptotic stability of linear differential equations on Banach spaces, Studia Math.**88** (1988), 37-42.

[LV2] YU. I. LYUBICH AND VŨ QUÔC PHÓNG, On the spectral mapping theorem for one-parameter groups of operators, J. Soviet Math. **61** (1992), 2035-2037.

[Ml] E. MARSCHALL, On the functional calculus of non-quasianalytic groups of operators and cosine functions, Rend. Circ. Mat. Palermo **35** (1986), 58-81.

[MM] J. MARTINEZ AND J. MAZON, $C_0$−Semigroups norm continuous at infinity, Semigroup Forum **52** (1996), 213-224.

[MN] P. MEYER-NIEBERG, *Banach Lattices*, Springer-Verlag, Berlin-Heidelberg-New York (1991).

[Mo] S. MONTGOMERY-SMITH, Stability and dichotomy of positive semigroups on $L_p$, to appear in: Proc. Am. Math. Soc.

[Mü] V. MÜLLER, Local spectral radius formula for operators in Banach spaces, Czech. Math. J. **38** (1988), 726-729.

[Na] R. NAGEL (Ed.), *One-parameter Semigroups of Positive Operators*, Springer Lect. Notes in Math. 1184 (1986).

[NH] R. NAGEL AND S.Z. HUANG, Spectral mapping theorems for $C_0$−groups satisfying non-quasianalytic growth conditions, Math. Nachr.**169** (1994), 207-218.

[NR] R. NAGEL AND F. RÄBIGER, Superstable operators on Banach spaces, Israel J. Math.**81** (1993), 213-226.

[NF] B. SZ.-NAGY AND C. FOIAS, *Harmonic Analysis of Operators on Hilbert Space*, North-Holland, Amsterdam (1970).

[Ne1] J.M.A.M. VAN NEERVEN, *The Adjoint of a Semigroup on Linear Operators*, Springer Lect. Notes in Math. 1529, Springer-Verlag (1992).

[Ne2] J.M.A.M. VAN NEERVEN, Exponential stability of operators and operator semigroups, J. Func. Anal. **130** (1995), 293-309.

[Ne3] J.M.A.M. VAN NEERVEN, On the orbits of an operator with spectral radius one, Czech. Math. J. **45** (**120**) (1995), 495-502.

[Ne4] J.M.A.M. VAN NEERVEN, Characterization of exponential stability of a semigroup of operators in terms of its action by convolution on vector-valued function spaces over $\mathbb{R}_+$, J. Diff. Eq. **124** (1996), 324-342.

[Ne5] J.M.A.M. VAN NEERVEN, Individual stability of $C_0$−semigroups and uniform boundedness of the local resolvent, to appear in: Semigroup Forum.

[Ne6] J.M.A.M. VAN NEERVEN, Inequality of spectral bound and growth bound for positive semigroups in rearrangement invariant Banach function spaces, to appear in: Arch. Math. (Basel).

[Ne7] J.M.A.M. VAN NEERVEN, The norm of a complex Banach lattice, submitted.

[Ne8] J.M.A.M. VAN NEERVEN, Elementary operator-theoretic proof of Wiener's Tauberian theorem, submitted.

[NS] J.M.A.M. VAN NEERVEN AND B. STRAUB, On the existence and growth of mild solutions of the abstract Cauchy problem for operators with polynomially bounded resolvents, submitted.

[NSW] J.M.A.M. VAN NEERVEN, B. STRAUB, AND L. WEIS, On the asymptotic behaviour of a semigroup of linear operators, Indag. Math., N.S. **6** (1995), 453-476.

[Nb1] F. NEUBRANDER, Well-posedness of higher order abstract Cauchy problems, Trans. Am. Math. Soc. **295** (1986), 257-290.

[Nb2] F. NEUBRANDER, Laplace transform and asymptotic behavior of strongly continuous semigroups, Houston J. Math. **12** (1986), 549-561.

[Nm] D.J. NEWMAN, Simple analytic proof of the prime number theorem, Am. Math. Monthly **87** (1980) 693-696.

[Pa] B. DE PAGTER, A characterization of sun-reflexivity, Math. Ann. **283** (1989), 511-518.

[PW] R.E.A.C. PALEY AND N. WIENER, *Fourier Transforms in the Complex Domain*, Am. Math. Soc. Colloq. Publ. 19, New York (1934).

[Pz] A. PAZY, *Semigroups of Linear Operators and Applications to Partial Differential Equations*, Springer-Verlag (1983).

[Pe] J. PEETRE, Sur la transformation de Fourier des fonctions à valeurs vectorielles, Rend. Sem. Mat. Univ. Padova **42** (1969), 15-26.

[Pi] G. PISIER, *Factorisation of Linear Operators and Geometry of Banach Spaces*, CBMS, Regional Conference Series, No. 60, AMS, Providence (1986).

[Po] H. POLLARD, Harmonic analysis of bounded functions, Duke Math. J. **20** (1953), 499-512.

[PZ] A.J. PRITCHARD AND J. ZABCZYK, Stability and stabilizability of infinite-dimensional systems, SIAM Rev. **23** (1981), 25-52.

[Pr1] J. PRÜSS, Equilibrium solutions of age-specific population dynamics of several species, J. Math. Biology **11** (1981), 65-84.

[Pr2] J. PRÜSS, On the spectrum of $C_0$-semigroups, Trans. Am. Math. Soc. **284** (1984), 847-857.

[Pl] K.M. PRZYLUSKI, On a discrete time version of a problem of A.J. Pritchard and J. Zabczyk, Proc. Roy. Soc. Edinburgh, Sect A, **101** (1985), 159-161.

[RS] F. RÄBIGER AND R. SCHNAUBELT, The spectral mapping theorem for evolution semigroups on spaces of vector-valued functions, Semigroup Forum **52** (1996), 225-240.

[RW] F. RÄBIGER AND M. WOLFF, Superstable semigroups of operators, Indag. Math., N.S. **6** (1995), 481-494.

[Ra] R. RAU, Hyperbolic evolution semigroups on vector valued function spaces, Semigroup Forum **48** (1994), 107-118.

[Ro] S. ROLEWICZ, On uniform $N$-equistability, J. Math. Anal. Appl. **115** (1986), 434-441.

[Ru] W. RUDIN, *Real and Complex Analysis*, McGraw-Hill, 3rd ed. (1986).

[RV] W.M. RUESS AND VŨ QUÔC PHÓNG, Asymptotically almost periodic solutions of evolution equations in Banach spaces, J. Diff. Eq. **122** (1995), 282-301.

[Sf] H. H. SCHAEFER, *Banach Lattices and Positive Operators*, Springer-Verlag, Berlin-Heidelberg-New York (1974).

[Si] B. SIMON, Schrödinger semigroups, Bull. Am. Math. Soc. **7** (1982), 447-526.

[SS] G.M. SKYLAR AND V.YA. SHIRMAN, On the asymptotic stability of a linear differential equation in a Banach space, Teor. Funktsii Funktsional Anal. Prilozhen **37** (1982), 127-132.

[Sl] M. SLEMROD, Asymptotic behavior of $C_0$-semigroups as determined by the spectrum of the generator, Indiana Univ. Math. J. **25** (1976), 783-792.

[Tr] H. TRIEBEL, *Interpolation Theory, Function Spaces, Differential Operators*, Berlin VEB (1978).

[Tg1] R. TRIGGIANI, A sharp result on the exponential operator-norm decay of a family $T_h(t)$ of strongly continuous semigroups uniformly in $h$, in: Lect. Notes in Pure and Appl. Math. 160, Marcel Dekker, New York (1994), 325-335.

[Tg2] R. TRIGGIANI, A sharp result on the exponential operator-norm decay of a family of strongly continuous semigroups, Semigroup Forum **49** (1994), 387-395.

[Vo] J. VOIGT, Interpolation for positive $C_0$-semigroups on $L^p$-spaces, Math. Z. **188** (1985), 283-286.

[Vu1] VŨ QUÔC PHÓNG, Theorems of Katznelson-Tzafriri type for semigroups of operators, J. Func. Anal. **103** (1992), 74-84.

[Vu2] VŨ QUÔC PHÓNG, On the spectrum, complete trajectories and asymptotic stability of linear semi-dynamical systems, J. Diff. Eq. **105** (1993), 30-45.

[Vu3] VŨ QUÔC PHÓNG, Semigroups with nonquasianalytic growth, Studia Math. **104** (1993), 229-241.

[Vu4] VŨ QUÔC PHÓNG, Stability of semigroups commuting with a compact operator, to appear in: Proc. Am. Math. Soc.

[Vu5] VŨ QUÔC PHÓNG, Almost periodic and strongly stable semigroups of operators, to appear in: Banach Center Publications.

[VL] VŨ QUÔC PHÓNG AND YU. I. LYUBICH, A spectral criterion for almost periodicity of representations of abelian semigroups, J. Soviet Math. **48** (1990), 644-647.

[Wb] G. WEBB, An operator theoretic formulation of asynchronous exponential growth, Trans. Am. Math. Soc. **303** (1987), 751-763.

[We1] L. WEIS, Integral operators and changes of density, Indiana Univ. Math. J. **31** (1982), 83-96.

[We2] L. WEIS, The stability of positive semigroups on $L_p$ spaces, Proc. Am. Math. Soc. **123** (1995), 3089-3094.

[WW] L. WEIS AND V. WROBEL, Asymptotic behavior of $C_0$-semigroups in Banach spaces, preprint.

[Ws1] G. WEISS, Weak $L^p$-stability of a linear semigroup on a Hilbert space implies exponential stability, J. Diff. Eq. **76** (1988), 269-285.

[Ws2] G. WEISS, Weakly $l^p$-stable linear operators are power stable, Int. J. Systems Sci. **20** (1989) 2323-2328.

[Ws3] G. WEISS, The resolvent growth assumption for semigroups on Hilbert spaces, J. Math. Anal. Appl. **145** (1990), 154-171.

[Wi] D.V. WIDDER, *The Laplace Transform*, Princeton University Press, Princeton (1946).

[Wo] M. WOLFF, A remark on the spectral bound of the generator of semigroups of positive operators with applications to stability theory, in: *Functional Analysis and Approximation*, Proc. Conf. Oberwolfach 1980, Birkhäuser (1981), 39-50.

[Wr] V. WROBEL, Asymptotic behavior of $C_0$−semigroups in $B$-convex spaces, Indiana Univ. Math. J. **38** (1989), 101-114.

[Yo] P. YOU, Characteristic conditions for a $C_0$−semigroup with continuity in the uniform operator topology for $t > 0$, Proc. Am. Math. Soc. **116** (1992) 991-997.

[Za1] A.C. ZAANEN, *Integration*, 2nd ed., North Holland (1967).

[Za2] A.C. ZAANEN, *Riesz Spaces II*, North Holland (1983).

[Zb1] A. ZABCZYK, Remarks on the control of discrete-time distributed parameter systems, SIAM J. Control **12** (1974), 721-735.

[Zb2] A. ZABCZYK, A note on $C_0$−semigroups, Bull. Acad. Polon. Sci. **23** (1975), 895-898.

[Zh] Q. ZHENG, The exponential stability and the perturbation problem of linear evolution systems in Banach spaces, J. Sichuan Univ. **25** (1988), 401-411 (in Chinese).

[Zh] Q. ZHENG, *Strongly Continuous Semigroups of Linear Operators*, Huazhong University of Science and Technology Press, Wuhan (1994) (in Chinese).

# Index

abscissa, of
   improper convergence 15
   uniform boundedness
     of the resolvent 14
abstract Cauchy problem 2
algebraically simple 107
almost periodic
   function 93, 179
   semigroup 201
analytic RNP 129
approximate eigenvector 27
associate space 222

$B$-convex 127
Banach function space 75, 221
   rearrangement invariant 80, 87, 137, 222
Beurling algebra 44
boundedly locally dense 91

Calkin algebra 106
Cesàro means 16
complete orbit 182
complexification 220

$\Delta_2$-condition 82

exponentially stable 4

Fejér kernel 164
Fourier transform 16
Fourier type 116
fractional power
fundamental function 222

generator 3
$C_0$-group 21
growth bound 9

fractional 114
uniform 8

Hölder's inequality 222

individual stability, 121ff, 154ff
isometry 150

Laplace transform 15
   complex inversion, 18
left invariant mean, 161, 178, 184
limit isometric (semi)group 152
local resolvent 1

non-quasianalytic 42, 212

order continuous norm 220
Orlicz space 81, 139, 223

$\odot$-reflexive 163
representation 212
resolvent 1
   identity 6
   local 1
   set 1

Schrödinger operator 187, 211
sectorial 215
semigroup,
   adjoint 28, 34, 163
   bounded away from 0, 180
   $C_0-$ 3
   irreducible 109
   isometric 150
   positive 20
   quotient 7

scalarly integrable 88, 140
topological 200
uniformly continuous 35
    at infinity 37
solution,
    mild 2
    classical 2
spectral,
    bound 2, 8
    inclusion theorem 26
    mapping theorem,
        Gearhart 34
        Greiner 31
        Latushkin-Montgomery-
            Smith 66, 83, 95
        Martinez-Mazon 38
        for non-quasianalytic
            groups 48
        for point spectrum 27
        for residual spectrum 29
    projection 48
    radius 2
    set 163, 177
    synthesis 163, 177
spectrum, 2, 212
    approximate point 27
    Arveson 176
    essential 106
    peripheral 2
    point 26
    residual 27
    unitary 2
stable,
    exponentially 4
    uniformly 4
    uniformly exponentially 4

strictly dominant 107
Sushkevich kernel, 200

theorem,
    Arendt-Batty-Lyubich-Vũ
        153, 178, 185
    Datko-Pazy 77, 80, 104
    Evans 70, 176
    Fejér 16, 31
    Gearhart 34, 53, 68, 126
    Glicksberg-DeLeeuw 206
    Hille-Yosida 5
    Katznelson-Tzafriri 165, 181
    Krein-Rutnam 104
    Loomis 179
    Müller 77
    Phragmen-Lindelöf 59, 85,
        135
    Pringsheim-Landau 19, 20,
        126, 136
    Rolewicz 81, 138
    Vitali 83, 102, 155
    Weis 99, 126
    Weis-Wrobel 114, 119, 142

uniformly exponentially stable
    4
unimodular 203

weakly almost periodic 161,
    201
weight 42
well-posed 3

Yosida approximation 95

# Symbols

$\langle \cdot, \cdot \rangle$    2
$\|\cdot\|_{D(A)}$    4
$(-A)^\alpha$    215
$AP(\mathbb{R}_+, X)$    93
$\mathcal{A}_{\mathbf{T}}$    181
$B_p$    64
$BUC(\mathbb{R}_+, X)$    90
$\Gamma$    2
$(C, 1)$    16
$C_{00}(\mathbb{R}_+, X)$    68
$\mathcal{C}(X)$    106
$\chi$    2
$D(A)$    3
$E'$    222
$\hat{f}$    16
$\hat{f}(\mathbf{T})$    46, 175
$\hat{f}(\mathbf{T})x$    163
$\mathcal{F}$    16
$\overline{G}^{wo}$    201
$H^p(D, X)$    129
$H^p_\omega$    60
$I_{\mathbf{T}}$    176
$j : X \to X^{\odot\odot}$    163
$j_E$    177
$J_E$    177
$k(E)$    177
$\mathcal{K}(X)$    106
$\mathcal{L}f$    15
$\mathcal{L}(X)$    2
$\mathcal{L}_0(X)$    165
$L_\omega(\mathbb{R})$    44
$L^p \cap L^q$    22
$L^\Phi$    223
$\mu_g$    96
$\omega_0(f)$    16
$\omega_1(f)$    15

$\omega_0(\mathbf{T})$    8
$\omega_0^{ess}(\mathbf{T})$    106
$\omega_1(\mathbf{T})$    9
$\omega_\alpha(\mathbf{T})$    114
$\varphi_E$    222
$\varrho(A)$    1
$R(\lambda, A)$    1
$r(T)$    2
$\sigma(A)$    2
$\sigma_a(A)$    27
$\sigma_p(A)$    26
$\sigma_r(A)$    27
$s(A)$    2, 8
$s_0(A)$    14
$s_\beta(A)$    146
$\mathbf{S}_p$    64
$\text{Sp}(\mathbf{T})$    176
$\mathbf{T}$    3
$\mathbf{T}_{(\alpha)}$    15
$\mathbf{T}_\omega$    15
$\mathbf{T}^*$    28
$\mathbf{T}^\odot$    28
$\mathbf{T}_0^\infty$    35
$\mathbf{T}_{x_0}$    154
$\mathbf{T}_{X/Y}$    7
$\hat{\mathbf{T}}$    36
$\mathcal{T}$    165
$X^*$    2
$X^\odot$    28
$X_{x_0}$    154
$X_0 \cap X_1$    216
$X_0 + X_1$    216
$[X_0, X_1]_\theta$    217
$(X_0, X_1)_{\theta,\infty}$    217
$\xi$    181
$(Y, \pi, \mathbf{U})$    152

Titles previously published in the series

# OPERATOR THEORY: ADVANCES AND APPLICATIONS
BIRKHÄUSER VERLAG

Edited by **I. Gohberg,**
School of Mathematical Sciences, Tel-Aviv University, Ramat Aviv, Israel

78. **M. Demuth, B.-W. Schulze** (Eds): Partial Differential Operators and Mathematical Physics: International Conference in Holzhau (Germany), July 3-9, 1994, 1995 (ISBN 3-7643-5208-6)

79. **I. Gohberg, M.A. Kaashoek, F. van Schagen**: Partially Specified Matrices and Operators: Classification, Completion, Applications, 1995 (ISBN 3-7643-5259-0)

80. **I. Gohberg, H. Langer** (Eds): Operator Theory and Boundary Eigenvalue Problems International Workshop in Vienna, July 27-30, 1993, 1995 (ISBN 3-7643-5259-0)

81. **H. Upmeier**: Toeplitz Operators and Index Theory in Several Complex Variables, 1995 (ISBN 3-7643-5280-5)

82. **T. Constantinescu**: Schur Parameters, Factorization and Dilation Problems, 1996 (ISBN 3-7643-5285-X)

83. **A.B. Antonevich**: Linear Functional Equations. Operator Approach, 1995 (ISBN 3-7643-2931-9)

84. **L.A. Sakhnovich**: Integral Equations with Difference Kernels on Finite Intervals, 1996 (ISBN 3-7643-5267-1)

85. **Y.M. Berezansky, G.F. Us, Z.G. Sheftel**: Functional Analysis, **Vol. I**, 1996 (ISBN 3-7643-5344-9)

86. **Y.M. Berezansky, G.F. Us, Z.G. Sheftel**: Functional Analysis, **Vol. II**, 1996 (ISBN 4-7643-5345-7)

87. **I. Gohberg / P. Lancaster / P.N. Shivakumar** (Eds): Recent Developments in Operator Theory and Its Applications, International Conference in Winnipeg, October 2–6, 1994, 1996 (ISBN 3-7643-5413-5)

# MATHEMATICS

A. Lunardi, Università di Parma, Italy

## Analytic Semigroups and Optimal Regularity in Parabolic Problems

**PNLDE 16**
Progress in Nonlinear Differential
Equations and Their Applications

1995. 442. Hardcover
ISBN 3-7643-5172-1

The book shows how the abstract methods of analytic semigroups and evolution equations in Banach spaces can be fruitfully applied to the study of parabolic problems.

Particular attention is paid to optimal regularity results in linear equations. Furthermore, these results are used to study several other problems, especially fully nonlinear ones.

Owing to the new unified approach chosen, known theorems are presented from a novel perspective and new results are derived.

The book is self-contained. It is addressed to PhD students and researchers interested in abstract evolution equations and in parabolic partial differential equations and systems. It gives a comprehensive overview on the present state of the art in the field, teaching at the same time how to exploit its basic techniques.

Please order through your bookseller or write to:
Birkhäuser Verlag AG
P.O. Box 133
CH-4010 Basel / Switzerland
FAX: ++41 / 61 / 205 07 92
e-mail: farnik@birkhauser.ch

For orders originating in the USA or Canada:
Birkhäuser
333 Meadowlands Parkway
USA-Secaucus, NJ 07094-2491
FAX: ++1 201 348 4033
e-mail: orders@birkhauser.com

**BIRKHÄUSER   BASEL • BOSTON • BERLIN**

# MATHEMATICS

**J. Prüss**, Universität-GH Paderborn, Germany

## Evolutionary Integral Equations and Applications

**MMA 87**
Monographs in Mathematics

1993. 392 pages. Hardcover
ISBN 3-7643-2876-2

This book deals with evolutionary systems whose equation of state can be formulated as a linear Volterra equation in a Banach space. The main feature of the kernels involved is that they consist of unbounded linear operators. The aim is a coherent presentation of the state of art of the theory including detailed proofs and its applications to problems from mathematical physics, such as viscoelasticity, heat conduction, and electrodynamics with memory. The importance of evolutionary integral equations - which form a larger class than do evolution equations - stems from such applications and therefore special emphasis is placed on these. A number of models are derived and, by means of the developed theory, discussed thoroughly. An annotated bibliography containing 450 entries increases the book's value as an incisive reference text.

"...The book under review deserves to be a valuable source and reference for at least two kinds of researchers: (a) mathematicians ..., (b) people coming from the side of concrete applications ... Birkhäuser Verlag is to be congratulated for having made this text available to the mathematical community."

ZAMM 75, 282, 1994

| | |
|---|---|
| Please order through your bookseller or write to:<br>Birkhäuser Verlag AG<br>P.O. Box 133<br>CH-4010 Basel / Switzerland<br>FAX: ++41 / 61 / 205 07 92<br>e-mail: farnik@birkhauser.ch | For orders originating in the USA or Canada:<br>Birkhäuser<br>333 Meadowlands Parkway<br>USA-Secaucus, NJ 07094-2491<br>FAX: ++1 201 348 4033<br>e-mail: orders@birkhauser.com |

**BIRKHÄUSER BASEL • BOSTON • BERLIN**

# MATHEMATICS

H. Amann, University of Zürich, Switzerland

## Linear and Quasilinear Parabolic Problems
### Volume I, Abstract Linear Theory

**MMA 89**  
Monographs in Mathematics

1995. 372 pages. Hardcover  
ISBN 3-7643-5114-4

This treatise gives an exposition of the functional analytical approach to quasilinear parabolic evolution equations, developed to a large extent by the author during the last 10 years. This approach is based on the theory of linear nonautonomous parabolic evolution equations and on interpolation-extrapolation techniques. It is the only general method that applies to noncoercive quasilinear parabolic systems under nonlinear boundary conditions.

The present first volume is devoted to a detailed study of nonautonomous linear parabolic evolution equations in general Banach spaces. It contains a careful exposition of the constant domain case, leading to some improvements of the classical Sobolevskii-Tanabe results. The second volume will be concerned with concrete representations of interpolation-extrapolation spaces and with linear parabolic systems of arbitrary order and under general boundary conditions.

Please order through your bookseller or write to:  
Birkhäuser Verlag AG  
P.O. Box 133  
CH-4010 Basel / Switzerland  
FAX: ++41 / 61 / 205 07 92  
e-mail: farnik@birkhauser.ch

For orders originating in the USA or Canada:  
Birkhäuser  
333 Meadowlands Parkway  
USA-Secaucus, NJ 07094-2491  
FAX: ++1 201 348 4033  
e-mail: orders@birkhauser.com

**BIRKHÄUSER BASEL • BOSTON • BERLIN**